AF334642

Hair in Toxicology
An Important Bio-monitor

Issues in Toxicology

Series Editors

Professor Diana Anderson, *University of Bradford, UK*
Dr Michael D Waters, *National Institute of Environmental Health Science, N Carolina, USA*
Dr Timothy C Marrs, *Food Standards Agency, London, UK*

This Series is devoted to coverage of modern toxicology and assessment of risk and is responding to the resurgence in interest in these areas of scientific investigation.

Ideal as a reference and guide to investigations in the biomedical, biochemical and pharmaceutical sciences at the graduate and post graduate level.

Titles in the series:

Hair in Toxicology: An Important Bio-Monitor
Edited by Desmond John Tobin, *University of Bradford*

Visit our website on www.rsc.org/issuesintoxicology

For further information please contact:

Sales and Customer Services
Royal Society of Chemistry
Thomas Graham House
Science Park, Milton Road
Cambridge CB4 0WF
Telephone +44 (0)1223 432360, Fax +44 (0)1223 426017, Email sales@rsc.org

Hair in Toxicology
An Important Bio-monitor

Edited by

Desmond John Tobin
University of Bradford, UK

ISBN 0-85404-587-2

A catalogue record for this book is available from the British Library

Published by The Royal Society of Chemistry,
Thomas Graham House, Science Park, Milton Road, Cambridge CB4 0WF, UK

Registered Charity Number 207890

For further information see our web site at www.rsc.org

Typeset by Alden Bookset, Northampton, UK
Printed by Athenaeum Press Ltd, Gateshead, Tyne and Wear, UK

Dedication

I dedicate this book to my former mentors who were instrumental in providing me with the opportunities and support to lay the foundations of my academic career: the late Aodán Breathnach who took me under his wing while he was Emeritus Professor at St. Thomas's Hospital, London, during my PhD student days; Jean-Claude Bystryn, who patiently and generously facilitated my development as an independent scientist while a post-doctoral fellow in his laboratory at NYU Medical Center, New York; and Ralf Paus, who through his unique combination of fun, fairness and frenetic activity, inspired me to attempt cultivation of these same qualities.

This book is also dedicated to my parents and family in Ireland, who from the start encouraged, supported and indulged me – *Go raibh míle maith agaibh go léir – foai scáth a chéile a mhaireann na daoine! [English: Thank you all very much – people depend on each other in life!]*

Preface

While the estate agent may chant 'location, location, location!', and the stock exchange trader shout "Buy, sell, buy!", the editor of a book devoted to hair has to engage in a little bit of buying, selling and unmasking of the hair follicle as the dream location for unrivalled access to myriad processes underlying much of modern and ancient human life and lifestyle.

This book, entitled *Hair in Toxicology – An Important Bio-Monitor,* is part of a new book series *Issues in Toxicology* from The Royal Society of Chemistry. It is the first book of this kind to be devoted exclusively to the hair follicle and its shaft as an important tool in modern toxicology. This multi-author book will serve as both a reference and a guide to investigations in the biomedical, biochemical and pharmaceutical sciences. Written by investigators from the fields of chemistry, biochemistry and biology, the authors have first-hand knowledge from their chosen sub-specialities and are active contributors to the peer-reviewed scientific literature.

Hair in Toxicology – An Important Bio-monitor is divided into four cognate sections, reflecting the range of interest in the exploitation of this bio-resource to provide valuable information about the world we live in today and have lived in yesterday. The book's 'womb to tomb/cradle to grave' scope begins with a section reviewing our current knowledge of the biology of hair growth: how it cycles, how it generates a fibre, and the value of its component pigment. This section sets the scene by explaining why the hair should serve as a unique bio-resource in toxicology and is followed by a section that considers issues that emerge from the hair's ability to capture snap-shots of our diverse interactions with the environment, both during and after life, and is contributed by experts in the forensic and environmental sciences.

From there we proceed to a section that concentrates on several important toxicological issues to emerge from current strategies for the 'personal care' of hair. Finally, the book draws to a close with an assessment of the hair fibre's contribution to our understanding of human life from archeological and historical investigations. A final perspective on future directions in the use of hair in toxicology closes the book.

The increasing sophistication required of forensics, toxicology, and health and personal care product development requires the exploitation of novel non-invasive approaches to source accurate and vital information. The hair fibre and follicle stand alone in fulfilling these criteria. Thus, there is no better time to shout 'location, location, location!' for the hair and its follicle within the mammalian system, and even 'Buy, sell, buy' for those who make their business from this wondrous mini-organ. But for even the most detached observer, we hope that a perusal of these chapters may instill some sense of the 'Wow!' factor that is the hair follicle.

Contents

Chapter 2   The Human Hair Fibre   34

Desmond J. Tobin

Chapter 3   Pigmentation of Human Hair 57

Desmond J. Tobin

Chapter 6   Hair and Metal Toxicity   125

Stefanos N. Kales and David C. Christiani

Part 3 – Chemistry and Toxicology of Personal Hair Care Products

Thomas Clausen and Wolfgang Balzer

G. Frank Gerberick and Cindy A. Ryan

Chapter 11   Hair Colorant Use and Associated Pathology – Cancer?   229

Tongzhang Zheng, Yawei Zhang, Yong Zhu and Lindsay Morton

Chapter 12   The Chemistry of Hair Care Products: Potential Toxicological Issues for Shampoos, Hair Conditioners, Fixatives, Permanent Waves, Relaxers and Depilatories   286

Janusz Jachowicz

Chapter 13   Hair Care Products – Regulatory Issues   311

P. Raniero De Stasio

Part 4 – Hair in Archaeology

Editor

Desmond J. Tobin, PhD, was born in Ireland in 1965. After graduating with a BSc. (Biology, Chemistry and Maths) from the National University of Ireland, Maynooth in 1986, and a graduateship of the Institute of Biology in Immunology from NESCOT, Surrey, England in 1988, he joined the Division of Biochemsitry at St.Thomas's Hospital Medical School, London where he obtained his PhD degree in 1991. His post-doctoral training was conducted at the Dept. of Dermatology, New York University Medical School, which was followed by a period as Assistant Research Professor in this Department until 1996. Returning to Europe in 1996 for a Lectureship in Biomedical Sciences at the University of Bradford, England, Desmond Tobin is currently Professor of Cell Biology at this Department. Prof.Tobin has authored or co-authored over 100 papers, reviews and chapters.

Contributors

Wolfgang Balzer PhD
Corporate Vice President R&D
Colorants,
Wella AG,
Berliner Allee 65,
D – 64274 Darmstadt,
Germany.
WBalzer@wella.de

Bruce A. Benner PhD
Analytical Chemistry Division,
Chemical Science and Technology
Laboratory,
National Institute of Standards and
Technology,
Gaithersburg,
MD20899,
USA.
bruce.benner@nist.gov

Vladimir Bencko, MD, PhD
Professor and Head,
Institute of Hygiene & Epidemiology,
First Faculty of Medicine,
Charles University of Prague,
Studnickova 7,
CZ 128 00 Prague 2
Czech Republic.
vbencko@lf1.cuni.cz

Prof. David C. Christiani, MD
Professor and Director,
Occupational Health Program
& Professor of Occupational
Medicine & Epidemiology,
Harvard School of Public
Health,
Professor of Medicine,
Harvard Medical School,
Cambridge, MA, USA.
dchris@hohp.harvard.edu

Thomas Clausen PhD
Vice President R&D,
Wella AG,
Berliner Allee 65,
D – 64274
Darmstadt,
Germany.

P. Raniero De Stasio PhD
Proctor & Gamble,
Rusham Park Technical
Centre,
Whitehall Lane,
Egham,
Surrey TW20 9NW
UK.
destasio.r@pg.com

G. Frank Gerberick, PhD
The Procter & Gamble Company
Miami Valley Laboratories
P.O. Box 538707
Cincinnati,
OH 45253–8707
USA.
gerberick.gf@pg.com

Janusz Jachowicz PhD
1361 Alps Rd.,
Wayne,
NJ 07470,
USA.
jjachowicz@ispcorp.com

Stefanos N. Kales, MD
Professor and Medical Director,
Employee Health & Industrial
Medicine,
Cambridge Health Alliance
& Assistant Professor,
Harvard Medical School & Harvard
School of Public Health,
Dept. of Environmental Health,
Macht Bldg., Room 427,
The Cambridge Hospital,
1493 Cambridge Street, Cambridge,
MA 02139,
USA.
skales@challiance.org

Pascal Kintz PhD
Laboratoire ChemTox,
3 rue Gruninger, F-67400 Illkirch
France.
pascal.kintz@wanadoo.fr

Barbara C. Levin
Analytical Chemistry Division,
Chemical Science and Technology
Laboratory,
National Institute of Standards and
Technolgy,
Gaithersburg,
MD 20899,
USA.

Lindsay Morton
Division of Environmental Health
Sciences,
Yale School of Public Health
60 College Street,
New Haven C7 065 20,
USA.

Tamsin O'Connell PhD
McDonald Institute for Archaeological
Research,
University of Cambridge,
Downing Street,
Cambridge CB2 3ER,
UK.
tco21@cam.ac.uk

Cindy A. Ryan PhD
The Proctor & Gamble
Company,
Miami Valley Laboratories,
P.O. Box 538707
Cincinnati,
OH 45253–8707
USA.

Desmond J. Tobin PhD.
Professor of Cell Biology,
Biomedical Sciences,
School of Life Sciences,
University of Bradford
Bradford, West Yorkshire,
BD7 1DP, UK.
dtobin@bradford.ac.uk

Marion Villain PhD
Laboratoire ChemTox,
3 rue Gruinger, F-67400
Illkirch
France.
pascal.kintz@wanadoo.fr

Andrew S. Wilson PhD
Wellcome Trust Research Fellow
in Bioarchaeology,
Dept. of Archaeological Sciences &
Dept. of Biomedical Sciences,
University of Bradford,
Bradford, West Yorkshire,
BD7 1DP, UK.
A.S.Wilson2@bradford.ac.uk

Tongzhang Zheng, BMed, ScD, ScM
Associate Professor and Head,
Division of Environmental Health
Sciences,
Yale School of Public Health,
60 College Street,
New Haven, CT 06520,
USA.
tongzhang.zheng@yale.edu

Yawei Zhang
Division of Environmental Health
Sciences
Yale school of Public
Health
60 College Street
New Haven, CT 06520,
USA.

Yong Zhu
Division of Environmental Health
Sciences
Yale school of Public Health
60 College Street
New Haven, CT 06520,
USA.

Part 1
Biology of Hair

The Biogenesis and Growth of Human Hair

DESMOND J. TOBIN

1.1 The Hair Follicle Mini-organ

1.1.1 Introduction

As social beings we communicate significantly *via* our physical appearance and so together with epidermal pigmentation the hair fibre-producing mini-organ accounts for most of the variation in the phenotype of different mammals and between different human population groupings. Although commonly dismissed as being of superficial importance, the hair follicle(s) (HF) is truly one of human biology's most fascinating structures.[1] Hair growth, one of only two uniquely mammalian traits (in addition to mammary glands), serves several important functions. These include thermal insulation, camouflage (melanin affords significant protective value, *e.g.* change of coat colour in the arctic fox with season), social and sexual communication (involving visual stimuli, odorant dispersal *etc.*), sensory perception (*e.g.* whiskers), and protection against trauma, noxious insults, insects, *etc.* These features have clearly facilitated evolutionary success in animals, but it is not immediately clear how these may have proved critical for human survival. That said, one should not diminish the role of hair in social and sexual communications among humans. Because of our relative nakedness most attention and study is focused on scalp hair that, uniquely amongst primates, can be very thick, very long and very pigmented. Conversely, its absence from the human scalp can result in significant psychological trauma[2], *e.g.* in cases of androgenetic alopecia, alopecia areata and chemotherapy-induced alopecia. Our ancient pre-occupation with hair is further heightened today as our increasing longevity inevitably fuels our desire to extend youthfulness. This increasing attention to hair-care is reflected in the unremitting growth of the hair-care market, already a multi-billion euro enterprise world-wide (*Euromonitor*).

Unlike most other mammals, we humans have all but lost our ability to grow hair synchronously or as a wave. Instead, our hair grows in a mosaic pattern where significant autonomy of growth and pigmentation resides in individual HF.

The evolutionary selective pressure for why humans developed such a luxurious growth of pigmented scalp hair is more perplexing. One possible explanation may relate to the hair fibre as a dispersal conduit for pigment – melanin is an avid binder of a broad range of toxins and metals (for further discussion see Chapter 3 in this volume). This view, advanced by Hardy, derives from the evolution of early humans along riverbanks and seacoasts.[3] As such, a diet rich in fish, concentrators of heavy metals, could have had significant health implications. A mechanism to quickly remove these toxic metals, thereby preventing their build-up in the body, may have been exploited by melanin's capacity to bind these compounds into a rapidly dividing tissue that ultimately keratinises to form the hair fibre. The hair bulb exhibits the body's second highest rate of proliferation (after hematopoietic tissue) and so could swiftly incorporate metals and toxins into a pigmented and cornified hair shaft,[4] and in this way limit their access to the living tissue of the highly vascularised scalp.

1.1.2 A Unique Mammalian Epithelial-Mesenchymal-Neuroectodermal Interactive System

The HF or more accurately the 'pilo-sebaceous unit' encapsulates all the important physiological processes found in the human body, namely controlled cell growth/death, interactions between cells of different histologic type, cell differentiation and migration, hormone responsivity *etc*. Thus, the value of the HF as a model for biological scientific research goes way beyond its scope for cutaneous biology or dermatology alone. Indeed, the recent and dramatic upturn in interest in HF biology has focused principally on the pursuit of two of biology's holy grails: post-embryonic morphogenesis and control of cyclical tissue activity.

The HF mini-organ is formed from a bewilderingly complex set of interactions involving ectodermal, mesodermal and neuroectodermal components, which go to elaborate five or six concentric cylinders of at least fifteen distinct interacting cell sub-populations. These together provide a truly exceptional tissue[5] that rivals the vertebrate limb-bud[6] as a model for studies of the genetic regulation of development. An important consideration for the remit of this book is that the formation of the HF product, its fibre, occurs in a highly time-resolved manner and so *locks in* a snap-shot of the individual's physiology and chemistry at the time of the hair fibre's formation. Thus, the hair fibre does not undergo further biogenic change.

1.1.3 Comparison with Other Keratinised Skin Appendages – The Nail

Hair, scales, feathers, claws, horns and nails are all derived from skin and so all consist of keratinised modified epidermal cells.[7] Like the hair fibre, the biological and chemical composition of the nail is not altered by changes in the blood chemistry or by exposure to toxins, chemicals *etc*. occurring after these structures were formed (*i.e.* no post-biogenic change). Therefore, both hair and nails are of major interest to toxicologists and to those interested in forensic and medico-legal investigations. The slower growth of nail (toenail, 0.05 mm per day: finger nail, 0.1 mm per day) compared with human scalp hair fibres ($\approx$0.35 mm per day), and the fact that nails (especially of the foot) are not normally exposed to external contaminants, make the nail particularly useful for retrospective analysis. For a discussion of this topic, the reader is directed to an excellent clinical review by Daniel *et al.*[8]

1.2 Embryology of the Hair Follicle

1.2.1 Hair Follicle Induction

Hair follicles are skin appendages that develop from the human epidermis around the end of the third gestational month. The events surrounding the early stages of HF development (Figure 1.1) are currently the topic of intense research with several excellent reviews available to the interested reader.[5,9–11] In brief, these 'inductive' events rely on signals between specialised mesenchymal cells of the dermis that lie below the superficially-arranged epidermal cells (*i.e.* the epidermis), which direct specialised outcomes in these epithelial (keratinocytes) and dermal cells (fibroblasts). Dermal 'first' signals are thought to determine where along the overlying epidermis a HF is likely to be produced (so-called 'patterning' controls), while the epithelium's 'first' signal is thought to determine and direct the clustering of the dermal fibroblasts to form the hair growth 'inducer' component – the so-called follicular dermal papilla. There are also 'second' signals to induce initial downgrowth of the epidermal 'placode' *via* a local proliferation of the HF epithelium. From these initial HF 'commitment' events, subsequent differentiation signals determine which cells will go on to form keratinocytes of the hair fibre and follicular sheaths, and which fibroblasts form the growth-inducing follicular papilla and dermal sheaths. The 'signals' referred to above are mediated by intercellular signalling molecules secreted from specific subpopulations of skin cells and which have multiple receptors/targets to transduce their effects.

Of note here is the exquisite specificity of the signalling events occurring between different cell sub-populations. For example, recombination experiments have shown that dermis taken from one body region when combined with epidermis from another body site (even in another individual) will direct the formation of HF that are characteristic of the dermis donor site. A test of this in humans was recently carried out whereby male follicular dermal tissue taken from the scalp directed the formation of terminal hairs when recombined with female arm epidermis.[12]

The identity of the 'first' dermal and epidermal signal(s) is currently a matter of some conjecture and several molecules have been implicated in these events. However, there is increasing evidence that these may be associated with the WNT and β-catenin signalling pathway.[13] For example, WNT signalling molecules, fibroblast growth factors, transforming growth factor, platelet-derived growth factor and others are thought to lead the promotion of placode formation and early HF development and addition of some of these factors can induce the formation of

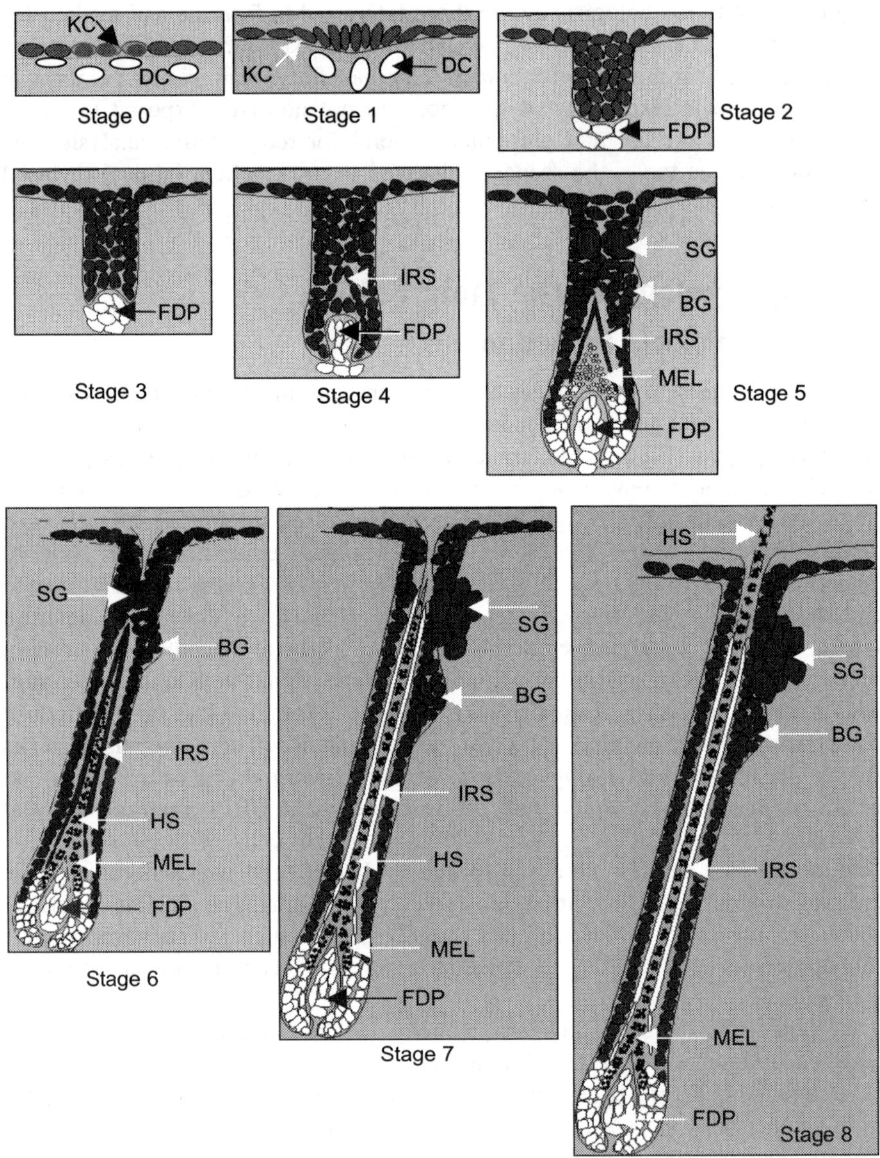

Figure 1.1 *Schematic representation of murine hair follicle morphogenesis. Staging 0–8 according to Paus et al.[14] Key: KC, keratinocytes; DC, dermal cells; FDP, Follicular papilla; IRS, inner root sheath; SG, sebaceous gland; BG, bulge; MEL, melanin; HS, hair shaft*

HF at ectopic sites. Conversely, loss of the ability to produce these factors (*e.g. via* mutation) can result in defective HF development.[10] The exotically-named secreted protein sonic hedgehog (SHH) engages in highly significant signalling events between HF epithelium and mesenchyme. These dictate whether HF development continues to completion once HF placode formation has occurred. In addition to epithelial and mesenchymal cells, the developing HF follicle also contains other cell populations, most conspicuously melanocytes that pigment the developing hair fibre. The reader is directed to Chapter 3 in this volume for a discussion on the development of the follicular pigmentary unit.

The assessment of morphologically recognisable stages of HF development has recently been revisited by Paus and colleagues and these authors have produced an elegant, user-friendly, guide to HF development using neonatal black mice as their model system.[14] I will refer to this guide when reviewing the morphological changes associated with the differentiation/development of the HF.

1.2.2 Hair Follicle Cell Differentiation

One of the enigmas of HF development is how a relatively undifferentiated cluster of epithelial and mesenchymal cells can give rise to such a large number of distinct cell lineages with variable different differentiated products (Figure 1.2).

HF morphogenesis is morphologically appreciable (Figure 1.3) only at Stage 3 when the forerunner of a HF 'bulb' becomes evident – a process reflecting the activation/induction of multiple keratinocyte differentiation pathways that lead to considerable structural change within the tissue. At Stage 3 the developing HF appears as a 'peg' of tissue consisting of an elongated column of concentrically-layered keratinocytes. The mesenchymal component of the HF, the follicular dermal papilla, is now located within a cavity of the developing epithelium bulb. The hair peg continues to elongate into the dermis of the skin during Stage 4 and there is evidence now of the formation of the inner root sheath (IRS), as it assumes a cone-shaped structure. The follicular papilla becomes progressively more invaginated by the enlarging hair bulb.

Significant additional morphological features can be distinguished in the Stage 5 developing HF. Not only does the IRS continue to develop and extend upwards within the HF, but several epithelial prominences or bulges also appear along the external wall of the developing HF called the outer root sheath (Figures 1.1 & 1.3). One of these 'bulges' will generate the future repository of the HF stems cells (*e.g.* for both epithelial cells and melanocytes). Another will become a site of specialised lipid-forming epithelial cells that will form the holocrine *sebaceous gland*. Sebum flow from this gland will coat/lubricate the hair shaft surface. Also around this time melanocytes that have migrated from the embryonic neural crest through the dermis and subsequently to the epidermis, will now distribute within the developing HF. Some of these cells will localise to the epithelial hair bulb matrix just above the follicular dermal papilla and begin to produce the pigment melanin (Figure 1.3). The first melanin granules are evident in pre-cortical keratinocytes at this stage. However, the forming fibre must now start to pass through the HF core in order to exit the skin surface. Its passage through intact follicular epithelium is facilitated by the formation of a 'hair canal', which is constructed *via* focal cell death or apoptosis.

The developing HF continues to extend deeper and deeper into the skin until its proximal bulbar end is situated within the adipose-rich (*i.e.* fat cell) subcutis.

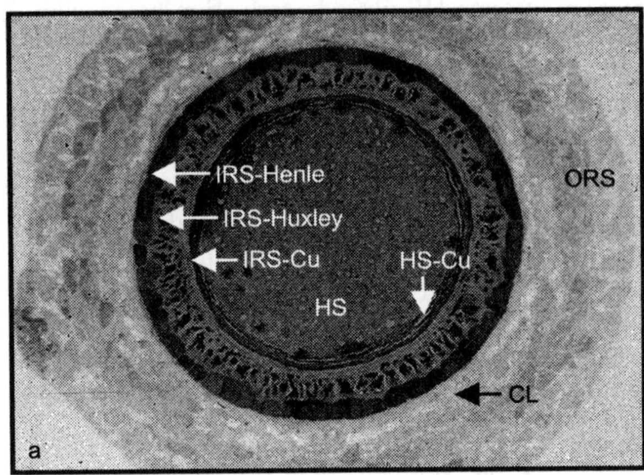

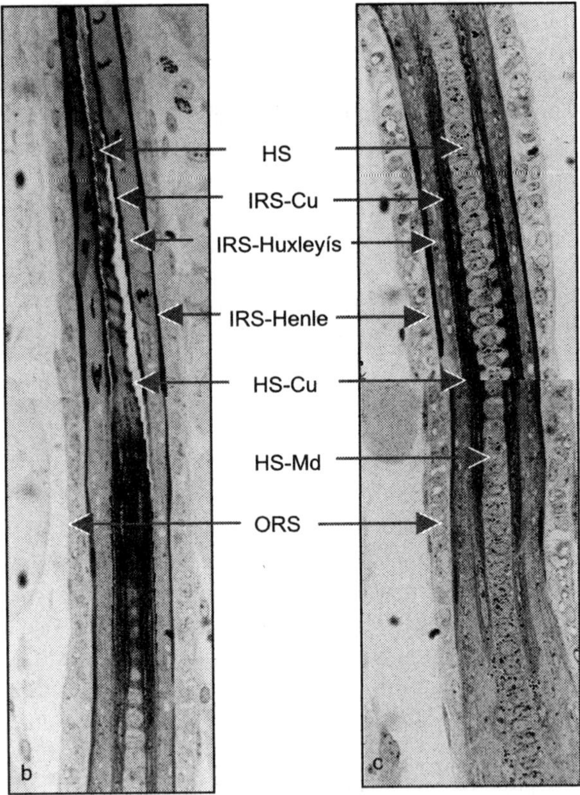

Figure 1.2   (a) *High resolution light micrograph of a transverse (horizontal) section of a human anagen hair follicle taken at the supra-bulbar level.* (b) *Longitudinal (vertical) section of the upper murine hair follicle.* (c) *Longitudinal section of the mid to lower murine hair follicle*
Key: HS, hair shaft; HS-Cu, hair shaft cuticle; IRS-Cu, inner root sheath cuticle; IRS-Huxley, inner root sheath Huxley's layer; IRS-Henle, inner root sheath Henle's layer; ORS, outer root sheath; CL, Companion layer

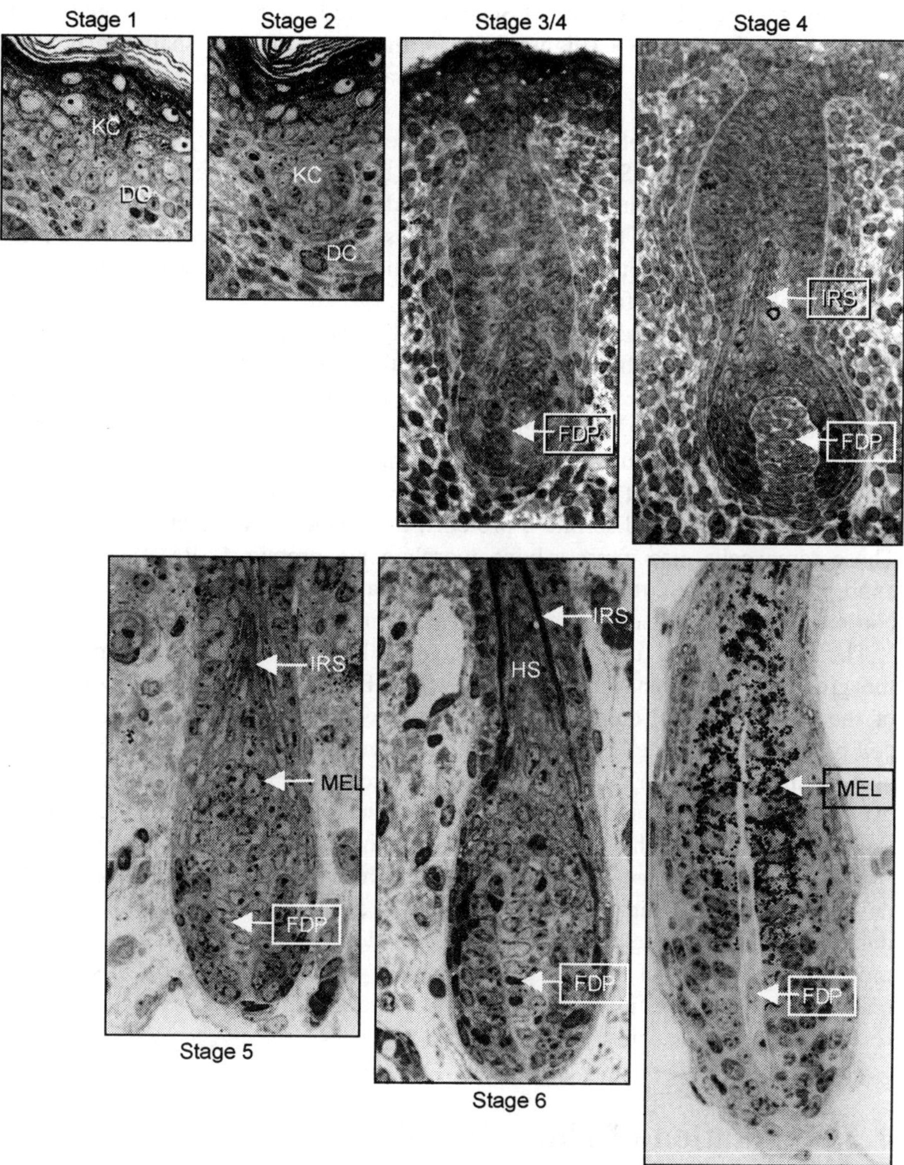

Figure 1.3 *High resolution light micrographs of a longitudinal (vertical) resin sections taken throughout murine hair follicle morphogenesis. Staging 1–8 according to Paus et al.[14] Key: KC, keratinocytes; DC, dermal cells; FDP, follicular papilla; IRS, inner root sheath; MEL, melanin; HS, hair shaft*

Anatomic features characteristic of Stage 6 HF include an increasing complexity of the now multilayered IRS (Figure 1.3). Furthermore, the hair shaft can now be visualised within the hair canal and melanin granules can be seen within its cortical keratinocytes.

The final two stages of HF development are characterised by the growth of the hair shaft through the IRS and hair canal until the tip of the fibre emerges from the surface of the skin (Figures 1.1 & 1.3). In addition, the aspect of the sebaceous gland changes relative to the HF horizontal axis at this stage. Finally, the HF attains its maximal length and bulk during Stage 8 when the distal hair shaft is positioned well free of the skin surface (Figures 1.1 & 1.3).

It should be stressed that the above are purely morphological descriptions, which do not do justice to the enormous molecular complexity that underpins the regulation of cell lineage commitment or cell fate induction that ultimately yields the keratinised hair fibre. Considerable research efforts are currently underway to dissect the molecular pathways and mechanisms involved. For example, cells in the hair bulb matrix that are destined to become hair shaft cortical keratinocytes express the serrate 1 and serrate 2 proteins, the ligands of the Notch 1 receptor.[10] These same cells also express bone morphogenic protein 4 (BMP4), while the expression of Noggin, the inhibitor of BMP4, can disrupt the differentiation of the cortical keratinocytes needed for hair shaft formation.

The anatomy of the fully developed Stage 8 HF is very similar to the anatomy of the growing/cycling anagen HF in the adult. Epithelial and mesenchymal cells of the developing HF therefore contrive to produce the mature HF *via* massive cell proliferation and cell differentiation. However, considerable tissue sculpting is also required to fashion this complex multilayered 3-D mini-organ. Such sculpting events also require intermittent and highly localised programmed cell death (apoptosis).[15] Only in this way can the HF assume its full hair shaft-forming status.

Cessation of hair fibre production, *i.e.* hair shaft growth, during Stage 8 of HF development signals entry into the 'hair growth cycle', from which the HF usually does not escape during the life of the individual. A detailed discussion of the molecular regulation of the hair growth cycle is beyond the scope of this volume; readers interested in a full treatment of this topic are directed to a recent excellent review by Stenn and Paus.[16]

1.3 Regulation of Hair Growth

1.3.1 The Hair Growth Cycle

1.3.1.1 Introduction

The second episode in the life of the HF, *i.e.* its entry into the *hair growth cycle*, involves a paradoxical 'phoenix from the ashes' scenario. In brief, this process begins with the precipitation of the fully-formed and hair fibre-producing HF into a regression phase characterised by massive apoptosis (the *first* catagen). Upon completion of this catagen phase, the HF is reduced to about 30% of its original Stage 8 tissue mass. The HF thereafter enters a period of relative rest (*i.e.* telogen) and remains therein until the third and final episode in the HF life history, *i.e.* its entry into the anagen phase of the *first* cycle. Thereafter the HF continues with life-long cyclical activity (Figures 1.4 & 1.5).

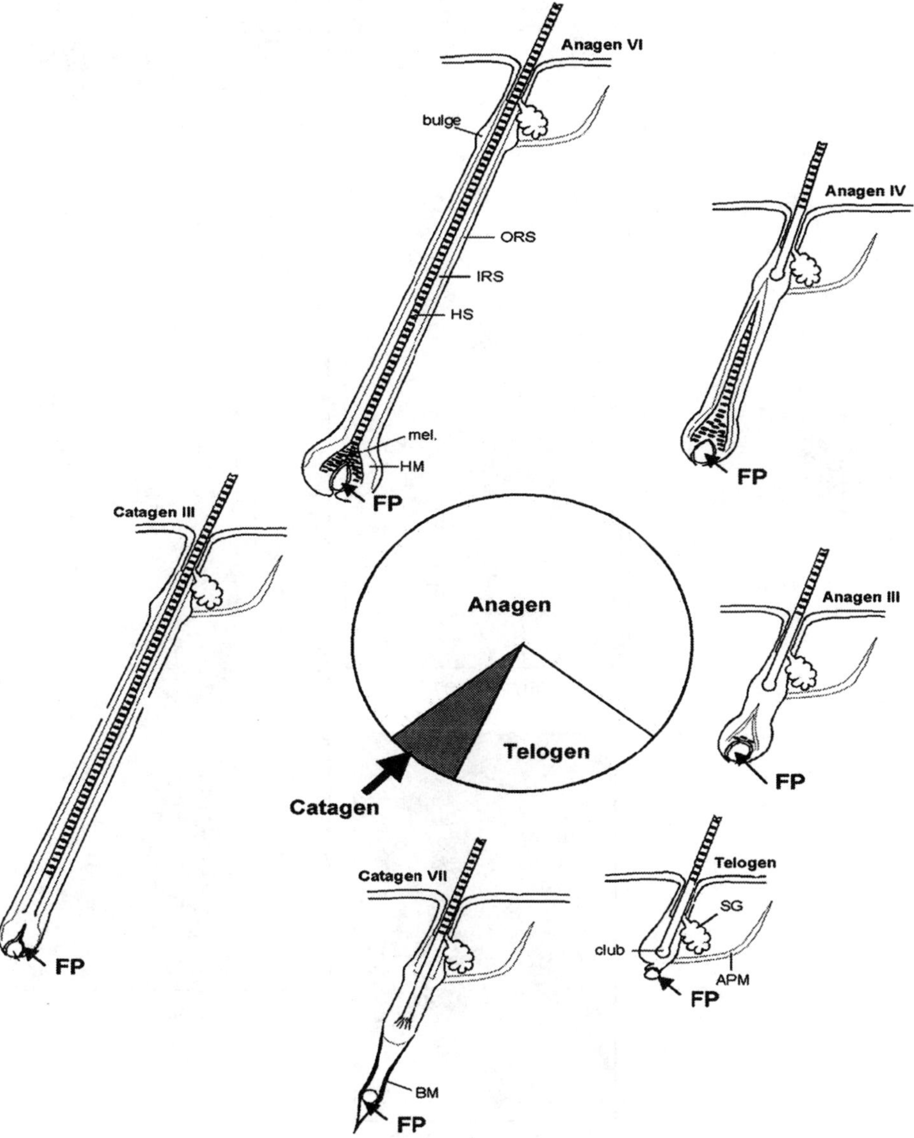

Figure 1.4 *Schematic representation of murine hair growth cycle. Staging according to Paus et al.*[24] *Key: FP, follicular papilla; IRS, inner root sheath; SG, sebaceous gland; BM, basement membrane; mel, melanin; HS, hair shaft; APM, arrector pili muscle*
(Reprinted from *Experimental Gerontology*, Tobin DJ, Paus R, Graying: Gerontology of the hair follicle pigmentary unit, **36**(37), 2001, with permission from Elsevier.)

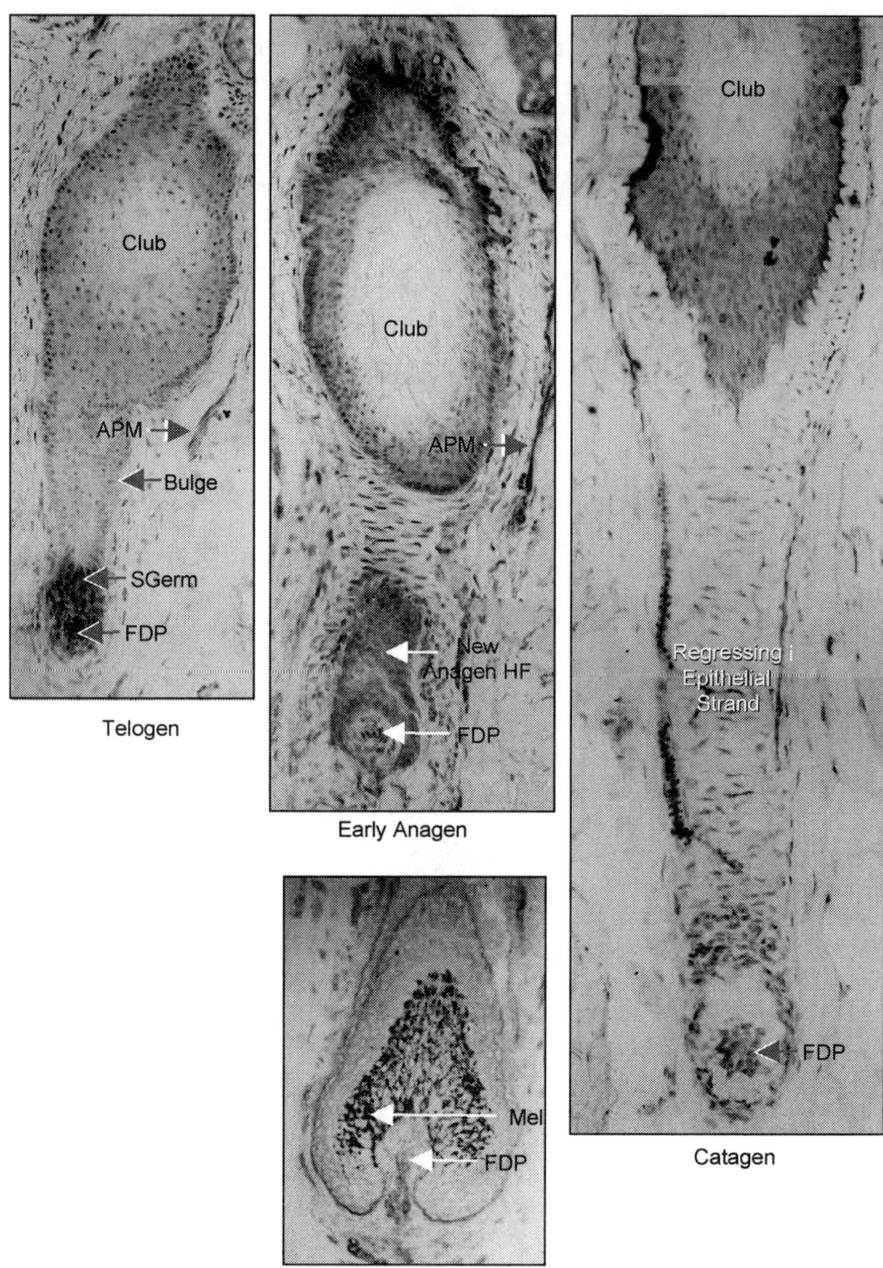

Figure 1.5 *Light micrographs of a longitudinal (vertical) cryosections taken throughout human hair growth cycle. Staging 1–8 according to Paus et al.[14] Key: FDP, follicular papilla; SGerm, secondary germ; AMP, arrector pili muscle; mel, melanin*

This first anagen of the hair growth cycle morphologically resembles several aspects of HF development *in utero*, so much so that the life-long cyclical activity of HF appears to recapitulate, at least in part, several embryologic events involved in HF morphogenesis.[1] In this way signalling pathways (see below) that were active during morphogenesis are re-used during in the hair growth cycle – again underpinned by signalling events between dermal and epithelial cells.[17]

While hair growth cycle dynamics vary somewhat between different mammalian species, between different body sites and even between different HF types in the same body site, there exists an intrinsic rhythmic activity that can be modulated not only by systemic factors but more importantly also by the HF's own locally-directed activities. Indeed, individual HF retain a large degree of autonomy, as evidenced by the prolonged 'donor-site dominance' in hair transplantation surgery[18]/grafting experiments and by hair growth mosaicism.

1.3.1.2 Catagen

Catagen represents a partial organ 'suicide', whereby the growing HF is reduced by more than two-thirds of its growth size (Figure 1.1). This massive cell death is highly controlled and very rapid (about 2–3 weeks on scalp) and involves substantial remodelling of the HF. The mechanism of HF regression during catagen is principally by an energy-dependent programmed cell death called apoptosis.[19] However, catagen also involves coordinated cell differentiation and extracellular remodelling events. Less than one in every thousand scalp HF will be in catagen at any one time.

The signal(s) for induction of catagen remains elusive. However, it has recently been shown in animal studies that severe psycho-social stress can precipitate the HF into catagen.[20] In addition, environmental factors including chemicals, trauma and experimentally-administered hormones can also induce catagen. It is likely that catagen may be induced here by changes in the levels of cytokines/hormones that are critical for continued cell proliferation. For example, inhibition of the receptor for insulin-like growth factor and altered levels of neurotrophins can induce catagen.[16]

The first morphological clue of impending HF regression (catagen) is the cessation of HF melanogenesis,[21] also involving melanocyte apoptosis.[22] Keratinocyte proliferation slows down at this stage, though some hair shaft production continues without incorporation of melanin. As a result the most proximal end of shed hair fibres are usually pigment-free. Thereafter, a massive apoptosis of keratinocytes occurs in the lower HF. However, not all keratinocytes are similarly affected and there is evidence that some bulbar keratinocytes form part of the secondary germ of the telogen HF (see below).

While keratinocytes and melanocytes of the lower HF are vulnerable to the catagen/apoptosis signal, fibroblasts of the follicular dermal papilla appear to be uniquely resistant. While the HF epithelium shortens dramatically, the papilla does not become stranded in the deep dermis/sub-cutis at the level of the former anagen hair bulb cavity. Instead, it moves, or is drawn upwards, toward the epidermis by maintaining direct association with the shrinking HF epithelial strand until it comes to rest at the hair 'bulge'. Maintenance of close follicular dermal papilla/epithelium contact is critical for continued cycling,[23] without which hair growth and cycling will fail. Tissue change associated with the transition from full growth (anagen) to rest (telogen) is conventionally divided into the characteristic morphologically-defined stages catagen I-VIII.[24]

1.3.1.3 Telogen

When the HF involution is complete at catagen VIII, the HF enters a period of relative rest called telogen. Keratinocytes of the epithelial sac moor the hair club/hair shaft tightly in the telogen HF. At this stage the HF appears paltry by comparison to its former majestic growing state. However, it remains vascularised, innervated and contains all the cell types needed to produce the next fully developed anagen HF. This region of the HF is thought to contain a range of different stem cell populations, including epithelial stem cells.[25] Stem melanocytes/melanoblasts are also thought to be located in the bulge and surrounding outer root sheath.[26] Furthermore, it is likely that mesenchymal stem cells for replenishment of the follicular dermal papilla and dermal sheath also reside close to the base of the telogen HF[27] (Figures 1.4 & 1.5). At any one time, approximately one in every ten scalp HF will be in the telogen phase of the hair growth cycle.

The completion of the telogen phase has sometimes mistakenly been associated with the passive shedding of the old hair shaft in humans. Instead the process of hair shedding, termed *exogen*, constitutes an independent active event.[28] Unfortunately this stage of the hair growth cycle has received little attention to date. This situation may be surprising, perhaps, to many readers, as hair shedding can be a preoccupation for many people/patients, not only for those with a recognised hair loss disorder but also for interpreting amounts of hair removed during normal grooming. Exogen appears to be independent of hair cycling controls and shedding also does not occur due to some motive force exerted by the tip of anagen hair shaft below, as is seen with the extrusion of milk teeth by adult permanent teeth. Exogen usually occurs during anagen IV rather than later during anagen V when the hair fibre tip reaches the base of the telogen club hair. Thus, shedding may appear to be the result of altered protease/protease inhibitor levels/activity, perhaps involving the lysosomal protease cathepsin L.[29]

Although there is considerable deviation between individuals and also between seasons, a daily shed rate of 100–200 is considered normal.

1.3.1.4 Anagen

Human scalp HF remains in the 'resting' telogen phase for approximately three months until they are 'awoken' by some, as yet unspecified, signal that initiates the new growth/anagen phase. Several fundamental questions remain regarding the regulation of this critical transition point of the hair growth cycle. For example it is not yet clear which cell type is responsible for initiating entry into anagen, what this stimulus is nor how it influences its target. Moreover, there are many other systems including vascular, neural, pigmentary, extracellular matrix *etc.* that also exhibit remarkably tight coupling to the hair growth cycle. Such questions are fundamental to many aspects of human biology (*e.g.* for wound healing and tissue remodelling), and so the highly accessible HF provides a very useful model system for understanding several fundamental biological phenomena.[16]

The cells that initiate anagen growth must have certain features to enable them to direct and discharge such enormous biological activity. First, they must have stem cell-like characteristics (see below), with the biological capability to maintain a unique rhythm or periodicity of activity that is intrinsic to the individual HF itself (see below). But perhaps most importantly, these cells need to send inductive signals to the surrounding epithelium and/or mesenchyme. Morphologically, the telogen-to-anagen transition can be viewed as the addition of significant new 'cycling' epithelial tissue to the so-called 'permanent'/non-cycling portion of the HF – represented in large part by the HF during telogen (Figures 1.4 & 1.5). In this way, the HF transits from the 'permanent' shortened telogen HF to the full length growing HF as it passes through six characteristic morphologic stages of anagen (anagen I-VI).

The new anagen hair bulb emerges from the secondary germ at the base of the telogen HF. Researchers are particularly interested in dissecting the signals sent by the follicular dermal papilla to the epithelial stem cells or *vice versa*, which may lead to both the initiation and regulation of the massive proliferation of hair bulb keratinocytes and the more moderate proliferation of outer root sheath keratinocytes and melanocytes. Notably, this proliferative region of the HF dramatically exhibits a rare example of immune privilege during anagen only[30,31] (see below).

As previously mentioned anagen development resembles several of the morphological changes observed during HF development. Thus, it is not surprising that many soluble growth factors associated with HF morphogenesis have also been detected during anagen development. However, the process should not be regarded as a exact re-capitulation of HF morphogenical events, especially as some factors (*e.g.* NT-3 and TGFβ2) that are associated with tissue *development* during HF morphogenesis can induce tissue *involution* in the cycling HF.[9] It appears that levels of key regulatory molecules (*e.g.* WNT family members, sonic hedgehog, BMPs *etc.*) fluctuate during the hair growth cycle and that gradients/ratios of these stimulators and inhibitors can drive the process of cell proliferation, differentiation or involution during the hair growth cycle. For example, while sonic hedgehog does not appear to be required for anagen onset, this molecule is needed for subsequent events, including epithelial cell proliferation and downgrowth of the anagen HF into the dermis.[32]

Evidence gathered over the last few decades has clearly established that the follicular dermal papilla as the principal player in the regulation of the hair cycle 'clock'.[33] Therefore, it is perhaps not surprising that electrologists have also focused on delivering electrical current to this component of the growing HF in their efforts to permanently remove hair. The papilla's critical role in determining the developmental pathway of the overlying ectoderm during HF cycling was convincingly demonstrated in early tissue recombination and implantation studies.[34] Therefore, the follicular dermal papilla appears to exhibit minimal post-natal change. While there is some debate regarding its intrinsic proliferative capacity, it now appears very likely that follicular dermal papilla cell numbers do indeed increase/decrease during normal cycling (see below).

One hair cycle-associated follicular dermal papilla change that is very striking is the significant reduction in the anagen-associated extracellular matrix (ECM) at the end of hair growth. Significant attention has been paid to the nature of the follicular dermal papilla ECM as this differs strikingly from that in the dermis[35] and so there has been considerable speculation about the possible role of one or many of the papilla-associated proteoglycans in the control of the hair growth cycle. While convincing evidence has yet to be presented, it is clear that some components, particularly chondroitin proteoglycans, diminish during catagen and are absent in telogen. Emerging from these observations has been the suggestion that in anagen, these extracellular matrix proteins may confer on the HF a protective environment for maintenance and growth associated with maintaining immune privilege[30] (see below).

1.3.2 Hair Follicle Stem Cells

Until very recently the consensus view was that the follicular dermal papilla and the upper third of the adult HF (so-called 'permanent' portion) were relatively stable during post-natal life, and that only the lower two-thirds of the HF epithelium underwent dramatic cycle-associated regression in catagen. However, recent evidence now suggests that significant proliferation and apoptosis-driven remodelling of the upper HF, including sebaceous gland, also occurs during the so-called 'restful' telogen.[16] During catagen the follicular papilla is pulled up toward the "permanent" portion of the HF and particularly to a region called the 'bulge' – the permanent epithelium at the base of the telogen HF. Although there are several existing hypotheses to explain the hair cycle, none can yet fully explain all aspects of the hair growth cycle including the intrinsic periodicity of the cycle, the individual HF autonomy of these process and the persistence of cyclic activity throughout adult life. The 'bulge-activation' hypothesis[25,36] is currently favoured. This theory, for which there is considerable supporting evidence, suggests that the mesenchymal-epithelial interaction at the end of telogen involves the stimulation of stem cells in the 'bulge' by the papilla fibroblasts. This event is then followed by epithelial stem cell proliferation. The progeny of these stem cells (so-called 'transient amplifying' (TA) cells) reform the lower 'temporary' part of the HF in anagen.[25] Unlike the stem cell, the TA cell has a limited mitotic potential before undergoing differentiation. The bulge-activation hypothesis also suggests that the loss of proliferative potential by TA cells may be the cue for impending catagen. However, recent studies have shown that catagen induction may originate elsewhere in the epithelium, *e.g. via* the TGF-β family[37] or in the follicular papilla itself, perhaps *via* molecules including FGF-5.[38] It should be kept in mind throughout these discussions that most HF pathologies have their origin in HF cycle dysregulation, and thus some aspects of their pathology may be explained by the bulge-activation hypothesis.

The location of several types of stem cells within the highly regenerative and highly differentiated HF tissue has prompted several researchers to suggest that the HF may be exploited as a repository of primitive cell populations for replacement in other tissues and organs. This is perhaps not too surprising when we look at the sheer variety of cell types to emerge from the epithelial cell population of the primitive

Stage 3 hair peg during HF morphogenesis. These include not just the non-keratinising ORS and companion layer keratinocytes, but also an additional six cornifying keratinocyte lineages of the HF and hair shaft. Moreover, this epithelial tissue also generates multiple other epithelial cell sub-populations of the sebaceous and apocrine sweat glands. In addition to the epithelium the HF also generates neuroectodermal melanocytes as well as neural, vascular and muscle components (*i.e.* arrector pili). It has been reported recently that epithelial stem cells of the HF may transdifferentiate into other histological cell types including endothelial cells and that cells of the follicular papilla may transform into hematopoietic/blood cell adipocytes and bone cells.[39] We are indeed entering a very exiting period of a fuller realisation of this mini-organ's contribution to our understanding of human biology.

1.3.3 Systemic and Intrinsic Influences on Hair Growth

1.3.3.1 Introduction

While I have previously stressed the autonomy of individual HF, this mini-organ deserves further admiration for its ability to interconnect with additional systemic regulatory networks of the body. Remarkably, the HF can respond to most hormones known to biomedicine. Even more surprising is the HF capacity to produce its wide range of hormones *via* synthesis, conversions *etc.* Prominent examples include sex hormones, POMC peptides, CRH, prolactin *etc.*[40] Neuro-peptides/-transmitters/-hormones are also increasingly being implicated in mediating HF events particularly those that may be stress-related.[20,41]

Depilation of telogen hairs induces the rapid onset of anagen and is an excellent example of the effect of local factors on the regulation of the hair growth cycle. This cyclical activity has been ascribed to numerous causes including the reduction in the level of local growth inhibitors/chalones.[42] Further, there is increasing support for the involvement of local autocrine/paracrine factors, particularly cytokines, in the regulation of the hair growth.[43] Hair growth in mice for example, is associated with the sequential expression of TGF-β and its receptors during the different phases of the cycle. TGF-βs are potent growth inhibitors for most cell types, but also regulate cell differentiation, migration, extracellular matrix production and modulate immune function. Interestingly, the administration of large doses of EGF induces a catagen-like state both *in vivo* (*e.g.* in sheep) and *in vitro*. By contrast, the addition of IGF-1 or IGF-II maintains *in vivo* anagen growth rates *ex vivo* and HF can be precipitated into a premature catagen in their absence.

Thus, the HF represents a potent environmental responder/transducer of biologically-relevant information. Perhaps we could have expected this, given the enormous evolutionary selective pressure for furry mammals to maintain an adequate coat.

1.3.3.2 Hair Follicle Immunology

Another big surprise to biologists was the unique immunological status of the HF. The skin emerged as a so-called 'first-level lymphoid organ' in the 1970s.[44] Since then much work has identified both cellular and humoral constituents of the skin immune system. Skin cells important in *innate* skin immunity include keratinocytes, macrophages, monocytes, granulocytes and mast cells, while *acquired* immunity incorporates Langerhans cells, T & B cells, endothelial cells and tissue dendritic cells. The immune system's second arm, the humoral *innate* immunity in skin is supported by fibrolysins, anti-microbial peptides, complement peptides, eicosanoids, neuropeptides and cytokines, while *acquired* humoral immunity involves secretory immunoglobulins, interleukins, interferons, colony stimulating factors and other cytokines.

During all this fervent skin research effort, the HF was oddly neglected. Odd because there are many observations associated with hair growth and cycling that strongly suggest the involvement of the immune system. First, HF provide numerous ports of entry into the body for micro-organisms. Second, the outward movement of the new hair shaft and displacement of the old club hair appears to open an access route to the more proximal HF. Third, hair growth is affected by substances with immunomodulatory characteristics, *e.g.* cytokines, hormones, neuropeptides and some drugs. Fourth, HF regression during catagen is associated with dramatic alterations in the peri-follicular populations of both macrophages and mast cells.[45] Fifth, there are autoimmune diseases that damage the HF.[46] However, perhaps the most significant recent observation implicating the HF in the skin immune system is the curious lack of transplantation antigen (MHC class I) expression[30,47] in the proximal anagen hair bulb with an associated lack of Langerhans and T cells.[48]

An important early observation in human hair research was the observation that, in contrast to the epidermis, Langerhans cells were conspicuous by their absence in the human HF.[49] While Langerhans cells are present in large numbers in the HF infundibulum, very few occur below the level of the sebaceous gland and almost none are detected in the hair bulb. Langerhans cells are the main antigen-presenting cell in the epidermis and their localisation also in the upper distal HF suggests that they operate here as key components of the 'sentinel receptor pathway', where they can respond to noxious stimuli including invading microorganisms, chemicals and UV-B light, and tissue injury induced by mechanical trauma.

Human skin contains two populations of T cells: a rather mobile α/βTCR+ intra-epidermal lymphocytes and a very minor sub-population of $\gamma/\delta TCR+$.[50] These cells would exert an immunosurveillance function against cutaneous pathogens, damaged epidermal cells and also against the emergence of neoplastic cells that arise only from stem cells in the bulge, located at or above the level of the insertion of the arrector pili muscle. The reduction in the numbers of dendritic T cells below the level of the arrector pili muscle insertion site ($\approx$ bulge level) may reflect the decreasing exposure to environmental stressors, but may also reflect a relative paucity of T cell receptor stimulatory ligands in the more proximal HF. The distribution of these cells is modulated in a hair cycle-dependent manner.[45]

Skin mast cells are most usually found only in dermal tissues, where they display antigen-specific, IgE-mediated degranulation and secretary activities. Interestingly, mast cells appear to be strategically located very close to the HF, their vasculature and innervation and their numbers are greatest in hairy human skin. They are not seen within the normal human follicular epithelium although they can rarely be detected in the follicular papilla. The distribution of mast cells is also modulated in a hair cycle–dependent manner.[51]

Macrophages have antigen-presenting and immuno-modulatory functions within the skin immune system. They are detectable in the dermis including the perifollicular dermal sheath but usually not in the HF epithelium. Like mast cells, macrophages have a role in anti-microbial/parasitic defense, but also may secrete multiple immuno-modulatory cytokines during HF regression. The distribution of these cells is modulated in a hair cycle–dependent manner and macrophages are particularly implicated in the resorption of regressing epithelium curing catagen.[45]

The HF differs dramatically from the skin immune system in that the epithelium of the proximal anagen HF lacks classical MHC class Ia (*i.e.* HLA-A, B, C) antigen expression.[47] All other nucleated cells express MHC class I, with the rare exceptions of the testes, eye, parts of the brain and foetotrophoblast. It has been proposed that, like these sites, the HF may enjoy immune privilege. The proximal outer root sheath and inner root sheath and all hair bulb keratinocytes express no detectable MHC class I. Thus, for as long as they exist, the inner root sheath and the hair bulb matrix remain MHC class I negative. In contrast, peri-follicular dermal sheath fibroblasts and immunocytes are MHC class I positive. The lower HF also produces immuno-suppressive peptides derived from POMC including αMSH and ACTH.[52,53] There is some evidence that the ECM of the follicular papilla, and that encapsulating the HF in the dermal sheath may provide a protective role to the lower HF[35] as similar basement membrane matrix barriers surround other tissues with immune privilege.

Immune privilege appears to exist to prevent the inappropriate recognition of antigens that may result in the attack of cells presenting these antigens in the context of MHC. This is important in preventing the induction of autoimmunity if immuno-genic autoantigens were to be exposed. Thus, the hair bulb environment differs immunologically from the epidermis. Moreover, the HF is therefore a very attractive model system for studies on tolerance and immunosuppression in general, and provides an exceptionally accessible tissue to investigate the signal transduction events associated with immuno-modulatory drugs with known hair growth effects, *e.g.* glucocorticosteroids and immunosuppressive immunophilin ligands.

There is also some evidence that the HF uses a primitive epithelial defense system against bacteria. Normally, the hair canals contain a resident microflora of non-pathogenic bacteria including *Propionibacterium acnes, Staphylococcus aureus, Staph epidermidis, Demodex follicularum, Malassezia* species *etc.* The HF appears to have a very effective anti-infection capacity as can be seen by the very rare occurrence of folliculitis on the human scalp despite its approx. 100,000 individual HFs. Folliculitis in immuno-compromised individuals is common however, leading to a greatly increased risk of infection to the mammalian body through this entry port.

1.3.3.3 Hormones and Hair Growth

Systemic modulation of hair growth is occasionally seen in human hair growth where more local mechanisms appear to predominate in the regulation of the asynchronous growth. However, in some mammals synchronicity of hair growth is observed as life-long moult waves that are sensitive to environmental stimuli, *e.g.* photoperiod. It is here that the link between hair growth and endocrine systemic effects is especially clear.[54] Hair growth is sensitive to the transduction of environmental signals that occur *via* the pineal gland. The level of circulating prolactin is another factor important in the regulation of the moulting response in mammals. Modulation of the hair moult can also be seen as the result of multiple hormone effects peripherally. In rats, estradiol, testosterone and adrenal steroids delay anagen onset, while thyroid hormones accelerate HF activity. However, it is apparent that these hormones have different points of action, *e.g.* estradiol decreases anagen length while tyroxine has the opposite effect.

The observation that men castrated before puberty did not go bald or grow beards and the subsequent confirmatory finding that these individuals did so upon treatment with testosterone indicated a role of androgens in hair growth.[55] However, it is likely that the proposed 'central role' for androgens (*e.g.* androsterone, testosterone *etc.*) in the regulation of hair growth is somewhat overstated. Indeed, androgens are unlikely to be involved in the development of the HF *in utero*, and hair growth cycling is essentially normal in the scalps of individuals lacking functional androgen receptors. Thus, it is likely that androgen effects (like those of other hormones) are indirect and operate *via* modulation of other signalling molecules. Androgens enter the cell and thereafter undergo conversion (*e.g.* by 5α-reductase) before binding to a nuclear receptor. Subsequently, this complex binds to an androgen-responsive element in DNA to activate specific genes *via* transcription factors. The effect of androgens on hair growth is complicated and once in the target tissue they undergo considerable conversions.[56] One such conversion that is highly significant for hair growth is the metabolism of testosterone to the more potent 5α-dihydro-testosterone by the enzyme 5α-reductase (type II). Deficiency of this enzyme in the post-pubertal male is associated with female-distributed hair growth, poor beard growth and retention of frontal hairline. Thus, HF in different regions of the body respond differently to androgens. Inhibition of type II 5α-reductase, by finasteride, can induce some hair re-growth in some balding men (see below).

Androgens can alter HF size and thus hair shaft calibre and pigment level. They also can modulate the hair growth cycle in some cases. This effect can be observed during hair miniaturisation in androgenetic alopecia/male pattern baldness. Hair fibre size is dependent on follicular papilla volume, implicating the follicular papilla as a androgen target tissue in the HF (see below). Interestingly, cultured papilla fibroblasts from androgen-dependent HF (*e.g.* beard) contain more of these receptors than those from balding scalp.[56] It is likely that androgen-induced effects on hair growth are mediated through growth factors for follicular epithelium, *e.g.* *via* IGF-1 in a paracrine fashion.

Pregnancy is associated with a higher percentage of HF in the anagen growth phase and this may be followed by the simultaneous entry post-partum of many HFs into telogen (see below). Here, there appears to be a partial synchronisation of HF cycling stages due to increased estradiol levels during pregnancy followed by hair loss possibly due to a reduction in estradiol and thyroxine levels post-natally.

1.4 Hair Growth Pattern and Type

Although we sport over 5 million HF on our bodies it may shock the reader to learn that only about 2% of these are actually distributed on our scalps.[57] In fact, the great majority of the other 4.9 million HF produce only the finest of hair fibers. Therefore, we can reasonably be called the 'naked ape'. Fortunately, the duration of the hair growth cycle also varies significantly in HF on different body sites – the thought of having to plait one's eye-lashes is too scary for words! In this way, a 2–8 year long anagen in scalp HF contrasts strikingly with a 2–3 month anagen of terminal arm HF and up to 6 months on terminal leg HF. Vellus hairs by dint of their design grow only for short periods, typically less than three months.

There is considerable variation in the density of HF in different body regions. This is likely to reflect the variable growth rates of tissues in these body sites during development from the neonate to adulthood. An approximation for adult Caucasians follows: adult cheek skin ($880+/- 60$ HF cm^{-2}); forehead: $770+/- 60$ HF cm^{-2}); forearm: $100+/-50$ HF cm^{-2}); upper arm: $40+/-10$ HF cm^{-2}).[58]

1.5 Hair Growth Rate and Fibre Diameter/Calibre

Scalp terminal hair grows at approximately 0.35 mm per day or about 1 cm per month. There is some evidence that hair elsewhere on the body may grow faster (*e.g.* androgen-stimulated beard growth ~ 0.4 mm per day).[59] These growth rates appear to reflect an intrinsic clock within individual HF and there is little evidence to suggest that this rate can be increased by external factors *e.g.* grooming, shaving *etc.*

It has long been recognised that the volumetric ratio of follicular papilla to hair matrix keratinocytes is important in determining the size of the hair produced by a HF.[60] Elliott and colleagues recently confirmed this view in human HF and further concluded that the anagen-associated increase in the overall follicular papilla size was due more to increased cell numbers than to the increased volume of extracellular matrix produced.[61] The plasticity of the follicular papilla and dermal sheath is likely to be a critical element not only for hair cycle control, but also for HF transformations from vellus-to-terminal and terminal-to-vellus. Alterations in follicular papilla cell number lie at the heart of any attempt to explain clinically important increases and decreases in hair fibre size. The HF mesenchyme was widely believed to consist of very stable cell populations. New murine data however, indicate that follicular papilla cell numbers increase during anagen and that this increase is driven primarily by cell proliferation in the proximal dermal sheath.[27,62] Maximal follicular papilla cell numbers appear to be required only for the early part of anagen VI, and soon thereafter cells are 'lost' from the follicular papilla starting during full anagen and continuing through to complete hair follicle involution (catagen). The absence of intra-follicular papilla apoptosis suggests that emigration of cells from the follicular papilla back to the less apoptosis-resistant connective tissue sheath underlies this cell reduction. Such events may result from the modulation of follicular papilla cell migration to a less apoptosis-resistant connective tissue sheath throughout anagen VI to telogen, and *vice versa* during early anagen, and may be under the local control of regulatory molecules, *e.g.* androgens. In this way, the rapid changes in hair fibre volume seen during puberty and androgenetic alopecia may be facilitated within a single hair cycle[62–64] (see below).

However, hair shaft calibre also varies along its length (*i.e.* fine distal tip to thicker mid region to narrow proximal 'club') suggesting changing volumetric ratios of follicular papilla (and its secretary activity) to hair matrix keratinocyte number occur not only between the main stages of the hair cycle, but also during sub-stages of anagen.[65] The vellus-terminal hair transformation during puberty is associated however, with a significant increase in follicular papilla cells, and may be associated with the recruitment of additional cells from the lower dermal sheath.

1.6 Racial and Ethnic Characteristics of Hair Growth

Hair pattern shows striking variations as a function of age, endocrine (hormonal) status and genetic constitution. It appears that all humans are born with a defined number of HF and that this absolute number does not increase (*e.g. via* neo-folliculogenesis). While the adult brown-haired male is thought to have approximately 100,000 scalp HF,[57] blondes tend to have about 20% more and red-heads about 20% less. While many differences in hair growth pattern are likely to have a racial basis, data are very limited and then only available for a few body sites.

Before discussing this topic more fully it may be useful to deal first with the definition of 'excessive' hair growth. Although 'excessive' hair growth is of concern to some women, and indeed some men, it is often very difficult to describe this as 'abnormal' hair growth, as there exists strong compounding factors contributed by the ethnic or genetic background of the individual. It is often difficult to judge whether facial or body hair of a woman is within the range of normal variation or is excessive. Much has been written on excessive hair growth in women, either hirsutism or hypertrichosis. Despite this there is no rigid or standard criterion on what 'excessive' hair growth is. Hirsutism by definition occurs only in women and is caused by either an increased sensitivity of the HF to normal circulating levels of androgens or is due to increased androgen production by the endocrine glands. Hypertrichosis however, can occur in men as well as women and usually affects the entire body. Hairs are not usually increased in diameter nor are they restricted in women to a masculine pattern of hair growth. Hirsuitism is highly unusual in Asiatics (Chinese, Korean, Japanese *etc.*), despite the fact that levels of androgen may even be higher in Asiatic women.[58,66]

As mentioned above, scalp hair density is greatest among Caucasians with blonde hair and least among those with red hair. This is also reflected in a report by Danforth and Trotter, which concluded that body hair was greater in those Caucasian individuals with dark skin colour.[67] Again this was not linked directly to hair pigmentation, as blond Italian men were still 'hairier' than blond Scandinavians.[67] Thus, even amongst Caucasians, heavier hair growth was seen in individuals with Mediterranean ancestry *versus* Nordic ancestry.[58] There also appears to be some variations in HF density between humans of different ethnicities; Africans and Asians have less densely haired skin than Caucasians.

1.6.1 Body Site Variation in Hair Growth Patterns

The presence of 'excessive' hair growth at different body sites in women of different racial backgrounds has been defined by most authors as the presence of terminal hairs greater than 0.5 mm in length. Hirsutism studies tend to stress the importance of hair on the lip, chin, chest and upper back as sites, which differentiate those women who consider themselves to be hirsute from those women who do not. In one analysis[58] women of north-west European, Indian, Nordic and USA Caucasian stocks all reported the lower leg and lower arm as prominent sites of 'excessive' hair growth, while upper arm hair and breast/chest hair was reported more by the Nordic women than in the other groups. Excessive lip hair was only reported by north-west European women, while none of the Indian women reported chest hair growth as a concern. Given the ethnic backgrounds of these women, these figures are more likely to reflect actual hair growth differences rather than any increased visibility of the hair due to constitutive higher pigment levels.

In adult men the region of the body where hair growth shows the most conspicuous racial variation is the external ear or pinna. The dense growth of long coarse terminal hairs in this site is seen most commonly in Indian males after 20 years of age. Next most commonly affected include the Maltese and Israeli populations, though in these the top of the pinna is most affected. Nearly 85% of Caucasian males will have some terminal hair growth in this region especially from about 25 to 60 years of age. This type of hair growth appears to be androgen-related. By contrast, only 55% of Negroid males are affected with this type of hair growth.[68]

There is only limited racial distribution data available regarding chest hair growth. Genetic differences appear to regulate the potential to develop coarse terminal hairs in this site (*e.g.* greater concordance was seen with identical twins than in fraternal twins). Density of coarse chest hair is reported to be greater in whites (Caucasians) than in Mongoloids (*e.g.* Chinese, Japanese) or Blacks (Negroid). Differences between races reflected significant differences in hair number, hair length, hair breath, and hair weight. A small percentage of white women also grew coarse sternal/chest hairs and these hairs were usually few in number and very short. This may relate to ovarian activities, as these often disappeared in post-menopausal women. In addition to hirsutism associated with endocrine disorders, idiopathic hirsutism is another diagnostic group of women with no detectable endocrine disease. However, increased androgen production may be a feature also in these women. Axilary hair growth in Caucasians is much denser than on Japanese of both sexes.[69]

1.7 Aging of the Hair Follicle and Human Scalp

While in our foetal state our skins sport a covering of fine hair called lanugo hair and this is usually already shed in the womb or very early after birth. We are born

Table 1.1 *Density of hair follicles during human development from neonatal to early adult life*

Newborns	Average number of scalp HF $= 1135/cm^2$
1 year of age	Average number of scalp HF $= 795/cm^2$
20–30 years	Average number of scalp HF $= 615/cm^2$

Table 1.2 *Density of hair follicles during human aging from early to late adult life*

20–30 years	Average number of scalp HF $= 615/cm^2$
30–50 years	Average number of scalp HF $= 485/cm^2$
80–90 years	Average number of scalp HF $= 435/cm^2$

with our full complement of HFs. However, there is clearly a dramatic size difference between the skin coverage of the neonate and the fully-grown adult. Thus, the density of HF per unit area of our skin is greatly reduced during our journey to adult-hood.[66] An approximation for Caucasians appears above (Table 1.1).

There also appears to be a progressive loss of HF with age.[66] An approximation for Caucasians is presented in Table 1.2.

However, although a bald scalp may appear fully denuded, as many of 300 HF cm^{-2} can still be found – these HF produce only cosmetically useless vellus hair fibers (see below).

Scalp hair grows on average 0.3 mm per day and grows at its fastest between the ages of 15 and 30 in men, while hair growth in women is greatest between the ages of 16 and 24. Given that scalp hair typically grows continuously for up to eight years, about 10–20 hair growth cycles are possible during one's lifetime. Despite this, many animals, including humans, become increasingly telogenic with age, suggesting that telogen, with its hair shaft retained *in situ*, may even be the preferred HF status taking the mammals as a whole. This may appear reasonable, given that telogen would be associated with a reduced risk of malignant transformation in HF cells. Anagen-associated hair bulb matrix keratinocytes display one of the highest rates of proliferation in the body (higher than many malignant tumors!) and are exposed to melanogenesis-related reactive oxygen species generated only during active growth.

1.8 Common Disorders Affecting Human Hair Growth

1.8.1 Introduction

Many diseases affecting the HF can cause hair loss (alopecia) (Table 1.3). These can be roughly divided into: a) the relatively common *androgenetic alopecias* (including male pattern alopecia (MPA), also known as common baldness, and female androgenetic alopecias (FAA)): b) the non-scarring alopecias (including

Table 1.3 *Some hair disorders that can lead to alopecia (hair loss)*

Androgenetic	Diffuse alopecia	Alopecia areata	Cicatricial (Scarring) alopecia	Tricho-tillomania	Traction & Trauma	Infections & infestations	Hair-shaft Abnormality
Male pattern	Telogen Effluvium	Patchy AA	Pseudopelade, Follicular mucinosis	Hair pulling tic	Cosmetic	Tinea capitis (Trichophyton tonsurans)	Trichorrhexis nodosa
Female pattern	Anagen Effluvium	A. Totalis	Lichen planus/ Planopilaris		Accidents	Ring-worm fungi	Trichorrhexis invaginata
	Drugs & Chemicals	A. Universalis	Lupus		Tight ponytail	Kerion	Trichoclasis
	Thyroid Disorder	Ophiasis variant	Lichen sclerosus		Hair-Styling	Black piedra	Uncombable hair (pili trianguli)
	Iron Deficiency	Reticular variant	Lichen simplex		Dyeing	(Peri)folliculitis	Pili annulati
	Nutrition		Sarcoidosis		Pressure	2° Syphilis	Monilethrix
	Systemic		Scleroderma		Habit tics	Herpes zoster	Bubble hair
	Renal failure		Pemphigoid		Scratching	Boils/carbuncles	Pili torti & Woolly hair Trichonodosa etc.
	Hepatic failure etc.		Dermato-myositis etc.				

telogen effluvium and alopecia areata); c) the rare scarring alopecias (either lymphocytic, neutrophilic, mixed or non-specific); and d) rare hair shaft abnormalities (see Chapter 2 elsewhere in this volume).

Given the remit of this current book, I will restrict my focus to those conditions affecting a significant proportion of the population.

1.8.2 Male Pattern Alopecia

Male pattern alopecia (MPA) is by far the most common hair loss condition in humans, affecting about 60–70% of men by 70 years of age. The condition is easy to spot with several characteristic patterns of hair loss classified by Hammilton[70] and later modified by Norwood.[71] Where there is some doubt as to diagnosis, the condition should be differentiated from telogen effluvium and alopecia areata (see below). However, a single individual can be affected concurrently with more than one condition, *e.g.* balding man may get a patch of alopecia areata-associated hair loss.

The pathogenesis of MPA involves both genetic and hormonal components. Genetically, the inheritance is strong but complex and polygenic and can be inherited from either or both parents. However, not all men are affected even those who reach very old ages. In a recent study by Birch and Messenger[72] an increased frequency of balding was observed in the fathers of young bald men and there was a high relative risk of balding in young men if they had a balding father. Regarding the nature of the mode of inheritance, an analysis of the frequencies of balding/ non-balding in brothers of balding/non-balding elderly men (as categorised by their father's hair growth status) failed to show that either balding or non-balding is due to the action of a single gene.[72]

The hormonal component appears to implicate the involvement of dihydrotestosterone, the product of a 5α-reductase-catalysed conversion from testosterone. Interestingly the absence of 5α-reductase (Type II) protects the individual from developing MPA and levels of 5α-reductase (Type II) and dihydrotestosterone are higher in the balding *versus* non-balding scalp of men with MPA.[73]

Of particular clinical importance here is the terminal-to-vellus HF transition that is characteristic of MPA. The reader is referred to recent useful exploratory contributions to this topic.[61–64] These reports have already yielded the consensus that HF miniaturisation is most likely to occur *via* abrupt and marked reductions in follicular papilla and/or dermal sheath cell numbers. This contrasts markedly with the previously dominant view that hair shaft miniaturisation, in MPA at least, occurs *via* slow and gradual cycle-by-cycle change. There is convincing clinical support for the 'abrupt change' view, not least the rapid progression of MPA and the preponderance of fine hairs over intermediate hairs in balding scalp.[63] Perhaps more convincing still, is the reversal of the vellus-to-terminal transformations in finasteride or minoxidil-stimulated HF over a single hair cycle.[63]

Currently, treatment for MPA is limited, but improving. The US FDA has approved both minoxidil 5% and finasteride for the treatment of MPA[74] and both are able to slow hair loss and in some individuals to re-grow previously lost hair. However, sustained use of these products is necessary to prevent loss of any hair re-growth or even for maintenance of existing growth. Surgical treatments for MPA have also proved very popular and have become considerably more sophisticated during the last ten years.

1.8.3 Female Pattern Alopecia

Hair loss in women can also be androgenetic, but not all clinically similar cases will have an androgen abnormality. Where no hyperandrogenemia can be shown, a diagnosis of telogen effluvium or female pattern alopecia is more appropriate.[75,76] The pattern of type of hair loss in women is usually less clear than in men. However, the Ludwig classification system can be used. Overall, pattern hair loss in women is characterised by a more diffuse reduction in hair density than in men and this occurs principally over the crown and frontal scalp. There is a characteristic retention of the frontal hairline. As in MPA there is an increase in prevalence with advancing age. Recently the emerging view is that FPA may not be the female counterpart of MPA. This is due to that fact that the role of androgens is unclear in many cases. While loss of scalp hair occurs in female hyperandrogenism, many exhibiting FPA present with no clinical or biochemical evidence of androgen excess. Thus, the emerging view is that FPA is likely to be a multifactorial, though genetically determined, trait with both androgen-dependent and androgen-independent mechanisms involved.[77] Treatment options are limited, but 2% minoxidil can be prescribed. If indicated, anti-androgens can be given. In cases of marked patterned hair loss, transplantation surgery is also an option.

1.8.4 Alopecia Areata

Alopecia areata (AA) is a common cause of hair loss afflicting approx. 1.7% of the general human population.[78] It is manifested by patchy areas of hair loss on scalp and other body parts that can progress to complete loss of all body hair. AA results from selective, largely reversible, damage to anagen HF.[79] While not life threatening, the disease is nonetheless serious because it is disfiguring and in humans can cause severe psychological problems and loss of employment.

The etiology of AA is unknown, but an autoimmune pathogenesis is suspected.[80] AA appears to be a systemic disease because there is frequent involvement of organs other than the HF including the nails and eyes.[81,82] Thus, the defect may be extrinsic to HF. Nail changes occur in approximately one-third of patients. Moreover, the prognosis of AA appears to be worse when nail changes are present. Eye abnormalities are detected in up to 80% of patients with AA who otherwise have no ocular symptoms *versus* 30% in normal controls, though these changes are not severe enough to cause decreased visual acuity.

AA is associated with a number of immune abnormalities.[83] Some are non-selective while others are more specific and point to an immune abnormality selectively directed to a component of the HF when in anagen phase of the hair growth cycle. These include the presence of a peri- and intrafollicular mononuclear infiltrate, an increased expression of class I and II MHC antigens and of Langerhans cells in hair bulbs, deposits of immune reactants around HF, and the fact that effective therapies for AA have, as a common denominator, an immunosuppressive effect on immune cells in skin. Circulating antibodies to pigment and endothelial cells have be reported and more recently abnormal antibody responses specifically directed to HF antigens have been demonstrated in AA patients, in two new rodent models of the disease, and in dogs and horses.[84]

It has been proposed that AA may result from a breakdown in HF immune privilege during the early stages of anagen development[85] (see above) and there is loss of MHC class I/II negativity in AA anagen HF. Many autoimmune diseases have been shown to be associated with a particular MHC haplotype and certain haplotypes appear to predispose an individual to autoimmunity. Similarly, several studies suggest that AA is also associated with particular MHC class II haplotypes, which can separate AA of different types.[86]

The only AA-associated conditions for which there is sufficient firm data are atopy and Down's syndrome.[87] The incidence of AA appears to be increased in patients with Down's syndrome, a condition associated with functional deficiencies in T cell-mediated immune response and decreases in serum IgG levels. On average, 5% of patients with Down's syndrome have AA compared to approximately 0.1% of concurrent control mentally retarded patients. There is also increasing evidence of an increase in the prevalence of diabetes mellitus, especially type I insulin dependent diabetes, in relatives of patients with AA but not in the patients themselves. These findings suggest a genetic association between the two diseases whereby the expression of AA protects against the development of diabetes mellitus. It needs to be emphasised that the majority of patients with AA are in good health and have no other associated diseases.

Various organ-specific autoantibodies are said to occur with increased frequency in AA.[87] These associations remain for the most part controversial with the exception of thyroid autoantibodies. Of particular interest to this author is the observation of antibodies specifically directed to the HF in many mammalian species with AA.[84,88–92] These observations indicate that HFs express unique antigens that can stimulate autoimmune responses. Further studies have shown that AA antibodies target HF-specific keratins. The biological relevance of the HF antibodies in AA remains to be determined. However, their detection in two rodent models for AA and more recently in dogs and horses with AA, suggest that AA-like hair loss is widespread in mammals and that similar, highly conserved, antigens appear to be important disease targets. Still, a basic question is whether these antibodies are a cause or result of AA. Current evidence suggests that the presence of antibodies to HF may appear before the onset of clinically identifiable hair loss and so may not necessarily be produced secondarily as a response to HF damage associated with AA.[84,88–92] Finally, purified IgG from an AA-affected horses may adversely affect hair regrowth when passively transferred to normal mice.[91] This study, which needs to be confirmed, should be interpreted in light of an earlier study that reported the failure of passive transfer of serum from human patients with AA to inhibit hair growth in transplants of human scalp skin grafted onto nude mice.[92]

More recent research strongly supports the involvement of T-lymphocytes as the effector agents involved in the induction of AA, and maybe also its maintenance.

AA can be transferred to human scalp explants grafted onto SCID mice by the injection of T cells (isolated from lesional AA skin) into the grafts.[94] In addition, AA-like hair loss can be induced experimentally in the C3H/HeJ AA murine model using full-thickness skin grafts.[95] Here the suggestion is that the immune system itself is the regulator of AA rather than any intrinsic defect of the HF itself.[96]

1.8.5 Telogen Effluvium

Telogen effluvium is characterised by a perturbation of the hair growth cycle resulting in an abnormally high shedding rate of hairs in the telogen phase. First described by Kligman in 1961, it was considered that the HF responds in a similar way, *i.e.* by prematurely terminating anagen irrespective of the cause. The reason for such hair loss requires full examination of the patient's history including laboratory investigations, *e.g.* endocrine, nutritional or autoimmune status and perhaps also histological examination.[97] Several causes have been proposed to trigger this cyclical short-circuiting. These include parturition, illnesses with associated fever, surgery, psychological stress, crash diets, anticoagulant drugs etc.[98]

1.9 Conclusion

The HF has truly come of age in this biomedical sciences century. Indeed it could well be put: 'Ask not what bioscience can do for the hair follicle; rather what the hair follicle can do for bioscience'. The HF is truly a perfect scientific (both biological *via* the follicle and physical *via* its fibre) universe in which is represented all of life's major fundamental processes (microbial and eurkaryotic) required for the organisation of complex interactive tissues. The subsequent chapters in this volume will, I hope, provide ample evidence to justify these lofty statements.

1.10 Acknowledgements

The author would like to thank Dr Markus Magerl and Dr Eva M.J. Peters for assistance with the schematic drawings.

1.11 References

1. R. Paus and G. Cotsarelis, *N. Engl. J. Med.*, 1999, **341(7)**, 491.
2. I.M. Hadshiew, K. Foitzik, P.C. Arck and R. Paus, *J. Invest. Dermatol.*, 2004, **123(3)**, 455.
3. E. Morgan, in *The Ascent of Woman*, Souvenir Press, London, 1985.
4. A. Bertazzo, C. Costa, M. Biasiolo, G. Allegri, G. Cirrincione and G. Presti, *Biol. Trace. Elem. Res.*, 1996, **52**, 37.
5. M.H. Hardy, *Trends. Genet.*, 1992, **8(2)**, 55.
6. M. Philpott and R. Paus, Principles of hair follicle morphogenesis in *Molecular Basis of Epithelial Appendage Morphogenesis* Choung C.M. (ed), R.G. Landes Co. Austin, Texas, 1998, 75–110.

7. C.-M. Choung in *Molecular Basis of Epithelial Appendage Morphogenesis*, C.-M. Chuong (ed), R.G. Landes Co., Austin, Texas, 1998, 3.

8. C.R. Daniel III, B.M. Piraccini and A. Tosti, *J. Am. Acad. Dermatol*, 2004, **50(2)**, 258.

9. V.A. Botchkarev and R. Paus, *J. Exp. Zoolog. B. Mol. Dev. Evol.*, 2003, **298(1)**, 164.

10. S.E. Millar, *J. Invest. Dermatol.*, 2002, **118(2)**, 216.

11. E. Fuchs, B.J. Merrill, C. Jamora and R. DasGupta, *Dev. Cell.*, 2001, **1(1)**, 13.

12. A.J. Reynolds, C. Lawrence, P.B. Cserhalmi-Friedman, A.M. Christiano and C.A. Jahoda, *Nature*, 1999, **402(6757)**, 33.

13. S. Noramly, A. Freeman, B.A. Morgan, *Development*, 1999, **126(16)**, 3509.

14. R. Paus, S. Muller-Rover, C. Van Der Veen, M. Maurer, S. Eichmuller, G. Ling, U. Hofmann, K. Foitzik, L. Mecklenburg and B. Handjiski, *J. Invest. Dermatol.*, 1999, **113(4)**, 523.

15. M. Magerl, D.J. Tobin, S. Muller-Rover, E. Hagen, G. Lindner, I.A. McKay and R. Paus, *J. Invest. Dermatol.*, 2001, **116(6)**, 947.

16. K.S. Stenn and R. Paus, *Physiol. Rev.*, 2001, **81(1)**, 449.

17. R.F. Oliver and C.A. Jahoda, *Clin. Dermatol.*, 1988, **6(4)**, 74.

18. N. Orentreich, *N.Y. State. J. Med.*, 1972, **72(5)**, 578.

19. G. Lindner, V.A. Botchkarev, N.V. Botchkareva, G. Ling, C. van der Veen and R. Paus, *Am. J. Pathol.*, 1997, **151(6)**, 1601.

20. P.C. Arck, B. Handjiski, E.M. Peters, A.S. Peter, E. Hagen, A. Fischer, B.F. Klapp and R. Paus, *Am. J. Pathol.*, 2003, **162(3)**, 803.

21. D.J. Tobin, E. Hagen, V.A. Botchkarev and R. Paus, *J. Invest. Dermatol.*, 1998, **111(6)**, 941.

22. D.J. Tobin, A. Slominski, V. Botchkarev and R. Paus, *J. Investig. Dermatol. Symp. Proc.*, 1999, **4(3)**, 323.

23. A.A. Panteleyev, R. Paus and A.M. Christiano, *Am. J. Pathol.*, 2000, **157(4)**, 1071.

24. S. Muller-Rover, B. Handjiski, C. van der Veen, S. Eichmuller, K. Foitzik, I.A. McKay, K.S. Stenn and R. Paus, *J. Invest. Dermatol.*, 2001, **117(1)**, 3.

25. G. Cotsarelis, T.-T. Sun and R.M. Lavker, *Cell*, 1990, **61**, 1329.

26. E.K. Nishimura, S.A. Jordan, H. Oshima, H. Yoshida, M. Osawa, M. Moriyama, I.J. Jackson, Y. Barrandon, Y. Miyachi and S. Nishikawa, *Nature*, 2002, **416(6883)**, 854.

27. D.J. Tobin, A. Gunin, M. Magerl and R. Paus, *J. Investig. Dermatol. Symp. Proc.*, 2003, **8(1)**, 80.

28. Y. Milner, J. Sudnik, M. Filippi, M. Kizoulis, M. Kashgarian and K. Stenn, *J. Invest. Dermatol.*, 2002, **119(3)**, 639.

29. D.J. Tobin, K. Foitzik, T. Reinheckel, L. Mecklenburg, V.A. Botchkarev, C. Peters and R. Paus, *Am. J. Pathol.*, 2002, **160(5)**, 1807.

30. G.E. Westgate, R.I. Craggs and W.T. Gibson, *J. Invest. Dermatol.*, 1991, **97**, 417.

31. R. Paus, in *Skin Immune System* J.D. Bos (ed), CRC Press, Boca Raton, 1997.

32. L.C. Wang, Z.Y. Liu, L. Gambardella, A. Delacour, R. Shapiro, J. Yang, I. Sizing, P. Rayhorn, E.A. Garber, C.D. Benjamin, K.P. Williams, F.R. Taylor, Y. Barrandon, L. Ling and L.C. Burkly, *J. Invest. Dermatol.*, 2000, **114(5)**, 901.

33. C.A. Jahoda and A.J. Reynolds, *Dermatol. Clin.*, 1996, **4(4)**, 573.

34. R.F. Oliver, *J Embryol. Exp. Morphol.*, 1967, **18**, 43.

35. G.E. Westgate, A.G. Messenger, L.P. Watson and W.T. Gibson, *J. Invest. Dermatol.*, 1991, **96**, 191.

36. T.T. Sun, G. Cotsarelis and R.M. Lavker, *J. Invest. Dermatol.*, 1991, **96(5)**, 77S.

37. T. Soma, Y. Tsuji and T. Hibino, *J. Invest. Dermatol.*, 2002, **118(6)**, 993.

38. S. Suzuki, T. Kato, H. Takimoto, S. Masui, H. Oshima, K. Ozawa, S. Suzuki and T. Imamura, *J. Invest. Dermatol.*, 1998, **111**, 963.

39. C.A. Jahoda, J. Whitehouse, A.J. Reynolds and N. Hole, *Exp. Dermatol.*, 2003, **12(6)**, 849.

40. A. Slominski and J. Wortsman, *Endocr. Rev.*, 2000, **21(5)**, 457.

41. R. Paus, E.M. Peters, S. Eichmuller, V.A. Botchkarev, *J. Invest. Dermatol. Symp. Proc.*, 1997, **2(1)**, 61.

42. R. Paus, K.S. Stenn and R.E. Link, *Br. J. Dermatol.*, 1990, **122**, 777.

43. G.P. Moore, D.L. Du Cross, K. Issacs, P. Pisansarakit and P.C. Wynn, *Ann. NY. Acad. Sci.*, 1991, **642**, 308.

44. K.E. Fichtellius, O. Groth and S. Lidén, *Int. Arch. Allergy Appl. Immunol.*, 1970, **37**, 607.

45. R. Paus, T. Christoph and S. Muller-Rover, *J. Investig. Dermatol. Symp. Proc.*, 1999, **4(3)**, 226.

46. D.J. Tobin, D.A. Fenton and M.D. Kendall, *Amer. J. Dermatopathol.*, 1991, **13**, 248.

47. R. Paus, B.J. Nickoloff and T. Ito, *Trends Immunol.*, 2005, **26(1)**, 32.

48. T. Christoph, S. Muller-Rover, H. Audring, D.J. Tobin, B. Hermes, G. Cotsarelis, R. Ruckert and R. Paus, *Br. J. Dermatol.*, 2000, **142(5)**, 862.

49. A.S. Breathnach, *Histochem. J.*, 1963, **97**, 73.

50. R.E. Tigelaar and J.M. Lewis, *J. Invest. Dermatol.*, 1995, **105**, 43S.

51. M. Maurer, E. Fischer, B. Handjiski, E. von Stebut, B. Algermissen, A. Bavandi and R. Paus, *Lab. Invest.*, 1997, **77(4)**, 319.

52. S. Kauser, A.J. Thody, K.U. Schallreuter, C.L. Gummer and D.J. Tobin, *Endocrinology*, 2005, **146**(2), 532.

53. S. Kauser, A.J. Thody, K.U. Schallreuter, C.L. Gummer, D.J. Tobin, *J. Invest. Dermatol.*, 2004, **123(1)**, 184.

54. F.J.G. Ebling, in *Hair and Hair Diseases*, C.E. Orfanos, R. Happle (ed), Springer-Verlag, Berlin, 1990, 267.

55. J.B. Hamilton, *Am. J. Anat.*, 1942, **71**, 451.

56. V.A. Randall, N.A. Hibberts, M.J. Thornton, K. Hamada, A.E. Merrick, S. Kato, T.J. Jenner, I. De Oliveira and A.G. Messenger, *Horm. Res.*, 2000, **54**, 243.

57. G. Szabo, in *The Biology of Hair Growth*, W. Montagna and R.A. Ellis (ed), Academic, New York, 1958, 33.

58. R.B. Greenblatt, in *Hirsuitism and Virilism*, V.B. Mahesh and R.B. Greenblatt (ed), John Wright PSG, Boston. 1983, 1.

59. W. Nagl, *Br. J. Dermatol.*, 1995. **132**, 94.
60. E.J. Van Scott and T.M. Ekel, *J. Invest. Dermatol.*, 1958, **31**, 281.
61. K. Elliott, T.J. Stephenson and A.G. Messenger, *J. Invest. Dermatol.*, 1999, **113**, 873.
62. D.J. Tobin, A. Gunin, M. Magerl, B. Handijski and R. Paus, *J. Invest. Dermatol.*, 2003, **120,** 895.
63. D.A. Whiting, *J. Am. Acad. Dermatol.*, 2001, **45**, S81.
64. C.A. Jahoda, *Exp. Dermatol.*, 1998, **7(5)**, 235.
65. L. Ibrahim and E.A. Wright, *J. Embryol. Exp. Morphol.*, 1982, **72**, 209.
66. L. Giacometti, in *Aging*, Pergamon, London, 1965, 97.
67. C.H. Danforth and M. Trotter, *Amer. J. Phys. Anthropol.*, 1922, **5**, 259.
68. J.B. Hamilton, in *Advances in the Biology of Skin*, W. Montagna and R.L. Dobson (ed), Pergamon Press, New York, 1967, 129.
69. J.B. Hamilton and H. Terada, *J. Clin. Endocrinol.*, 1963, **7**, 465.
70. J.B. Hamilton, *Ann. NY. Acad. Sci.*, 1951, **68**, 1359.
71. O. Norwood, *South Med. J.*, 1993, **68,** 1359.
72. M.P. Birch and A.G. Messenger, *Eur. J. Dermatol.*, 2001, **11**, 309.
73. M.E. Sawaya and V.H. Price, *J. Invest. Dermatol.*, 1997, **109**, 296.
74. J. Shapiro, M. Wiseman and H. Lui, *Can. Fam. Physician.*, 2000, **46**, 1469.
75. E.A. Olsen, M. Hordinsky, J.L. Roberts and D.A. Whiting, *J. Am. Acad. Dermatol.*, 2002, **47(5)**, 795.
76. V.H. Price, J.L. Roberts, M. Hordinsky, E.A. Olsen, R. Savin, W. Bergfeld, V. Fiedler, A. Lucky, D.A. Whiting, F. Pappas, J. Culbertson, P. Kotey, A. Meehan and J. Waldstreicher, *J. Am. Acad. Dermatol.*, 2000, **43(5 Pt 1)**, 768.
77. M.P. Birch, S.C. Lalla and A.G. Messenger, *Clin. Exp. Dermatol.*, 2002, **27(5)**, 383.
78. K.H. Safavi, S.A. Muller, V.J. Suman and A.N. Moshell, L.J. Melton III, *Mayo Clin. Proc.*, 1995, **70**, 628.
79. D.J. Tobin, *Microsc. Res. Tech.*, 1997, **38**, 443.
80. K.J. McElwee, D.J. Tobin, J.C. Bystryn, L.E. King Jr. and J.P. Sundberg, *Exp. Dermatol.*, 1999, **8(5)**, 371.
81. A. Tosti, S. Colombati, G.M. Caponeri, C. Ciliberti, G. Tosti, M. Bosi and S. Veronesi, *Dermatologica*, 1985, **170(2)**, 69.
82. A. Tosti, R. Morelli, F. Bardazzi and A.M. Peluso, *Pediatr. Dermatol.*, 1994, **11(2)**, 112.
83. D.J. Tobin and J.-C. Bystryn, in *Hair and its disorders: Biology, Pathology and Management*, F.M. Camacho, V.A. Randall and V.H. Price (eds), Martin Dunitz, London, 2000, 187.
84. K.J. McElwee, D. Boggess, T. Olivry, R.F. Oliver, D. Whiting, D.J. Tobin, J.C. Bystryn, L.E. King Jr. and J.P. Sundberg, *Pathobiology*, 1998, **66(2)**, 90.
85. T. Ito, N. Ito, A. Bettermann, Y. Tokura, M. Takigawa and R. Paus, *Am. J. Pathol.*, 2004, **164(2)**, 623.
86. B.W. Colombe, V.H. Price, E.L. Khoury, M.R. Garovoy and C.D. Lou, *J. Am. Acad. Dermatol.*, 1995, **33**, 757.
87. A.J. McDonagh and A.G. Messenger, *J. Dermatol. Sci.*, 1994, **7**, S125.

88. D.J. Tobin, D.A. Fenton, N. Orentreich and J.-C. Bystryn, *J. Invest. Dermatol.*, 1994, **102**, 721.
89. D.J. Tobin, J.P. Sundberg, L.E. King, Jr., D. Boggess and J.-C. Bystryn, *J. Invest. Dermatol.*, 1997, **109**, 329.
90. D.J. Tobin, T. Olivry and J.-C. Bystryn, *Adv. Vet. Dermatol.*, 1998, **3**, 355.
91. D.J. Tobin, Z. Alhaidari and T. Olivry, *Exp. Dermatol.*, 1987, **5**, 289.
92. D.J. Tobin, S.K. Hann, M.S. Song and J.-C. Bystryn, *Arch. Dermatol.*, 1997, **133**, 57.
93. A. Gilhar, T. Pillar, B. Assy and M. David, *Br. J. Dermatol.*, 1992, **126**, 166.
94. A. Gilhar, Y. Ullmann, T. Berkutzki, B. Assy and R.S. Khalish, *J. Clin. Invest.*, 1998, **101**, 62.
95. K.J. McElwee, D. Boggess, L.E. King Jr. and J.P. Sundberg, *J. Invest. Dermatol.*, 1998, **111**, 797.
96. M. Zoller, K.J. McElwee, M. Vitacolonna and R. Hoffmann, *Exp. Dermatol.*, 2004, **13**, 435.
97. S. Harrison and R. Sinclair, *Clin. Exp. Dermatol.*, 2002, **27(5)**, 389.
98. W.D. Steck, *Cutis.*, 1978, **21(4)**, 543.

CHAPTER 2

The Human Hair Fibre

DESMOND J. TOBIN

2.1 Psychological Issues Surrounding Hair

As social and highly visual beings we humans place great value on our visual appearance, in large part derived from our skin and hair. Although we are relatively 'naked' by comparison with other primates, our highly sophisticated communication skills depend disproportionately on focused attention to our head and face – both body sites which can exhibit dramatic and luxuriant hair growth.[1] Thus, both the form and colour of our hair play significant roles as signals for social and sexual communication, *e.g.* as indicators of age, sex, status, values *etc.* Despite the above, hair has little real biological significance for humans, apart from some minor cushioning of the cranium, protection from direct sunlight, and as protective filters for eyes, nose and ears.[1]

In our modern, highly mediated world, hair is increasingly becoming a central part of our self-identity/body image, and due to the hair fibre's structure, it is one phenotypic trait that we can modify relatively easily. Indeed, while our pre-occupation with hair, and particularly its colour, is truly ancient, this is further intensified today as increasing longevity fuels an insatiable desire to delay loss of youthfulness. Unremitting exposure to advertising of hair care/colour products has successfully seduced most of us in the economically developed and developing world to participate in a multi-billion euro global hair colour market. Indeed, current estimates indicate that up to 60% of the adult western population (both men and women) use hair colouring products and that 80% of British 'blondes' are well . . . not!

Given the potency of the hair 'signal', it is perhaps unsurprising that significant psychological trauma is associated with hair 'problems' from too much to too little or being of the 'wrong' type and colour *etc.* in between. Common examples of hair loss include androgenetic alopecia (male pattern baldness) and telogen effluvium, the latter associated with stressors including childbirth, fever, dieting, surgery, medications *etc.* Alopecia areata is the most common pathologic form of hair loss. Furthermore, iatrogenic chemotherapy-induced alopecia is of major concern as it is the often the most difficult consequence of cancer treatment for the patient to cope with. It is this dramatic visual signal of ill-health that compounds feelings of loss of control over one's own life.[2] There is also a general gender difference in psychological/psychiatric problems associated with hair loss. Woman who experience diffuse hair loss tend to experience even great distress than an equivalent degree of hair loss in men.[3–5] For a small fraction of individuals, even mild hair loss can trigger a somatoform disorder called dysmorphphobia.[4] Some psychiatric conditions may themselves manifest in hair-pulling tics, *e.g.* trichotillomania.

Increased hair growth can also be a source of considerable distress, whether due to *hirsutism* – androgen-mediated, male-pattern hair growth in women–or to *hypertrichosis*, a non-androgen-dependent form of hair growth deemed excessive by the individual.

2.2 Anatomy of the Human Scalp Hair Fibre

The hair shaft is the principle product of the hair follicle and is of practical interest to an increasing band of scientists and technologists, reflected by the range of contributions to this volume. It is the unique growth characteristics of hair that permits such enormous scope for examination, from study of interactions with topical modifications (cosmetic scientists) to assessment of individuals' medical status (dermatologists, trichologists) to yielding medico-legal (forensic scientists) and historical (bio-archeologists) information. As in most aspects of human hair follicle biological research, we owe much to investigators of hair growth in other mammal systems and, for the hair fibre, the wool literature has been most informative.

The terminal hair shaft, as exemplified by human scalp hair, consists of an outer cuticle ensheathing an inner cortex. All hairs, irrespective of their calibre (*i.e.* diameter), length, colour or stage of growth within the hair follicle are constructed of a bulk cortex surrounded by a protective covering of flattened cuticle cells. However, a third component, the medulla, is commonly located in the centre, especially or large terminal hair fibres (Figure 2.1). The relative size of each of these three components differs throughout the mammalian class. This variation is, in large part, dependent on their associated requirements for thermoregulation. Hair fibres can contain significant water content – up to 30% by weight, with the remainder consisting of proteins.

Before examining each of these layers in detail, it is perhaps appropriate to review some general terminology applied in descriptions of hair fibre morphology. The cylindrical hair fibre exhibits an elliptical transverse sectional shape, the extent of which can vary considerably between different racial groups (see below), so much so that hair cross-section in specific groups may appear round, oval or flattened. As a result, hair fibres have a major and minor axis. However, for ease of comparison, an average 'equivalent circular diameter' (ECD) may be quoted. The average Caucasian ECD value for scalp hairs is approximately 70 μm (though can be up to 180 μm).[6] Specific structural traits can also generate hairs that vary between straight, wavy or crimped.

Fine hairs commonly average less than 30 μm and tend to be termed 'vellus' if less than 1 mm in length, though vellus hairs can be as fine as 4 μm across. It is of

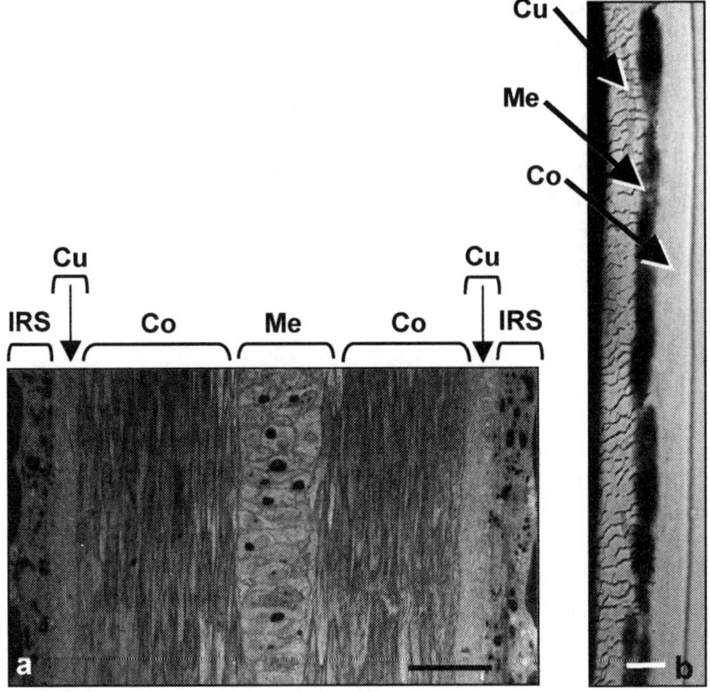

Figure 2.1   (a) *Light micrograph of a transverse section of the forming hair fibre within the anagen scalp hair follicle. IRS: inner root sheath of hair follicle; Cu: cuticle; Co: Cortex; Md: Medulla. (b) Surface view of a scalp hair fibre whole mount showing cuticle (Cu), cortex (Co) and medulla (Me) components. Scale bars a = 40 μm; b = 80 μm*

note that normal human scalp contains hairs throughout this calibre range. Furthermore, different types of hair fibre can be produced during the life cycle of the same hair follicle as observed in the vellus-to-terminal transformation of hairs on the face of the post-pubertal male and terminal-to-vellus miniaturisation of hair on the balding scalp (see below).

2.2.1 Hair Fibre Surface – The Cuticle

The surface of the hair fibre consists of the hair cuticle, which forms an overlapping, cortex-encircling sheath of flattened cells (often called scales) arranged in a longitudinal-circumferential manner with free ends of the scales oriented in a root-to-tip direction (Figure 2.2). It should be remembered that the surface of the cuticle is likely to reflect the significant interactions experienced during its 'moulding' within the hair follicle's *second* cuticular layer provided by one of the three component layers of the inner root sheath (IRS) of the hair follicle. Indeed, the hair fibre remains tightly anchored in the skin *via* reciprocal *hair cuticle-on-IRS cuticle* interactions deep within the growing hair follicle. Fine striations on the surface of

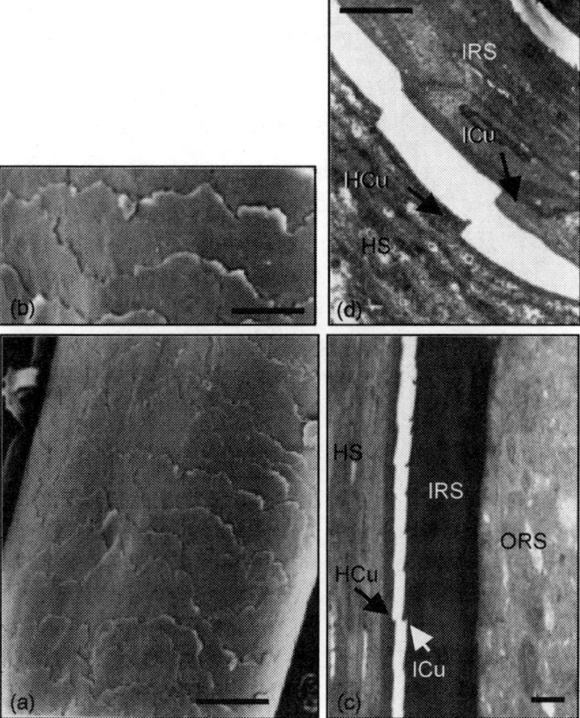

Figure 2.2 (a) *Scanning electron micrograph view of a scalp hair shaft. Note the root-to-tip orientation of the cuticle cells. (b) Higher power view of (a). (c) Imbricated nature of the hair shaft (HS) cuticular layers (HCu) and of the inner root sheath (IRS) cuticular layers (ICu). Scale bars a = 30 μm; b = 15 μm; c = 80 μm; d = 5 μm*

the cuticle cell, present even after removal of bound fatty acids (see below), suggest that these reflect contacts between the hair cuticle as it moves through the upwardly-moving IRS cuticle. Thus, hair fibres of each mammal species exhibit their own peculiarities, which facilitate easy identification.

Each cuticle cell is a multi-laminated, translucent and pigment-free structure measuring approximately 0.3–0.5 μm thick and 60 μm long.[7] The cells are slightly overlapping and create a 4° angle with the long axis of the hair fibre. Cuticle cells are at their most intact as the hair fibre emerges from the follicle at the skin surface, and here there may be as many a ten cuticle cells ensheathing the hair cortex. However, as one moves distally along the hair fibre, natural wear-and-tear forces, *e.g.* grooming, reduce this number. Moreover, cuticle quality is further modified by chipping of small fragments at the scale edges. The multi-laminated sub-structure of a cuticle cell can be best appreciated by transmission electron microscopy (TEM) (Figure 2.3). Using specific stains, each cuticle cell is observed to contain an external surface or epicuticle, followed by the A-layer and exocuticle, followed by the endocuticle, inner layer and cell membrane complex.[6]

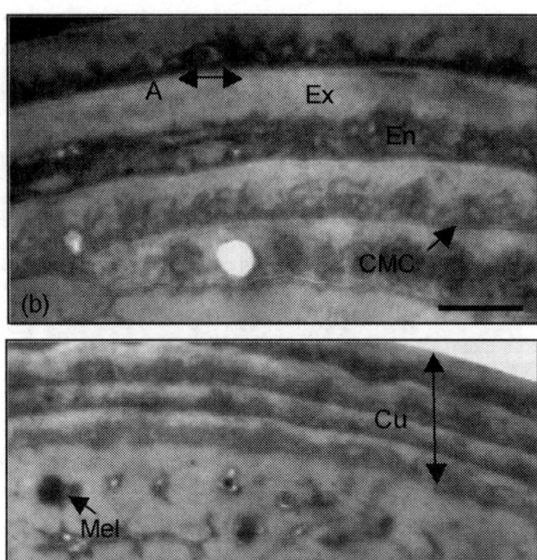

Figure 2.3 *Transmission electron micrograph of transverse section of a human scalp hair fibre stained with uranyl acetate and lead citrate. (a) Low power view of hair fibre cuticle (Cu) and hair cortex (Co). (b) Higher power view of the hair cuticle structure where the A-layer (A), exocuticle (Ex), endocuticle (En), and cell membrane complex (CMC) can be seen. Scale bars: a = 500 nm; b = 250 nm*

The *epicuticle* or 'fibre cuticle surface membrane' is approximately 13 nm thick, is visible with specialised staining and is derived from the original plasma membrane of the cuticle cell before hardening. Immediately subjacent to this layer is the *exocuticle* consisting of a narrow dense *'A layer'* and a broader less densely-stainable layer below. The A layer, which measures approximately 100 nm across, is a mostly proteinaceous layer containing ultra-high cysteine-rich sulfur proteins. Most of its component cysteines are involved in disulfide bond formation and so contribute a barrier function, protecting the fibre from attack. The remainder of the exocuticle is less intensely stained, reflecting its composition of sulfur-rich proteinaceous material, and measures approximately 300 nm across. This layer also contributes to the overall protective function of the cuticle.

In contrast to the exocuticle, the *endocuticle* stains weakly and heterogeneously. Of variable thickness (20–300 nm), the endocuticle layer contains less sulfur-rich proteins than the other cuticle layers and so is the weakest/softest part of the cuticle. Its heterogeneous staining pattern is due to its composition of remnants of cytoplasmic organelle material. Importantly, this region of the cuticle is accessible to dyes (see below), is most 'responsive' for external modification and can swell in water, unlike other, more rigorous, highly cross-linked parts of the cuticle.[8] Immediately internal to the endocuticle is a narrow layer (~20 nm), similar in staining pattern to the exocuticle, called the *inner-layer* or I-layer.

Individual cuticle cells connect to one another at intercellular membrane complexes (approx. 25 nm across), which contain a tri-laminar sub-structure of a central dense δ-band (approx. 18 nm), sandwiched between two lightly or non-staining narrower β-bands[9] (Figure 2.3).

The latter β-bands (approx. 5 nm thick) represent the former plasma membranes of the cuticle cells. The precise composition of these bands is unclear, though it appears that they contain high amounts of glue-like carbohydrates and glycoproteins with cohesive properties. Under ideal staining conditions, the dense δ-band itself consists of further laminar components.

2.2.2 Hair Fibre – The Cortex

Without the cuticle, the hair fibre cortex would fray and fall apart. The cortex, which forms the bulk of the fibre mass, is composed of elongated (50–100 μm long) spindle-shaped (3–6 μm across) cortical keratinocytes with fine finger-like projections. These highly sulfur-containing cells are arranged symmetrically (unlike in wool) and contain both fibrillar and non-fibrillar material arranged in a mostly orthocortical pattern.[6,10,11] Those cortex cells next to the cuticle exhibit a more flattened morphology and contain lower levels of sulfur proteins than elsewhere in the cortex. Each cortical cell usually contains recognisable features of its pre-hardened life, *e.g.* centrally-located nuclear remnants and pigment/ melanin granules. In the absence of pigment granules, *e.g.* in senile white hair, the entire cortex appears translucent with a yellow-ish tinge – the intrinsic color of keratin. In transverse section, the packing of individual cortical cells can be seen to depend on the presence of intercellular membrane complexes, which appear tortuous due to incorporating the fluted intersections of the cells' finger-like projections. Non-keratinaceous membrane complexes consist not only of the two cell membranes, but also contain a glue-like material that confers adhesion to the cortical cells.

The repeating structural unit of the hair cortex is the cysteine-rich macrofibril unit (40–200 μm across) that extends the full length of the cell. Hair cortex, where individual macrofibril units are positioned very close to one another (Figure 2.4), has been termed paracortex, while macrofibril units interspersed with non-keratin inter-macrofibrillar matrix has been termed orthocortex.[6]

Under high-power transmission electron microscopy, individual keratinaceous macrofibrils can be seen to consist of a composite of spirally-arranged α-keratin intermediate filaments (α-KIF) (7.5 nm across and forming a spiral 40 nm across) termed *microfibrils* and matrix, consisting of KIF-associated protein (KAP), in a ratio generally less than 1.0.[11]

2.2.3 Hair Fibre – The Medulla

Very few hairs at birth contain a medulla (see below under *Aging and Hair Fibre Form*). Even in the adult, only hair fibres of a particular diameter will contain this

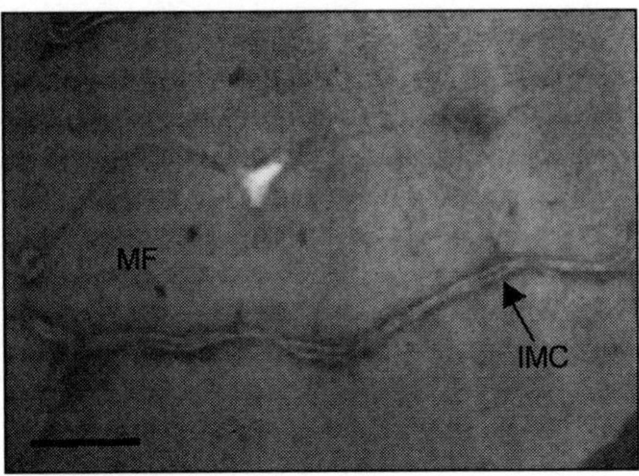

Figure 2.4 *Transmission electron micrograph of a transverse section of human scalp hair stained with uranyl acetate and lead citrate. Note the close juxtaposition of the macrofibril (MF) units of keratin microfibrils and the intercellular membrane complexes (IMC). Scale bar = 200* nm

additional component. All scalp hairs, excepting the most fine or miniaturised, contain a medulla (Figure 2.1) – though unlike in mammals with very coarse hairs (*e.g.* horse) this contributes little to the overall mass of the hair fibre. The human medulla may be continuous (especially in high calibre fibres), discontinuous or fragmented in finer hairs.

The medullary cell is a relatively poorly described, spherical and mostly hollow cell within the hairs haft core, which becomes highly vacuolated during the process of dehydration during formation. However, its presence and degree of vacuolation can influence the physical properties of the fibre, particularly with regard to reflection of incident light, and so may contribute to sheen and subtle colour tone. For animals, the air-trapping characteristic of the medulla importantly provides a degree of insulation to the pelage. During medulla cell formation, distinctive large electron-dense granules can be seen by TEM. The granules contain trichohyalin, a non-filamentous material that lacks cysteine but contains citrulline residues with isopeptide cross-links of ε-(γ-glutamyl)-lysine, accounting for the low solubility of this fibre component.

2.3 Biosynthesis of the Hair Fibre

The hair fibre is constructed as a highly integrated system of several components including in order of decreasing amount: keratins, water, lipids, pigment and trace elements. The biosynthesis of hair proteins begins in the bulb of the anagen hair follicle and ceases in the keratogeneous zone, approximately 500 μm above the level of the zone of maximal keratinocyte proliferation (so-called sub-Auber's matrix) in scalp terminal hair follicles. Low-sulfur and high-sulfur proteins are synthesised simultaneously, though the synthesis of the latter peaks later. Despite the significant variability in hair form between humans of different ethnicities, *e.g.* with regard to environment, diet, and hair texture, the chemical composition of hair protein across the ethnic groups is remarkably uniform.[11] Thus, there are no significant differences in amino acid composition of hair of different ethnicities (see below).

2.3.1 Hair Keratins

The highly complex field of hair or 'hard' α-keratins (*i.e.* hair, nail, claw *etc.*) has benefited much from recent advances in molecular cloning (*e.g.* the screening of human scalp cDNA libraries) yielding unexpected new information about keratin genes and their expression as a function of hair growth and cycling. Hair keratins are the products of several gene families, each with closely related family members and with structural similarities shared across different families. Recently, there have been attempts to simplify the rather confusing nomenclatures for classifying keratins within the KIF gene system and to apply a similar system to the one used in naming keratin-associated proteins (KAPs). The reader is directed to excellent and comprehensive treatments of this topic.[12–14]

Hair keratins have been grouped into two major classes: a) low sulfur, high molecular weight α-helical keratins (45–58 kDa) and b) high sulfur, low molecular weight (10–20 kDa) random-coil keratins. The former, the fibrillar fraction, is termed 'α-keratin intermediate filaments' and is embedded in matrix protein of the latter. Thus, the hair fibre exhibits very high sulfur content (5% in humans). KIF can be further sub-grouped into Type I (acidic) and Type II (basic to neutral) keratins, with nine hair keratin Type I genes and six Type II members currently recognised. Both gene families exhibit great similarities of physical organisation. Type I hair keratin genes can be further sub-classified by gene organisation and sequence homology into *subgroups A, B, C*, where A and B are highly related and C keratins are less related. The six Type II genes are also organised into subgroups that are counterparts of subgroups A and C in Type I.

Broadly speaking, highly related Type I and II group A genes are all expressed in the cortex of terminal scalp hair, while unrelated Type I and II genes appear to be differentially expressed in hair matrix, cuticle or cortex. Indeed, some specific keratin gene products are expressed in only singly scattered cortical keratinocytes (*e.g.* Type I group B keratin hHa8). Recently, an exceptional Type I group B hair keratin (hHa7), previously undetectable in terminal scalp hair, was found to be expressed in sexual hair.[15]

The keratin intermediate filament unit (8–10 nm across) differs from other intermediate filaments in that it is a heterodimeric unit formed as a pair of equimolar amounts of one Type I and one Type II keratin. Furthermore, these units are expressed in a highly tissue-specific and differentiation status-specific manner. Type I hair keratins are acidic and range in size from 392–416 amino acids, with mostly highly conserved N-terminal sequences. Type II hair keratins are basic to neutral and range in size from 479–506 amino acids. Both type I and II hair keratins have similar levels of cysteine residues.

The 'hard' hair keratins (7–8 nm across) can be distinguished from epidermal 'soft' keratins by their lack of extended runs of glycine residues; instead 'hard' hair keratins contain many cysteine residues (particularly at the N- and C-terminal domains) that enable them to form extensive disulfide bond cross-linking with other cysteine-rich proteins. Due to the presence of non-polar amino acid residues (leucine and valine) within a repeating heptad sequence, the filaments take up a coiled-coil conformation whereby right-handed α-helices coil around each other in a left-handed coil to form a rod structure.[16] A pair of these coiled rods (heterodimer α-helices) aligns in an anti-parallel manner to form a tetrameric unit – the KIF subunit.

2.3.1.1 Keratin-associated Proteins

Keratin-associated proteins (KAPs) are essential for the formation of a rigid and resistant hair shaft because they form extensive disulfide bonds that cross-link with the abundant cysteine residues of the hair keratins. Using various analytical tools *e.g.* solubilisation, chromatography, electrophoresis, amino acid sequencing *etc.*, hair proteins have been fractionated to reveal an increasingly complex group of proteins. In the mid 1990s, Rogers and Powell described eight families of KAPs and classified these on the basis of their amino acid composition as: a) high sulfur (16–30 mol% cysteine), b) ultra-high sulfur (<30 mol% cysteine) and c) high glycine/tyrosine proteins.[12] These were described mostly in wool but also in rabbit fur. Since then there has been considerable growth in this field with at least 23 KAP families now known.[17] These have been grouped as follows: high sulfur KAPs – family 1–3, 10–16 and 23; ultra-high sulfur – family 4, 5, 9 and 17; and high glycine/tyrosine KAPs – 6–8 and 18–22. However, research from the last couple of years has lead to even greater complexity, with large clusters of novel human KAP genes located.[17–19]

Keratin intermediate filaments (KIFs) and the inter-filamentous matrix of the cortex composed of KAPs form a complex group of interactive proteins in the hair cortex and cuticle. Hard (hair) keratins contain many cysteine residues (particularly at the N- and C-terminal domains) that enable them to form extensive disulfide bond cross-linking (both intra-chain and inter-chain) with other cysteine-rich proteins of the matrix. Indeed, some KAP families may contain up to 70 cysteine residues (out of approximately 180 amino acids) that would facilitate many covalent interactions within and between proteins. Thus, there is a potential for a bewilderingly complex set of interactions given the sheer numbers of proteins involved and their potential for interaction. It is of some relief to note from *in situ* hybridisation studies that genes for different KIFs and KAPs are sequentially activated as a function of hair growth. For example, Type II KIF genes are expressed sequentially in pre-cortical keratinocytes in separate stages. Moreover, expression for hard hair and soft epidermal KIF genes is usually mutually exclusive during hair growth. These observations may suggest a degree of order for sets of preferential protein interactions during hair fibre formation. There is good evidence for linkage between KIFs and the keratinocyte plasma membrane *via* desmosomes,[20] suggesting the desmosomes play an important role in hair keratinocyte differentiation.

Unlike the pre-cortical keratinocyte, cuticle cells contain a very different protein composition of mostly cysteine-rich proteins that form cytoplasmic granules.

While, these cells do contain some KIFs, only very rarely are filamentous structures detectable. Thus, protein-protein interactions here are mostly to permit the formation of lamellar structures – the product of fusion between these granules. This difference is reflective of the cuticle's greater protective function.

The medulla of the hair fibre contains a major arginine-rich multi-repeat matrix protein of 200–220 kDa, called trichohyalin seen most dramatically as a differentiation-associated protein in the inner root sheath (IRS). Indeed, trichohyalin serves as an excellent marker of hair follicle differentiation as it is expressed earlier than KIF mRNA. However, the formation and functionality of trichohyalin in the IRS and medulla are every different, perhaps reflective of the non-obligatory involvement of the medulla in hair growth. In inner root sheath cells trichohyalin forms as granules of matrix protein linking arrays of KIF. These then fuse to form multiple trichohyalin-KIF granules in citrulline-rich lamellar arrays. By contrast, medullary cells synthesise small granules of trichohyalin that later fuse to form large mature medullary granules, though without any apparent involvement of KIFs. The unusual conversion of arginine to citrulline with IRS and medullary cell differentiation is catalysed, at least in part, by a peptidylarginine deiminase.[21]

2.3.1.2 Biochemical Aspects of Hair Keratinisation

The dynamics of assembly of keratin intermediate filaments in epidermis and in hair fibres is very different. Whereas these can disassemble during mitosis in epidermal keratinocytes, they are the product of non-viable keratin-producing cortical keratinocytes in the keratogeneous zone of the hair follicle.

2.3.1.3 Nutrition and Hair Fibre Proteins

Cysteine is a crucial amino acid in protein synthesis and hair follicle growth. While the hair fibre does not contain methionine, the amino acid can be converted to cysteine and is also needed in the diet. However, it is not at all clear whether any 'normal' change in diet, including the provision of hair protein amino acids or their precursors, would significantly alter the nature of composition of hair proteins. A similar view may also apply to the provision of these reagents topically. That said, there is a dearth of rigorous data regarding the effects of nutrients on hair structure, although the wool literature contains several reports suggesting that correction of deficiencies in copper and zinc can improve wool fibre growth.[22] Here, copper is known to play important roles in cysteine to cystine oxidation during hair fibre growth, while zinc is important in cell growth and division. For further discussion in the use of the hair fibre in dietary assessment please see Chapter 8 elsewhere in this volume.

2.3.1.4 Genetic Variation in Hair Fibre Proteins

Several mouse mutations have been described that map to KIF or KAP genes.[23] For example, the nude phenotype (*nu/nu*) in mice is due to reduced content of cysteine-rich proteins resulting in weakened hair fibres. A similar mutation in the winged-helix-nude (WHN) gene is associated with a hairless phenotype in humans.[24] The congenital hair disorder monilethrix is caused by mutations in the hair keratin *hHb1* or *hHb6* genes[25] (see below), while copper deficiency in Menkes' syndrome with kinked hair results in defective oxidation of sulfhydryl groups in disulfide bonds.[26] Reduction in sulfur content also leads to the increased fibre fragility seen in the tricho-thiodystrophies.

2.4 Physical Properties of the Hair Fibre

2.4.1 Hair Fibre Lipids

Lipids on the hair fibre surface (*e.g.* cuticle) are likely to be functionally relevant as they provide a hydrophobic interface to protect the hair cortex from a hostile wet/dry environment. This field is relatively young and it was not until the mid 1980s when authors described a lipidic layer on the surface of hair fibres termed the F-layer.[27] A marked increase in the wet-ability or hydophilicity of the hair was observed when hair fibres were treated with an alkaline solution (potassium tertiary butoxide) in tertiary butanol due to the removal of fatty acids from the surface of wool fibre. This fraction contained a large amounts (58% of total) of a methyl branched 21-carbon fatty acid. This 21-carbon saturated fatty acid (exclusive to the hair/wool cuticle) is present in truly exceptionally high amounts, and has been identified by mass spectroscopy as 18-methyl-eicosanoic acid.[28] In humans the major fatty acids in hair fibre lipids include 16:0 (17%), 18:0 (10%), 18:1 (5%) and 21:0 (48%). It is of some considerable interest that in marked contrast to inter-species variability in sebaceous gland lipids, hair cuticle lipids are highly conserved.

Evidence that these lipids are surface-bound is inferred from data showing that increasing hair fibre diameter is associated with a decrease in total bound fatty acid. Thus, it is now well appreciated that hair fibres have approximately a 3 nm coating (although some estimate this may be up to 30 nm) of long chain fatty acids covalently bonded to the protein membrane of the epicuticle. Recent technologic advances, particularly in atomic force microscopy, have allowed exceptionally high observational power views of the surface of the hair fibre surface, and how this is altered by changes in temperature, hydration, pH, lipid layer removers, topically applied cosmetic products *etc.* The main function of the branched methyl-eicosanoic acid is currently unknown, but it may be involved with increasing the degree of hydrophobicity over straight-chain fatty acids and/or altering the frictional quality of the fibre.[29]

Small amounts of cholesterol sulfate, cholesterol and fatty alcohol are also associated with the F-layer, though the nature of bonding (*e.g. via* thioester linkages) is unclear.

2.4.2 Mechanical Properties of the Hair Fibre

What our eyes see when we look at hair is the interplay of many physical characteristics. The obvious physical parameters include the hair fibre's geometry consisting of its curvature, its colour/shade and its shine/lustre. These attributes however, can be sub-categorised further to include fibre diameter, tensile strength, torsion, swelling, friction *etc.*[11]

While we have stressed throughout this chapter that the cuticle is most important for protecting the hair fibre, the actual strength and elasticity of the hair fibre is very much derived from the nature of the KIF and KAP interactions within the hair cortex. Thus, stretching the hair fibre can cause the cuticle, particularly the endocuticle and cell membrane complexes, to split. Although keratin biosynthesis occurs in an aqueous environment within the hair follicle, the process of hydration-dehydration of the hair fibre reveals an anisotropism (non-uniform hardness) when wet. However, despite the relatively high affinity of hair keratin for water and the relative ease of water to permeate the hair fibre, this process is highly selective whereby water molecules are variably excluded from certain molecular structures within the hair (see below). Whereas the matrix material of the hair cortex absorbs water and then softens, the microfibrilar components themselves remain water-impenetrable.[11] Keratin fibers are significantly weakened when wet.

The mechanical behaviour of hair fibres displays a linear stress-strain curve with three distinct regions. For the first 2% of increase in extension the hair fibre responds stiffly. If extension continues the hair fibre enters a more yielding state (*i.e.* with little increase in stress force) where the fibre can be extended to about 30%. If the stress is released at this point, the hair will return to its original 'resting' length. However, if stress force is raised still higher a post-yield region of the curve is reached where the fibre responds stiffly and ultimately snaps. This remarkable ability to extend by almost a third is due to properties of the KIF rather than reflecting any contribution of the KAPs.

The radial swelling of hair fibres that occurs when wet results from an uptake of water by the KAP matrix material and this can be demonstrated by increased X-ray spacings. Several models have been proposed to explain the remarkable stress-strain curve of hair fibres in terms of either the KIFs alone or KIF-KAP combined.[11] Only when more complete information is available, on the molecular aspects of keratin and keratin-associated interactions (*e.g.* where exactly the disulfide bonds are located, whether they are within and between the KAPs *etc.*) and the precise contributions of the elastic and adhesive inter-cortical cell membrane complexes, will this issue be fully resolved. It should be noted, however, that matrix proteins do have the effect of excluding water from the inter-cortical KIF space. Thus, the higher the content of KAPs the stiffer (and less 'swell-able') the fibre is likely to be.

Torsional properties of the hair fibre rely on mechanical properties of hair fibre components that are very different from those for tensile strength. Whereas the matrix material confers little to the tensile strength of the hair fibre *i.e.* the *tensile modulus* of hair decreases by only a factor of ≈ 3 when wet (with the matrix absorbing the water), there is a 15-fold corresponding decrease in the *torsion* or *rigidity modulus*. This is explained by the shear stress associated with torsion change being imparted only to the softened matrix, not to the keratin filaments themselves. Finally, while the cuticle is not considered to play a significant role in the mechanical properties of the fibre, the uptake of water by the endocuticle is likely to absorb some of the deformation stress applied to the hair fibre.[11]

There are some important differences in the mechanical properties of hair fibres from individuals of different ethnic background. Caucasian hair, like for so many other hair fibre parameters (*e.g.* ellipticity) fall somewhere in the middle of the spectrum with a narrow maximum of 60–75 µm.[11] Furthermore, the tensile properties of hair fibres from Caucasian and Asian individuals are broadly similar. There is some evidence that axial twisting of African hair can provide zones of weakness to breaking stresses upon extension. Other interesting differences include the dramatic variation in 'comb-ability' stress in African hair. While combing of Caucasian and Asian dry hair induces hair shaft parallelism, this does not occur in curly African hair. Moreover, a surprising reversal occurs when the comb-ability/combing stress of wet hair is compared. Here, wet African hair becomes much more easily combed than wet hair in Asians and Caucasians. This difference is largely due to the configuration change of hair from Asians and Caucasians when wet compared to when dry. Little configuration change accompanies the wetting of African hair fibres. See below for further discussion on ethnic variations in hair structure.

2.4.2.1 Hair Fibre Curvature

It is well known that hair fibres can assume a non-circular profile *via* twists, curls, crimps and indentations. In some animals seasonality in the level of crimp may be apparent, usually with the summer coat exhibiting less crimp. The presumption here is that the additional volume of air trapped in highly crimped hair acts as a thermal insulator. Despite our familiarity with these phenotypes, and our ability to technically induce these forms chemically, little is known of the actual physical explanations for natural curl and crimp. One possibility is that the IRS may 'guide' or 'mould' the still malleable hair fibre as it grows through the follicular wall. Any such moulding influence is likely to be greatest at the keratogeneous zone of the hair follicle. In this manner, a curved follicular wall of the IRS should cause the emerging and hardening hair fibre to assume the corresponding shape. Perhaps more convincingly, variations in hair curl and crimp may reflect a variable degree of asymmetry in the hair fibre itself such that keratinisation occurs asymmetrically on one side of the fibre causing it to exhibit short and long longitudinal axes. Asymmetrical keratinisation also could also affect how the hair shaft responds with variable levels of moisture. There is ample evidence from the wool literature for asymmetry in cortical keratinocyte differentiation such that different types of cells are formed.[30,31] When high-crimp and low-crimp merino wools were compared, cells on one side of the crimp expressed KIFs in a whorled pattern (so-called *orthocortex*) while cells on the opposite side of the fibre lacked this whorled pattern termed *paracortex* in high-crimp fibres and *mesocortex* in low-crimp fibres. Furthermore, as hair fibre diameter increases so also does the amount of orthocortex. The above description applies to ovine hair fibres, but this bi-, trilateral structure of hair cortex has been difficult to detect in humans. However, when Mongolian-type straight hair (consisting of paracortex) was compared with curly Caucasian hair, a small amount of orthocortex was found distributed at the cuticle-cortex junction in the latter.[7] Moreover, in high-crimp hair of some people of African origin, asymmetrical arrangements of ortho- and paracortex can be detected. It should be noted however, that significant curvature of the hair follicle itself could be observed in some of these individuals. In any event it appears that the composition of the matrix will be very important in determining the curve of the hair fibre.

2.4.3 Ethnic Variations in Hair Structure

Human hair is commonly grouped into just three main sub-types: Caucasian, Asian and African. Differences between these groups are usually determined with respect to a range of parameters including: hair fibre diameter and its cross-sectional form, overall fibre shape, mechanical properties (see above), comb-ability, shape, chemical makeup and moisture level. For many of these parameters Caucasian hair falls intermediate to the Asian and African extremes.

A recent, and long awaited, systematic examination of the protein structure of hairs from Asian, Caucasian and African individuals[31] revealed no differences by X-ray analysis in the structure of the hair keratin. This research team also examined the degree of radial swelling of hair of each ethnic grouping when placed in water. Here, African hair exhibited the lowest radial swelling rate and lowest maximum swelling. By contrast, both Asian and Caucasian gave similar statistically higher values, after normalisation for initial shaft diameters. An explanation for this finding remains elusive, particularly in light of the similar x-ray results on protein structure.

Asian hair has the greatest fibre diameter and exhibits a circular sectional profile with a mean ellipticity of approximately 90%, giving it an almost fully circular profile. In contrast, African hair exhibits high inter-individual variability with regard to diameter but also with respect to the degree of ellipticity of the hair fibre cross-section. The mean ellipticity value is closer to 60%, although there is also much variability along the length of the hair fibres. Importantly, one aspect of the cross-sectional profile may be prominently flattened. Together, these features impart an overall shape to African hair that resembles a twisted rod, but also with focal constrictions along the hair shaft. Caucasian hair fibres however, have an intermediate diameter with a cross-sectional shape that is less oval and with an ellipticity value of 75%.[31] As a result, Asian and Caucasian hair fibres are more cylindric than those of Africans.

The overall form of the hair fibre has implications for its mechanical properties, and there is some evidence that the twists and turns in African hair fibres reduce tensile strength at these turn sites. Thus, African hair may break more easily under certain circumstances than hair of Asians and Caucasians. Indeed, it has been shown that 'breaking stress' and 'breaking extension' values are lower in African hair fibres than in Caucasians or Asians. Meanwhile, despite the difference in diameters of Asian and Caucasian hairs, both exhibit similar behaviours during stress. Despite these gross ethnic differences in the form of the hair fibres, the chemical makeup of the hair keratin is remarkably similar throughout all humans. The hair protein compositional or structural features of hair protein in these ethnicities therefore do not provide immediate explanations for these behavioural differences, though it is probable that the flattening, twisting and cracking of African hair fibres are contributing factors.

2.4.4 Aging and Hair Fibre Form

Increasing age can leave its mark on several phenotypic properties of the hair fibre. The most visually apparent of these include: hair thinning, hair loss, reduction in the rate of growth, pigmentation loss[32–35] (see Chapter 3 elsewhere in this volume), and loss of hair fibre lubrication/moisturisation. Despite this, hypertrichosis can also occur in certain body sites of both men and women of advancing age, *e.g.* terminal hair growth on the upper lip and chin of post-menopausal women and on the ear pinae, nose and nasal vestibules in aging men. Hair begins to exhibit age-related change from the earliest period of hair follicle development *in utero* (see Chapter 1 elsewhere in the volume) where our first 'coat' of fine and sometimes pigmented lanugo hair is shed before birth to give way to still finer, shorter and unpigmented vellus hair. Great changes in circulating hormones (*e.g.* at puberty) fully 'awaken' the vellus hair follicle in several body zones to induce the vellus-to-terminal hair transformation. However, even the fine scalp hair of the growing child and adolescent exhibits striking changes with increasing age to mature adulthood, not only in colour (most typically a darkening of hair colour *e.g.* blond to brown), but also a coarsening of the hair fibres themselves. A discussion of the changes that are genetically controlled and involve various hormones and their cognate receptors is beyond the scope of this chapter. The reader is directed to a review of this topic.[36]

For most individuals however, the most dramatic age-related change in hair form (excluding loss of hair colour and loss of the hair itself) results from so-called hair follicle transformations.[37–39] Of particular clinical importance is the vellus-to-terminal transformation occurring during puberty, hirsutism and hypertrichosis, and the reverse terminal-to-vellus transition characteristic of androgenetic alopecia or male-pattern baldness (AGA).[37–39] It was long viewed that hair thinning was due to a gradual so-called 'miniaturisation' of the hair follicle occurring *via* slow and gradual hair-cycle-by-hair-cycle change. It is now thought however, that hair thinning/loss occurs *via* more abrupt and marked reductions in the numbers of cells of the follicular dermal papilla and connective tissue sheath sub-types.[39] There is convincing clinical support for the 'abrupt change' view, not least the rapid progression of AGA and the preponderance of fine hairs over intermediate hairs in the balding scalp.[38] Similarly, puberty-associated changes in hair growth are rapid. Perhaps more convincing still is the reversal of the vellus-to-terminal transformation that can occur over a single hair cycle in finasteride or minoxidil-stimulated hair follicles.[38]

In addition to hair follicle transformation to vellus or *invisible* hairs (the usual clinical appearance of 'hair loss'), there may also be a reduction in the absolute numbers of hair follicles, not only in the scalp but also throughout the body. The precise mechanism for hair follicle dropout is unclear, though it may mimic the hair follicle programmed organ deletion that can occur in mice,[40] and atrophic change and fibrosis may be indeed found in affected skin.

A reduction in the level of pigmentation of scalp hair in male pattern baldness is associated with the reduction in the calibre of these hairs (see Chapter 3 elsewhere in this volume). This is thought to be largely the result of the reduced capacity of smaller, finer hairs to accommodate large numbers of melanocytes. There is also a tendency for these 'miniaturising' hairs to be less medullated than terminal scalp hair. By contrast, loss of melanocytes from hair follicles producing hair fibres of normal calibre, *i.e.* during hair graying, may also result in a concomitant change in the structure of these hair fibres. This is perhaps not surprising given the close interaction between melanin granule-transferring melanocytes and hair shaft-forming/melanin-accepting pre-cortical keratinocytes. Therefore, it is likely that pigment-producing melanocytes in the hair bulb will influence cortical keratinocyte behaviour in several ways. For example, melanin transfer to cortical keratinocytes may hasten their terminal differentiation and cornification – a change that is mediated from increased levels of calcium, some of which may be transferred into the keratinocytes within the melanin granules. Indeed, there is evidence that white hair grows faster than pigmented hair both *in vivo*[35,41] and *in vitro* (Eva Peters, personal communication). It is likely that melanocytes influence neighbouring keratinocytes in many other ways including *via* the production of various cytokines, growth factors, eicosinoids, adhesion molecules and extracellular matrix.[42]

There is also some evidence that gray and white hair fibres exhibit different mechanical properties compared to adjacent pigmented hairs. Hollfelder provided some evidence that pigment-free hairs are not only coarser but also can be wavier than pigmented hairs.[42] Moreover, others have reported that the average diameter of white hair fibres is significantly greater that of pigmented hairs.[35] A more prominent development of the medulla in white, compared to pigmented, hair fibres, was also reported in this study. Interestingly, also described here was an age-related reduction in hair growth rate and in hair fibre diameter but this was broadly limited to pigmented hairs in these individuals. Thus, the implication here is that, counter-intuitively, white hairs may be partially spared these aging changes. The tensile strength of hair also decreases with age, having increased from birth to the second decade.[33,34]

In a similar manner to the observed change in lipid composition of sebum with advancing age, there is also an age-associated change in the chemical composition in the hair fibre. Metals involved in this change in hair fibres include cadmium, copper, zinc and strontium. There are also reductions in glutathione reductase, glutathione-S-transferase, glucose-6-phosphate dehydrogenase and gamma-glutamyl transpeptidase.[43–47]

The loss of hair shaft moisture and lubrication has been reported to occur as a function of increasing age in adults, especially in women. The hair exists within the context of the *pilo-sebaceous unit* – a term that implicates the sebum-producing gland in several aspects of hair biology. The open and interactive nature of the 'pilo-sebaceous unit' is facilitated by a duct that carries sebum from this holocrine tissue directly onto the hair fibre and from there to the skin surface. The activity of the sebaceous gland changes dramatically as a function of gender and age, from the relatively inactive pre-pubertal period, through a very active adolescence and young adulthood, to markedly reduced activity after the fourth decade of age, especially in females. In addition to crude overall changes in gland size and activity, more subtle changes involve modification of the composition of the lipids being produced by the gland at different times during our lifetimes. For example, sebum from children contains less squalene and cholesterol than sebum from adults.

2.5 Hair Shaft Abnormalities

2.5.1 Features of Damage in Normal Hairs

Extreme trauma or protracted mechanical stress can result in damage to the normal hair shaft and this needs to be distinguished from hair shaft abnormalities due to intrinsic defects, either genetic or nutritional. The location of the damage along, say, a 1 m length of hair is also highly relevant, as the 'oldest' fractions (*i.e.* distal sections towards and at the tip) will have been 'weathered' by repeated washings, cosmetic treatments, brushings and other manipulations that cause splitting and cuticular damage. Conversely, observing these features in short hair lengths suggests either the presence of more traumatic damage, *e.g.* aggressive grooming or cosmetic treatments (*e.g.* the appearance of trichorrhexis nodosa – focal splaying of the hair cortex giving the node the appearance of brush-on-brush junctions) or a underlying nutritional deficiency or intrinsic defect. Another source of structural damage to normal hair can be due to extreme heat stress. Cosmetic practices, for example those associated with straightening of African hair, may induce so-called 'bubble' hair where extreme heat induces bubble formation under the cuticle within the hair cortex.

The presence of numerous short broken hair 'stubs' on the scalp may neither reflect aggressive grooming/cosmetic practices nor a genetic or nutritional deficit. Alternatively, willful pulling of hair from the scalp can also lead to alopecia of broken hairs – as in trichotillomania. Affected individuals (mostly young girls) usually have some underlying psychological or emotional disorder. However, most practitioners now view trichotillomania similarly to nail biting or thumb sucking.

2.5.2 Hair Shaft Dysplasias

There are numerous hair shaft dysplasias with characteristic morphologies that result in malformations of the hair shaft. These can be either congenital (some of which may be hereditary) or acquired. In many cases alterations of the hair shaft reveals nothing more sinister that the result of excessive and aggressive grooming. However, they may also be important diagnostic clues for the presence of a specific disorder. Hair shaft dysplasias are generally grouped into those that result in increased hair fragility and those that don't. The interested reader is directed to an excellent review of these conditions.[26]

2.5.3 Hair Shaft Abnormalities with Increased Hair Fragility

This group includes rare disorders including trichorrhexis nodosa, monilethrix, pseudomonilethrix, pili torti (congenital and acquired forms), trichorrhexis invaginata (congenital and acquired forms; bamboo hair in Netherton's syndrome), trichothio-dystrophy (sulfur-deficient brittle hair), Bayonet hair and Menkes' syndrome (kinky hair), and common disorders like alopecia areata.

Trichorrhexis nodosa is the most common form of hair shaft dysplasia and refers purely to a morphological appearance that can occur in several disorders both congenital and acquired and can also occur in cases of excessive and aggressive grooming (see African hair above). These hairs feature a fracture zone with splaying out of cortical cells to form brush-on-brush ends. There is likely to be both a cuticular and cortical cell defect, though the relative contribution of these to the phenotype may vary by cause.

Monilethrix or beaded hair is a dominant hereditary defect that results from a mutation in a Type II hair cortex keratin. This change causes the hair to fracture at the characteristic constrictions close to the scalp surface. The fracture can result in trichorrhexis nodosa-like lesions.

Pili torti is another example of a dominantly inherited hair shaft dysplasia and is characterised by axial flattening and twisting that occurs periodically along the shaft. This defect renders the hair a dramatic glittery appearance in reflected light. This condition is associated with fragile and brittle hair that remains short due to fracture.

Pili torti with copper deficiency, also called *Menke's disease* or *trichopolio-dystrophy*, is an X-linked recessive disorder of copper metabolism. In addition to the hair shaft defect patients with this condition also have psychomotor defects, growth retardation and multiple other neurologic defects, which are usually lethal after about the first few years of life. Certain tissues in the affected individuals exhibit a marked reduction in copper (*e.g.* brain, liver) while other tissues have markedly increased copper concentrations (*e.g.* skin, gut, lung). The hair provides an important diagnostic clue and is sparse and lacking pigment, with an appearance of 'steel wool' standing-on-end. Individual hairs are twisted axially (pili torti).

The hair shaft abnormality in *bamboo hair* (trichorrhexis invaginata) is found in a rare ectodermal dermatosis with autosomal recessive inheritance. Hair shafts with nodes that resemble bamboo joints with a proximal cup portion and a distal ball portion forming the nodes, characterise this condition. There is also twisting of the shaft and alterations in hair fibre diameters and other structural abnormalities. Not all hair is similarly affected, however.

Trichothiodystrophy is a rare congenital disorder of sulfur deficiency that causes brittlness of the hair. The affected hair contains a reduced cystine content and affected individuals present with multiple neuroectodermal symptoms. These individuals are easily recognised as the hair on their scalp, eyebrows and eyelashes is sparse and, where present, is brittle and broken. Polarising microscopy reveals an alternating bright and dark banding pattern. Thiodystrophic hair has defects in the cuticle and cortical cells throughout the length of the hair shaft and in some cases the cuticle may be absent.

Bayonet hair is so called because these hyperpigmented hair fibres have a spindle-shaped swelling a few millimeters proximal from the tip. These hairs can also be found as a minor percentage of the normal scalp.

2.5.4 Hair Shaft Abnormalities without Increased Hair Fragility

This group includes pili annulati, pseudo pili annulati, wooly hair, spun-glass hair (uncombable hair, pili triangulati et canaliculi), pili bifurcati and alopecia areata.

Pili annulati is so called because hair exhibits bright and dark banding in reflected light with the bright bands due to light scattering from periodic air-filled cavities within the hair fibre. If familial, pili annulati is inherited in an autosomal dominant manner, though this condition can also be sporadic. Unlike the conditions mentioned above, the hair is strong and individuals do not report any clinical concerns.

Wooly hair is so called because of its characteristic crimp (*i.e.* like wool) and is usually restricted to description of Caucasian hair. The hair is of finer calibre than normal and is both curly and flattened. This hair can be associated with a diffuse or generalised state, or be part of a so-called *wooly hairy nevus*. There may also be acquired forms of wooly hair.

Spun-glass hair is another hair shaft dysplasia without associated increase in fragility. Here the flattened surfaces of the dry and rough hair shafts lie in multiple different orientations to reflect light in such a manner as to confer a spun-glass appearance. This dominant familial condition is most often seen in children with straight hair. The affected hair shafts contain one or more deep grooves, and the actual number of grooves will result in variations in fibre cross-sections from triangular and quadrangular to kidney-bean shaped *etc.*

Pili bifurcati refers to a condition where a single hair shaft exhibits bifurcation at several positions along its length. Within these intervals both rami of the parallel bifurcation have their own complete cuticle and there is a fusion point beyond the bifurcation where a single cuticle invests the entire hair shaft.

Alopecia areata is a presumptive autoimmune disease where an immune response raised against hair follicle-associated antigens results in a pathologic attack on growing hair follicles.[48] Targeted damage to the cortical keratinocyte population in the hair bulb interrupts normal keratinisation such that emerging hair shafts may break off close to the skin surface. The result is a so-called 'exclamation-mark' hair characterised by a fractured distal end. The proximal end tapers toward the scalp.[49] The brush-like fractured end consists of multiple separated cortical strands lacking inter-cortical cell adhesion and a covering cuticle (Figure 2.5).

2.5.5 Hair Shaft Abnormalities Due to Cosmetic Use

As briefly alluded to above, normal individuals can also exhibit hair shafts with structural abnormalities, though these are not the result of intrinsic defects. In most cases these are associated with poor grooming practice or the inappropriate use of cosmetic products. In the economically-developed world most adults will apply a range of cosmetic products to the scalp and hair from basic shampoos and conditioners to a plethora of more aggressive treatments including hair dyes, bleaches, straighteners, fixatives, relaxers, heat treatments, depilatories *etc* (see Chapters 9, 10, 12, 13 elsewhere in this volume and reference 49). Any cosmetic procedure if

Figure 2.5 *Scanning electron micrograph of the fractured distal tip of an 'exclamation-mark' hair in alopecia areata. Fibre breakage is due to focal weakness of the hair fibre at this point that is associated with loss of cohesion between cortical cells/strands and defective cuticle. Scale bar: a = 13 µm*

carried out carelessly and without following manufacturers' instructions can lead to significant hair fibre damage. One must emphasise that the hair is only in 'perfect' condition just as it emerges from the scalp epidermis – from then on the hair fibre is exposed to a battery of relatively adverse conditions.

The most devastating of negative outcomes from inappropriate product usage is hair loss. This can rarely, if ever, be linked directly to a cosmetic product used appropriately or even inappropriately – with the notable exceptions of the incorrect use of perms and relaxers. Matting and tangling occupies the other end of the outcome spectrum. Chemically treated hair is more likely to form tangles, as these chemicals can cause the cuticles to lift, resulting in increased roughening of the hair. The increased frictional properties of chemically treated hair will lead to the situation whereby normal grooming will increase the likelihood of hair breakage.

Between these two extremes are several other forms of hair shaft damage related to hair mis-treatment. Excessive heat (*e.g.* temperatures greater than 100 °C) applied to damp hair shafts is one way to cause the formation of 'bubble-hair'. This dry/brittle hair contains air cavities within the cortex of the fibre caused by boiling of water inside the shaft inducing keratin hydrolysis and local air expansion. Bubble hair is prone to breakage.

By far the greatest perceived danger to hair shaft integrity comes from the wide range of chemicals used in cosmetic products. These generally are permanent waving, bleaching, dyeing, highlighting and relaxing products. As these products first encounter the cuticle but also need to work deep in the cortex (*e.g.* to bleach melanin in the cortex), damage is first encountered at the hair surface. Chemicals used for permanent waving (*e.g.* thio-glycolates) and bleaching may cause the cuticle to lift or separate and these chemicals are associated with a reduction in the levels of cystine in the cystine-rich cuticle and beyond in the cortex and in a corresponding increase in the amounts of cysteic acid. These products rely on their ability to cleave disulfide bonds and depending on the concentrations used can result in damage, particularly to the superficial cuticle. The cuticle is thereafter even more susceptible to enhanced weathering. Lipids on the hair surface are also lost during chemical hair treatments and this may lead directly to reduction in the level of cohesiveness between cuticle cells.

Another change in the fibre during chemical processing and subsequent water rinsing/washing is the increase in fibre swelling. For example, if thioglycolates are used as the active chemical agent, the swelling of the water-rinsed hair can increase by as much as 200%. Hair treated in this way can exhibit trichorrhexis nodosa, especially Caucasian hair. Similarly treated African hair may present with longitudinal splits in the affected hair fibres.

2.6 Summary

The human hair fibre is a truly amazing natural composite with unique physical and chemical properties. The product of millennia of evolutionary selective pressures, the human species exhibits a wide variability in hair fibre forms – the scientific basis for the latter remains broadly enigmatic. Despite its robust nature, the hair shaft is not indestructible and lesions on/in the hair can provide us with a plethora of valuable information as a sensor of the external world we live in, on our grooming and cultural fetishes, and importantly, can be the most apparent diagnostic clue to intrinsic systemic biological disorders and deficiencies. We need to listen carefully to the messages coming 'down the wires' of our natural antennae!

2.7 References

1. D. Morris in *Body watching: A field guide to the human species*, Crown, New York 1985, 21.
2. K. Munstedt, N. Manthey, S. Sachsse and H. Vahrson, *Support Care Cancer*, 1997, **5**, 139.
3. T.F. Cash, V.H. Price, R.C. Savin, *J. Am. Acad. Dermatol.*, 1993, **29**, 568.
4. K.A. Phillips in *The broken mirror: Understanding and treating body dysmorphic disorder*, Oxford University Press, New York, 1996.

5. I.M. Hadshiew, K. Foitzik, P.C. Arck and R. Paus., *J. Invest. Dermatol.*, 2004, **123**, 455.

6. R.J. Randebrook, *J. Soc. Cosmet.*, 1969, **20**, 159.

7. J.A. Swift in *Formation and structure of human hair* P. Jolles, H. Zahn and H. Hocker (ed), Birkhauser Verlag, Basel, 1997, 149.

8. J.A. Swift, *J. Cosmet. Sci.*, 1999, **50**, 23.

9. J.D. Leeder, *Wool Sci. Rev.*, 1986, **63**, 3.

10. J.M. Gillespie *et al. Aust. J. Biol. Sci.*, 1964, **17**, 548.

11. L.J. Wolfram, *J. Am. Acad. Dermatol.*, 2003, **48**, S106.

12. B.C. Powell, G.E. Rogers in *Formation and structure of human hair* P. Jolles, H. Zahn, and H. Hocker (ed), Birkhauser Verlag, Basel, 1997, 59.

13. L. Langbein, M.A. Rogers, H. Winter, S. Praetzel and J. Schweizer, *J. Biol. Chem.*, 2001, **276**, 35123.

14. L. Langbein, M.A. Rogers, H. Winter, S. Praetzel, U. Beckhaus, H.R. Rackwitz and J. Schweizer, *J. Biol. Chem.*, 1999, **274**, 19874.

15. L.F. Jave-Suarez, L. Langbein, H. Winter, S. Praetzel, M.A. Rogers and J.J. Schweizer, *J. Invest. Dermatol.*, 2004, **122**, 555.

16. L.N. Jones, *Clinics in Dermatology*, 2001, **19**, 95.

17. Y. Shimomura, N. Aoki, J. Schweizer, L. Langbein, M.A. Rogers, H. Winter and M. Ito, *J. Biol. Chem.*, 2002, **277**, 45493.

18. M.A. Rogers, L. Langbein, H. Winter, C. Ehmann, S. Praetzel, B. Korn and J. Schweizer, *J. Biol. Chem.*, 2001, **276**, 19440.

19. Y. Shimomura, N. Aoki, M.A. Rogers, L. Langbein, J. Schweizer and M. Ito., *J. Investig. Dermatol. Symp. Proc.*, 2003, **8**, 96.

20. E.A. Smith and E.J. Fuchs., *J. Cell. Biol.*, 1998, **141**, 1229.

21. G.E. Rogers, H.W.J. Harding, I.J. Llewellyn-Smith, *Biochim. Biophys. Acta.*, 1977, **495**, 159.

22. J.M. Gillespie in *Biochemistry and Physiology of the skin*, L.A. Goldsmith (ed), Oxford University Press, New York, 1983, 475.

23. J.P. Sundberg in *Handbook of mouse mutations with skin and hair abnormalities*, J.P. Sundberg (ed), CRC Press, Boca Raton, 1994, 379.

24. J. Frank, C. Pignata. A.A. Panteleyev, D.M. Prowse, H. Baden, L. Weiner, L. Gaetaniello, W. Ahmad, N. Pozzi, P.B. Cserhalmi-Friedman, V.M. Aita, H. Uyttendaele, D. Gordon, J. Ott, J.L. Brissette and A.M. Christiano, *Nature*, 1999, **398**, 473.

25. H. Winter, M.A. Rogers, L. Langbein, H.P. Stevens, I.M. Leigh, C. Labreze, S. Roul, A. Taieb, T. Krieg and J. Schweizer, *Nat. Genet.*, 1997, **16**, 372.

26. V.H. Price in *Hair and Hair Disease*, C.E. Orfanos and R. Happle (ed), Springer-Verlag, New York, 1990, 363.

27. E.G. Bendit and M. Feughelman, in Encyclopedia of polymer science and technology, Wiley, New York, 1968, **8**, 1.

28. J.D. Leeder and J.A. Rippon, *J. Soc. Dyers Colourists*, 1985, **101**, 11.

29. P.W. Wertz and D.T. Downing, *Comp. Biochem. Physio. B: Comp Biochem.*, 1989, **92b**, 759.

30. L.N. Jones, D.E. Rivett and D.J. Tucker in *Handbook of Fiber Chemistry*, M. Lewin and E. M. Pearce (ed). Marcel Dekker, New York, 1988, 355.

31. A. Franbourg, P. Hallegot, F. Baltenneck, C. Toutain and F. Leroy, *J. Am. Acad. Dermatol.*, 2003, **48**, S115.
32. I.J. Kaplin, K.J. Whiteley, *Aust. J. Biol. Sci.*, 1978, **31**, 231.
33. M. Courtois, G. Loussouarn, C. Hourseau and J.F. Grollier, *Br. J. Dermatol.*, 1995, **132**, 86.
34. N. Naruse and T. Fujita, *J. Am. Geriatr. Soc.*, 1971, **19**, 308.
35. D. Van Neste, *Eur. J. Dermatol.*, 2004, **14**, 28.
36. M.E. Sawaya, *Ann. N. Y. Acad. Sci.*, 1991, **642**, 376.
37. C.A. Jahoda, *Exp. Dermatol.*, 1998, **7**, 235.
38. D.A. Whiting, *J. Am. Acad. Dermatol.*, 2001, **45**, S81.
39. D.J. Tobin, A. Gunin, M. Magerl, B. Handijski and R. Paus, *J. Invest. Dermatol.*, 2003, **120**, 895.
40. S. Eichmuller, C. van der Veen, I. Moll, B. Hermes, U. Hofmann, S. Muller-Rover and R. Paus, *J. Histochem. Cytochem.*, 1998, **46**, 361.
41. W. Nagl, *Br. J. Dermatol.*, 1995, **132**, 94.
42. A. Slominski, D.J. Tobin, S. Shibahara and J. Wortsman, *Physiol. Rev.*, 2004, **84**, 1155.
43. B. Hollfelder *et al.*, *Int. J. Cosmet. Sci.* 1974, **13**, 25.
44. P.R. Cohen and R.K. Scher in *Atlas of Hair and Nails*, M.K. Hordinsky, M.E. Sawaya and R.K. Scher, (ed) Churchill Livingstone, Philadelphia, 2000, 213.
45. M. Kermici, F. Pruche, R. Roguet abd M. Prunieras, *Mech. Ageing Dev.*, 1990, **31**, 73.
46. D.J. Cline, *Dermatol. Clin.*, 1988, **6**, 295.
47. A. Sturaro, G. Parvoli, L. Doretti, G. Allegri, C. Costa, *Biol. Trace Elem. Res.*, 1994, **40**, 1.
48. D.J. Tobin, D.A. Fenton and M.D. Kendall, *J. Cutan. Pathol.*, 1990, **17**, 348.
49. C.L. Gummer, in *Atlas of Hair and Nails*, M.K. Hordinsky, M.E. Sawaya and R.K. Scher (ed), Churchill Livingstone, Philadelphia, 2000, 173.

CHAPTER 3

Pigmentation of Human Hair

DESMOND J. TOBIN

3.1 Overview

Of all our phenotypic traits skin and hair colour communicate more immediate information to the observer than any other. Humans display a rich and varied palette of surface colour that not only highlights striking superficial variations between human sub-groups, but also underscores how we differ phenotypically from other mammalian species. Moreover, the colours of our 'crowing glory' can range from vivid reds and sun-bleached blondes to sober browns, raven blacks and with age, alas, to steel grays and snow whites.

Despite such variation, all hair colours are derived from the pigment *melanin*, synthesised *via* a phylogenetically ancient biochemical process termed melanogenesis. Synthesis occurs within melanosomes – specialised organelles unique to highly dendritic, neural crest-derived, cells called melanocytes.

Study of coat colour in mice has contributed hugely to our elucidation of the regulatory factors involved in the continuous pigmentation in the human epidermis. It is perhaps ironic therefore that, despite this knowledge base, we are only beginning to understand the mechanisms involved in regulating pigmentation in the human hair follicle.

In this overview, I focus on several critical issues relating to the biology of hair pigmentation including: possible reasons why scalp hair is pigmented and so variably in humans; how the hair pigmentary unit develops; how different follicular melanocyte sub-populations interact; the biochemistry of melanogenesis – and its intra-/ inter-ethnic variations; the positive and negative regulation of hair follicle pigmentation; the remodelling of the follicular melanocyte unit during hair cycling; the loss of hair pigment during chronological aging; and the affect of pigment loss on hair structure. This overview will hopefully provide a biological context for data presented elsewhere in this book (*e.g.* Chapters 4, 7, 8, 9, 14) where hair melanin also features.

3.2 Evolutionary Context for Hair Pigmentation

Hair growth – a particularly mammalian trait – has facilitated evolutionary success *via* contributions to thermal insulation, social and sexual communication (involving visual stimuli, odorant dispersal *etc.*) and sensory perception (*e.g.* whiskers). Additionally, hair colour provides camouflage (*e.g.* seasonal coat colour change in the arctic hare) and social and sexual communication (*e.g.* silver-backed gorilla). While humans do not rely on such adaptations for their actual survival, these still play a significant role in our social and sexual communication.

Often called the 'naked ape', our relative nakedness draws attention to our head hair – which, uniquely amongst primates, can be very thick, very long and very pigmented. This departure from the primate norm likely reflects significant selective pressure that exaggerated scalp and facial hair in humans. These are indeed potent communicative accompaniments for our already highly expressive faces. Pigmentation biologists have been at the vanguard of recent technological advances in molecular genetics that permit us to investigate the origin of various human sub-populations with a more objective eye. While, the 'naked' state of most of our skin highlights the need for epidermal pigment to filter harmful ultraviolet radiation (UVR), it is not immediately clear why we have developed such vigorous and luxuriant facial and scalp hair growth.

Recent evidence indicates that human development had its origins in the hot sunny African climes, where we had (and still have) paradoxically black skin and hair pigmentation. What possible advantage could there be in having dark skin and hair colour, where this acts not only as a trap for radiant heat but also acts as a good thermal insulator? One possibility is that deeply pigmented skin and hair may protect against sunstroke, due to melanin's very efficient and fast exchange of ion transport and efflux that facilitates adequate salt balance.[1] There may also have been other benefits to bi-pedal humans having a well-haired scalp epidermis. Akin to feather fluffing in birds, scalp hair acts as a cooling device by keeping the heat away from our bodies and when curly, can generate cooling air currents more efficiently than on exposed bald scalp. Further insights into possible benefits of scalp hair pigmentation in humans may be gleaned from the observation that a considerable portion of human evolution occurred along seacoasts and riverbanks. Therefore, relying for a significant part of their diet on fish,[2] early humans would have had to develop strategies for handling toxic heavy metals, which concentrate in many fish species. These metals, which selectively bind melanin[3] can be easily discharged *via* the hair fibre. This could have been the preferred route, given the hair follicle's very high proliferation rate and the subsequent concentration of hair melanin in a non-viable hair shaft. Further evolutionary value in having a deeply pigmented interface between the harsh external environment and the body's inner space containing delicate vital organs, may lie in the potent antibiotic properties of reactive quinone intermediates generated during melanin biosynthesis – of particular importance in warm humid climates where humans first emerged.[4,5]

One of the enduring enigmas of the modern human phenotype is the dramatic diversity of hair colours, particularly amongst Europeans. Molecular geneticists have started to unravel this enigma and have shown that skin and hair pigmentation phenotypes are linked to variations (so-called polymorphisms) in the melanocortin-1 receptor (MCR-1) gene. The resultant protein MCR-1 receptor acts as a 'lock' that is activated or engaged *via* binding a peptide 'key' called alpha-melanocyte stimulating hormone (αMSH). Most northern European individuals with red hair, for example, appear to be have inherited one of a limited number of structural variations in the MCR-1 gene.[6] It is possible that natural selection pressures, which ensured dark hair and skin in the tropics, were less critical as humans migrated northwards (*i.e.* to Europe), thus releasing a 'brake' that permitted opportunities for the MC1-R gene to mutate without hugely deleterious consequences. Some of these mutations in the MC1-R gene resulted in a reduced ability of the MCR-1 'lock' to fully engage the α-MSH, with consequential alteration in skin and hair colour.

3.3 Embryological Development of the Hair Follicle Pigmentary Unit

Cutaneous melanocytes (epidermis and pilosebaceous unit) originate from pluripotent cells that commit to the melanocyte lineage while in the embryonic neural crest. To reach the skin these cells, so-called melanoblasts, leave the neural tube and migrate dorso-laterally and differentiate along stereotypical routes from the closing neural tube, migrate between the derma-myotome of the somites and the overlying ectoderm, until they enter the dermis.[7] Much of our knowledge of the events involved in the development of melanocyte compartments within the skin and hair follicle derives from the analysis of gene mutations that effect differentiation, proliferation and migration of melanocyte precursors.[8] There are over 90 genes shown to affect hair colour in the mouse.[9] Importantly, human equivalents of each continue to be described, where mutations range from total loss of all-over pigmentation (*e.g.* types of albinism) to loss of pigment from specific body sites (*e.g.* piebaldism).[10]

After their long and tortuous migration from the neural crest during human development, melanocytes reach the human epidermis *via* the dermis at week 7 of intra-uterine life. Some of these cells leave the epidermis two weeks later (week 9) and enter the forming pilo-sebaceous units – the anatomic unit consisting of the hair follicle and associated sebaceous gland.[11] There, they appear to distribute randomly as either 'dopa-positive' or 'dopa-negative' melanocytes,[12] where dopa-positivity refers to their ability to produce the rate-limiting enzyme for melanin formation called *tyrosinase* and whether it can oxidise *dihydroxyphenylalanine* (Dopa) and produce the melanin precursor *dopaquinone*. As hair follicle morpho-genesis progresses to the stage of synthesising hair fibre, dopa-positive melanocytes cluster around the apex of the follicular dermal papilla in the hair bulb. Other amelanotic dopa-negative melanocytes occupy positions in the other root sheath (Figure 3.1).

3.4 Biology of the Mature Hair Follicle Pigmentary Unit

While follicular melanocytes are derived from epidermal melanocytes during hair follicle development, these pigment cell sub-populations diverge in many important ways as they distribute to their respective, though open, distinct compartments.

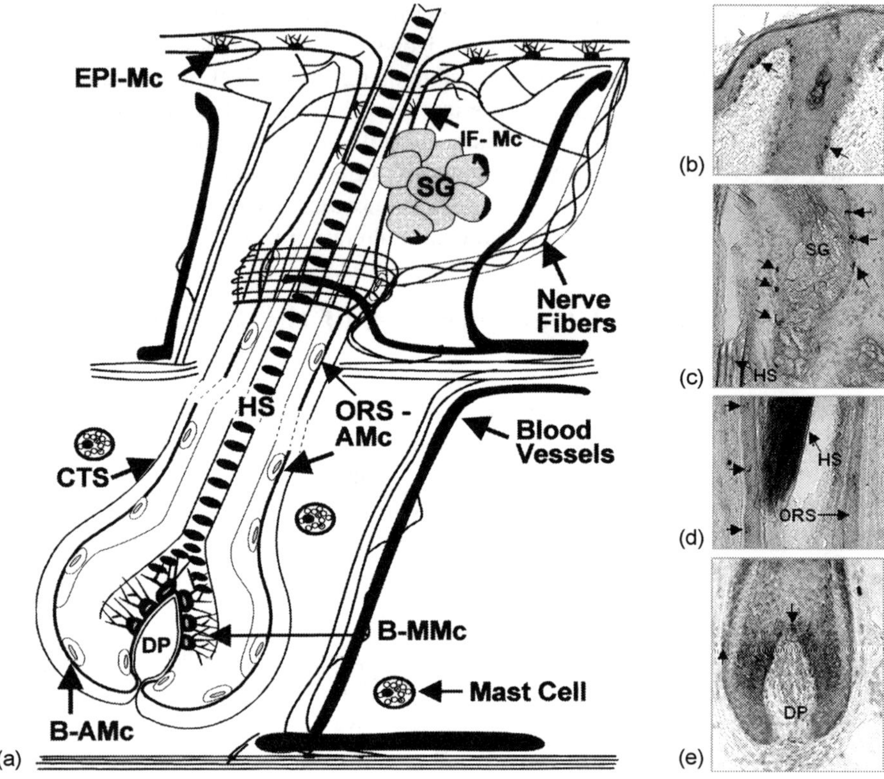

Figure 3.1 *Schematic* (a) *and immunohistologic representations* (b–e) *of the distribution of melanocytes in different regions of the human anagen scalp hair follicle. Melanocytes in frozen scalp sections were detected using a monoclonal antibody to the melanosomal antigen gp100. Key: EPI-Mc, epidermal melanocyte; IF–Mc, infundibulum melanocyte; SG, sebaceous gland; ORS-AMc, outer root sheath amelanogenic melanocyte; DP, dermal papilla; B-MMc, bulbar melanogenic melanocyte; B-AMc, bulbar amelanotic melanocyte; HS, hair shaft; CTS, connective tissue sheath*
(Reproduced with permission from Elsevier and taken from Ref. 13)

For example, hair bulb melanogenic melanocytes are larger, have longer and more extensive dendrites, contain more developed Golgi and rough endoplasmic reticulum, and produce melanosomes 2–4 times larger than those in epidermal melanocytes. While melanin degrades almost completely in the differentiating layers of the epidermis, melanin granules (especially eumelanin) transferred into hair cortical keratinocytes remain minimally digested.[13] In this way, some northern European Caucasians may have very black hair but also have fair, freckle-free skin and blue eyes (*e.g.* the so-called *'black' Irish*). Far the most striking difference between these two melanocyte sub-populations, and one with significant implications for the regulation of hair pigmentation, is the observation that the activity of the hair bulb melanocyte is tightly coupled to the hair growth cycle[14] (see below). Epidermal melanogenesis by contrast, appears to be continuous.[15]

Our knowledge of the cell biology of the hair follicle pigmentary apparatus owes much to the classic anatomic descriptions,[16] though the recent availability of melanocyte-specific probes (*e.g.* antibodies and molecular biology reagents) and genetically engineered murine models has added significant further refinements. As described in Chapter 1 of this volume, the growing anagen VI hair follicle is a useful starting position for a description of the cell biology of the hair follicle pigmentary unit. The growing pilosebaceous unit can be divided into five compartments on the basis of dopa, Masson silver and toluidine blue staining patterns for melanocytes. Dopa-positive melanotic melanocytes occur in three hair follicle locations, namely the basal layer/outer root sheath of the infundibulum, the sebaceous gland and in the hair bulb around the follicular papilla. Dopa-negative and amelanotic hair follicle melanocytes are distributed in the mid-to-lower outer root sheath and also in the peripheral and most proximal hair bulb (Figure 3.1). Some minor variations on this theme occur however, including rare outer root sheath distribution of melanogenically-active melanocytes in the scalp hair bulbs of Black[17] and Chinese[18] individuals among others. While dopa-oxidase activity of tyrosinase may not be detectable in amelanotic hair follicle melanocytes, the protein itself may be detected in some of these cells.[19] Moreover, amelanotic melanocytes may express c-kit and bcl-2,[20] though these cells do not tend to express the melanogenesis enzymes *tyrosinase related protein-1* and *-2* (TRP-1 & TRP-2).[21]

Although several melanocyte sub-populations are located in the hair follicle, the hair bulb is the only site of pigment production for the hair shaft. The melanogenically active melanocytes are restricted to the upper hair matrix of the anagen hair follicle, just below the pre-cortical keratinocyte population[22,23] (Figure 3.1 & Figure 3.2). This location correlates with the fact that melanin is transferred during anagen to the hair shaft cortex, less so to the medulla and, very rarely, to the hair cuticle.

3.4.1 Melanocyte–Keratinocyte Interactions in the Hair Follicle

In a manner akin to the *epidermal melanin unit* of the skin, melanogenically active melanocytes in the hair bulb form functional units with neighbouring immature pre-cortical keratinocytes. The *follicular melanin unit* resides in the proximal anagen bulb, and consists of approximately one melanocyte to five keratinocytes in the hair bulb as a whole and a 1:1 ratio in the basal layer of the hair bulb next to the follicular papilla. Melanogenic bulbar melanocytes interact closely with the follicular papilla, including *via* the thin/permeable basal lamina that separates them from the mesenchymal follicular papilla. By contrast, each epidermal melanocyte is associated with 36 'viable' keratinocytes in the epidermal-melanin unit.[24]

A subject of much current research focus is the mechanism (and its regulation) of melanin granule transfer from the melanocyte to the keratinocyte (see below). Transfer of melanin to cortical cells of the growing hair shaft is presumed to be similar to that in the epidermis.[25,26] It is widely thought to occur *via* cytophagy, where the keratinocyte, as active partner, phagocytoses the tips of melanocyte dendrites that contain mature stage IV melanosomes.[27] Whatever the exact mechanism(s), it is likely that melanocyte dendricity is critical in melanin transfer.[28,29]

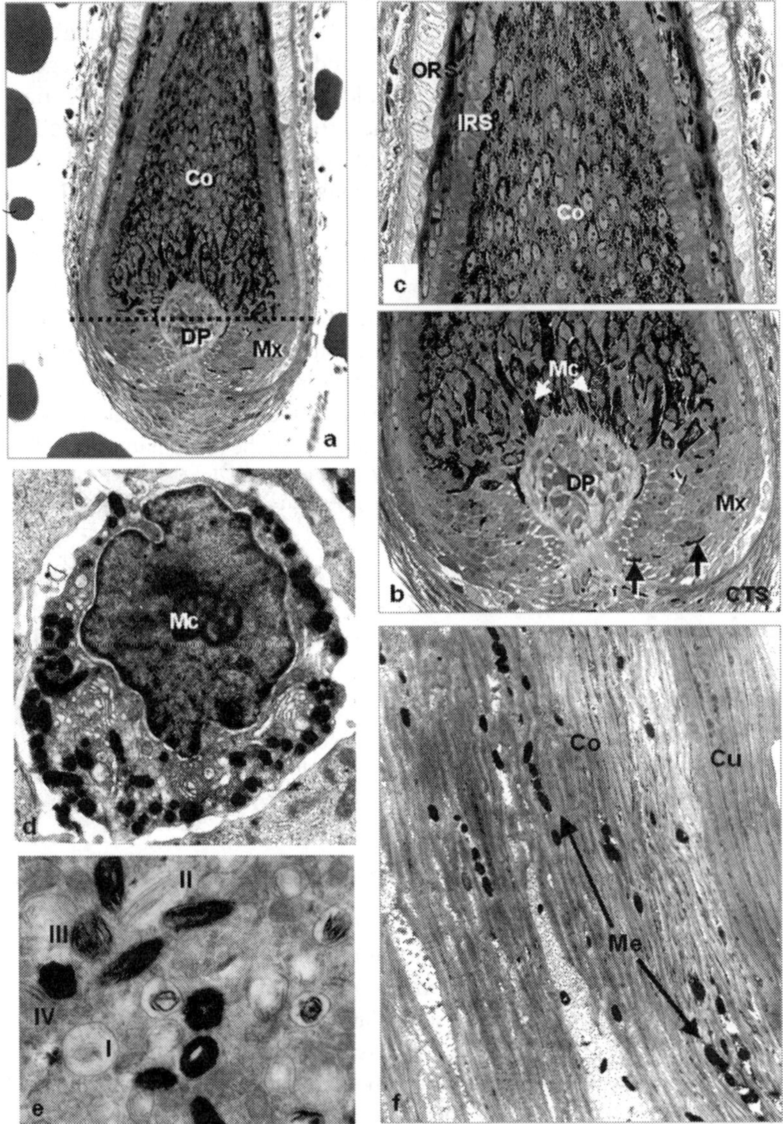

Figure 3.2 *Hair follicle melanocytes in the growing human scalp hair follicle. (a) Low power view of anagen hair bulb showing melanotic melanocytes restricted to above Auber's level of the hair bulb (dashed line). (b) Higher power view of melanogenic zone. Melanogenic melanocyte dendrites penetrate between most proximal matrix (Mx) keratinocytes. (c) Transfer of melanin from melanocytes to pre-cortical (Co) keratinocytes. Little or no melanin enters the inner root sheath (IRS), or outer root sheath (ORS). DP, dermal papilla; CTS, connective tissue sheath. (d&e) Transmission electron micrographs of hair bulb melanotic melanocyte. Note the presence of many melanosomes in different stages of maturation and the extensive Golgi apparatus. (f) Transmission electron micrograph of longitudinal section of hair shaft. Mature (stage IV) undigested melanin granules (Me) distribute between cortical macrofibrils (Co) but not in the hair cuticle (Cu)*

3.5 Biochemistry of Melanin Biosynthesis

The process of melanogenesis can be divided into: a) the formation of the melanosome – the morphologically and functionally unique organelle of melanocytes in which melanogenesis occurs and b) the biochemical pathway that converts L-tyrosine into melanin. Both processes are under complex genetic control that encodes a range of enzymes, structural proteins, transcription factors, receptors and growth factors. For a recent review of this topic, the reader to directed to Slominski, Tobin *et al.*[37]

3.5.1 Melanosome Organellogenesis

Although there is no evidence that melanosome biogenesis in follicular and epidermal melanocytes differs to a significant extent, we must keep an open mind. The formation and maturation of the eumelanosome (true brown/black melanosome) is currently the subject of intense research. Two views have emerged. The first indicates that the enzymatic elements required for melanogenesis are delivered *via* coated vesicles to pre-melanosomes that originate from the Golgi apparatus and endoplasmic reticulum respectively in the cell cytoplasm.[30] An alternate interpretation of melanosome biogenesis, based on the purification and analysis of early melanosomes, suggests that tyrosinase is sorted to early endosomes by the adaptor protein-3 system (from the trans Golgi network) and from there to late endosomes. The latter then fuse with stage I melanosomes.[31]

Melanosome development and maturation have been described morphologically as proceeding *via* four stages. While stage I melanosomes contain tyrosinase and other melanogenesis proteins, these are believed to remain catalytically inactive until subsequent protein cleavage events release them into the melanosome interior. Stage I in eumelanosomes corresponds to early matrix organisation, while in Stage II the matrix is already organised but without melanin formation. These events are associated with a change in melanosome shape, from spherical to ellipsoidal, and the formation of an intra-melanosomal fibrillar network (eumelanosomes). In Stage III melanosomes melanin is deposited, while it fills the melanosome at Stage IV. This orderly process can be deregulated in pathologic conditions. Tyrosinase may already be activated at stage I of some melanosomes, where 'granular type' melanin deposition may occur within organelles that lack either fibrillar or vesiculoglobular matrix.[32,33] However, in general, melanogenesis commences when tyrosinase and other relevant enzymes have been cleaved and this is dependent on an acidic environment that is provided by proton pumps. While tyrosinase is the rate-limiting enzyme of melanogenesis, several other proteins need to be recruited and transported to the melanosome for full functioning of these melanocyte-specific organelles. These include: tyrosinase-related protein-1 (TRP-1), tyrosinase-related protein-2 (TRP-2)/dopachrome tautomerase (DCT), gp100, p-protein and some of the members of the lysosome-associated membrane protein family (LAMP-1,-2 and -3).

Less is known about the events involved in the formation of the pheomelanosome that produces the red/yellow melanin. In contrast to the organised fibrillar network that characterises eumelanosomes, pheomelanosomes contain a vesiculoglobular matrix apparently derived from the fusion of vesiculoglobular bodies with the Stage I melanosomes. Tyrosinase activity appears earlier in these melanosomes such that pheomelanin is already deposited in Stage II melanosomes.[30]

3.5.1.1 Intra- and Inter-racial Variation in Human Scalp Hair Colour

Melanosome structure correlates with the type of melanin they produce. Melanocytes in black hair follicles contain the largest number of most highly electron-dense large melanosomes (eumelanosomes), while melanosomes in brown hair bulbs are somewhat smaller and in blonde hair melanosomes are poorly melanised, often with only the melanosomal matrix visible. Both eumelanogenic and pheomelanogenic melanosomes can co-exist in the same melanocyte,[34] but not within the same biochemical pathway, *e.g.* there is a switch committing melanosomes to either eu- or pheomelanin synthesis.[35] Albino hair melanosomes contain normal-appearing early pre-melanosomes but these fail to melanise.

3.5.2 Biosynthesis of Melanins (Eumelanin and Pheomelanin)

The biochemical reaction that converts the amino acid *L-phenylalanine*, *via* *L-tyrosine*, into a complex and heterogeneous group of compounds called *melanins* can be broadly divided into the following steps: a) the hydroxylation of L-phenylalanine/L-tyrosine to L-dihydroxy-phenylalanine (L-Dopa) – the limiting step in melanogenesis; b) the dehydrogenation/oxidation of L-Dopa – the precursor for both eu- and pheomelanins; c) the dehydrogenation of dihydrox-yindole (DHI) to yield melanin pigment. Eumelanogenesis and pheomelanogenesis both require the oxidation of Dopa to dopaquinone. Thereafter, the conversion of dopaquinone to leukodopachrome signals eumelanin production, while the addition of cysteine to dopaquinone to yield cysteinyldopa occurs in pheomelanin production (Figure 3.3). For eumelanogenesis, L-Dopa needs to be oxidised by tyrosinase to L-dopaquinone and again by tyrosinase from DHI to indole-5,6-quinone. Thus, tyrosinase with both tyrosinase hydroxylase and Dopa oxidase activities not only initiates the melanogenesis pathway but rapidly progresses it. Tyrosinase-related proteins (TRP-1 and DCT (TRP-2)) are important for maintaining the stability of tyrosinase at the melanosomal membrane. Eumelanogenesis is critically dependent on the velocity of the tyrosinase reaction, although this is also stimulated by TRP-1 and DCT. For pheomelanogenesis, the formation of cysteinyldopa is required.[36] Cysteinyldopa is further oxidised in multiple complex steps that may involve tyrosinase-dependent and tyrosinase-independent reactions as well as glutathione reductase and peroxidase activities, in order to form pheomelanin. The exact detail of the melanogenic biochemical pathway awaits full elucidation and is a matter of intense research beyond the scope of this chapter. The reader interested in further reading is directed to reference 37.

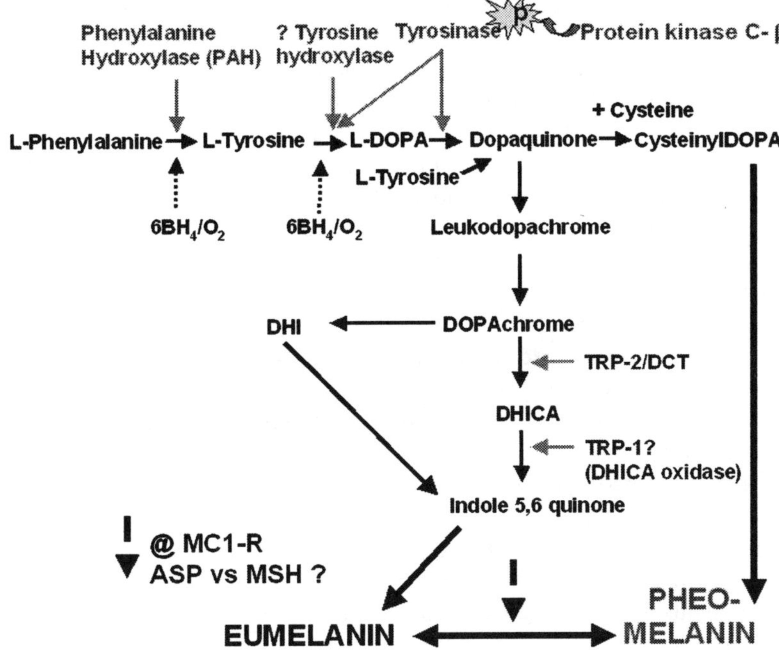

Figure 3.3 *Biosynthetic pathway of melanin. There is some debate about some of the finer details of this pathway, especially whether tyrosine hydroxylase can also act to convert L-tyrosine to L-dopa in the melanocyte, the precise role of 6-tetrahydrobiopterin (6BH₄) and whether PKC-β can be described as a rate-limited enzyme for melanogenesis via its ability to activate tyrosinase*

3.5.3 Physico-chemical Aspects of Hair Melanins

Melanins, the end products of multi-step transformations of L-tyrosine, are polymorphous and multifunctional biopolymers that include the cutaneous melanins: eumelanin, pheomelanin and mixed melanins (containing both eumelanin and pheomelanins),[38] and also neuromelanin.[39] Despite the broad association of eumelanin with brown/black hair and pheomelanins with red/blonde hair, relatively minor differences in melanin content can have significant effects on visible hair colour. Moreover, the blackest of Asian hair melanocytes can produce both eu- and pheomelanins within the same cell. Melanin in hair is tightly bound within a keratinaceous environment and its isolation requires degradation of keratin by strong chemicals.

Melanin pigments have a common arrangement of repeating units linked by carbon-carbon bonds, although they differ from each other with regard to their chemical, structural and physical properties.[39] Eumelanins are polymorphous nitrogenous biopolymers (mostly co-polymers of dihyroxyindole and dihyroxyindole carboxylic acid) and are black to brown in colour. They are insoluble in most solvents and are tightly associated with proteins through covalent bonds.

Eumelanins behave like polyanions and are capable of binding reversibly to cations, anions and polyamines – reactions facilitated by their high carboxyl group content. The semi-quinone units of eumelanin provide it with unique stable paramagnetic states.[40] Thus, the electron paramagnetic resonance (EPR) spectrum of eumelanin reflects its slightly asymmetric singlet which generates a free radical signal. The semiquinone units of eumelanin are also responsible for its redox status, with both reducing and oxidising capabilities towards oxygen radicals and other chemical redox systems.[41] Furthermore, eumelanin's physical structure and electrical properties are consistent with its behaviour as an amorphous semi-conductor.[42–44]

Pheomelanin, in contrast to eumelanin, has a backbone of benzothiazine units and exhibits a yellow to reddish-brown colour. It is alkali soluble,[45,46,39] and is tightly bound to protein – a chromoprotein – with high variability in nitrogen and sulfur content (C/N and C/S ratios). Like eumelanin, pheomelanin can act as a binding agent for drugs and chemicals[47,48] and contains semiquinones with their associated paramagnetic properties. However, pheomelanins also have additional semiquinoneimine centers.[49,50] Thus, their EPR spectra differ from that of eumelanin and so allow identification of melanin type and quantification. Pheomelanins are photo-labile with photolysis products including superoxide, hydroxyl radicals and hydrogen peroxide.[44,39]

3.6 Regulation of Hair Pigmentation

The pigmentation of hair fibres is affected by numerous intrinsic factors including hair-cycle dependent changes, body distribution, racial and gender differences, variable hormone-responsiveness, genetic defects and age-associated change.[37] Study of hair pigmentation may be further complicated by the effects of extrinsic variables including climate and season, infestations, pollutants, toxins and chemical exposure. Given that melanosome biogenesis and melanogenesis involves multiple steps, it is perhaps not surprising that positive and negative regulators of hair follicle melanogenesis will involve multiple biological factors. These include growth factors, cytokines, hormones, neuropeptides and neuro-transmitters, cyclic nucleotides, nutrient microelements and cations/anions.[15] These may act *via* autocrine, paracrine and endocrine mechanisms.[15,37] In stark contrast to its primary regulatory role in epidermal pigmentation, UVB radiation does not penetrate to the melanogenic cells of the anagen hair bulb. Thus, UVR is unlikely to influence the follicular melanin unit, at least directly. Examples of positive regulators of melanogenesis are α-MSH, ACTH, prostaglandin E and endothelin-1 and -3.[15,51,52] Negative regulation is provided by melatonin, IL-1, IL-6, TNF-α and TGF-β.[53–57]

3.6.1 Hair Growth Cycle Influences

Follicular pigmentation occurs only during the hair growth phase (anagen), which in human scalp hair can be very long (average three years, but can be up to ten

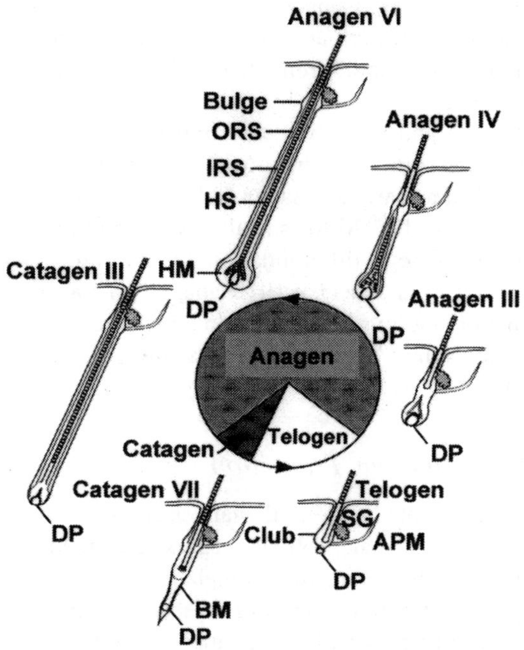

Figure 3.4 *Schematic representation of the hair growth cycle. Note that the melanogenic zone is present only during anagen III–VI. Key: ORS, outer root sheath; IRS, inner root sheath; DP, dermal papilla; HS, hair shaft; APM, arrector pili muscle, SG, sebaceous gland; BM, basement membrane; HM, hair matrix* (Reproduced with permission from Elsevier, taken from Ref. 13)

years) (Figure 3.4). The extended anagen phase of human scalp hair, together with its mosaic pattern of hair growth, hinders systematic analysis of melanocyte dynamics during the human hair cycle. By contrast, the short growth phase (15–17 days), synchronous hair growth pattern and restriction of melanogenically active truncal melanocytes to the hair follicles[58] all make black-haired mice an invaluable model for human hair pigmentation investigation.[59] For an in-depth treatment of this research model the reader is referred to reference 14. The junctions of the hair cycle phases exhibit striking change in the status of the follicular-melanin unit.

3.6.1.1 Telogen to Anagen Transition

The relatively quiescent telogen hair germ contains all cell precursors needed to reconstitute a fully developed anagen VI hair follicle, including the follicular melanin unit.[60] Telogen skin does not contain tyrosinase (either mRNA or protein), TRP-1 protein or melanin.[61] During the first one or two days of anagen induction tyrosinase mRNA and protein become just about detectable, and occasional tyrosinase-positive pre-melanosomes can be detected in amelanotic melanocytes by combined dopa-reaction cytochemistry and electron microscopy.[62] At this stage, the follicular papilla pools high concentrations of L-phenylalanine,[64] conditions that support the production of high amounts of L-tyrosine. In this way, the anagen-associated stimulation of undifferentiated telogen melanocytes/melanoblasts predates the melanogenic stimulus delivered during anagen III. This is followed by active melanogenesis and the subsequent transfer of mature melanosomes into keratinocytes of the pre-cortical matrix. Melanocytes in the S-phase of the cell cycle have been reported as early as anagen II[61,63,65] and significant proliferation is clearly apparent in anagen III.[66] Mitosis is also observed in melanogenically active cells, indicating that melanocyte differentiation does not preclude mitotic activity.[67] Bulbar melanocytes during the anagen III to anagen VI transition increase in their dendricity, develop more Golgi and rough endoplasmic reticulum, increase the size/number of their melanosomes[68] and begin to transfer mature melanosomes to pre-cortical keratinocytes.

3.6.1.2 Anagen to Catagen Transition

Even before catagen-associated structural changes are apparent in the hair bulb, the earliest signs of imminent hair follicle regression include the retraction of melanocyte dendrites and the attenuation of melanogenesis during late anagen VI.[68] Limited keratinocyte proliferation continues for a while, leaving the most proximal telogen hair shaft characteristically unpigmented. One can already detect a dramatic and rapid drop in levels of active tyrosinase beginning during late anagen VI itself. TRP-2 (DCT) and DHICA-CF activities also exhibit moderate reductions from mid to late anagen VI and are lowest during catagen. The termination of melanogenesis may reflect a swamping of a melanogenesis-dependent signalling system or the induction of melanogenesis inhibitory factors, *e.g.* IL-1, IL-6, INF-γ, TGF-β, TNF-α or corticosteroids.[37] Alternatively, the low supply of L-tyrosine in the now reducing hair bulb environment inhibits melanogenesis, until an increased production of GTP-cyclohydrolase-1, 6-tetrahydrobiopterin and phenylalanine hydroxylase occurs again during telogen in preparation for optimal melanogenesis conditions during the subsequent anagen phase.[64]

3.6.1.3 Fate of Pigmented Melanocytes During Catagen

An enduring enigma of both hair follicle and pigment biology concerns the fate of the hair bulb melanocytes during catagen (Figure 3.5). Particularly relevant questions are where formerly melanogenically-active melanocytes go during catagen and telogen, and where they originate from, when follicular melanogenesis is resumed during the next anagen phase. Although melanogenically active melanocytes are no longer detectable in the proximal hair follicle during catagen, their 'disappearance' is not unheralded and residual melanin generated during anagen can be seen 'deposited' in the follicular papilla.[16] A long held view in hair biology is that the hair bulb melanocyte system is a self-perpetuating arrangement, whereby melanocytes involved in the pigmentation of one hair generation are also involved in the pigmentation of the next.[69] While there is evidence of some plasticity in the hair follicle pigmentary unit, the level invoked by the

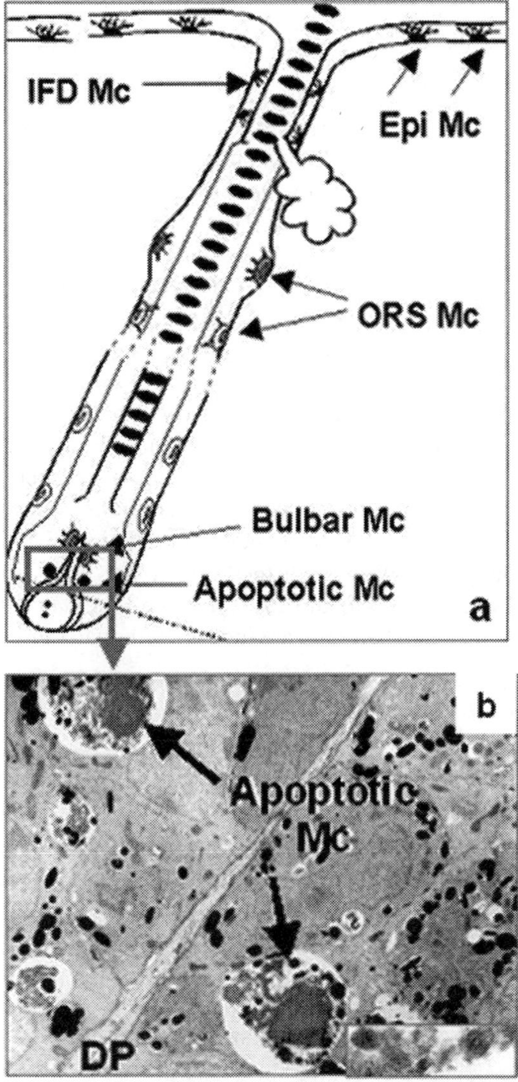

Figure 3.5 *Apoptosis of melanotic hair bulb melanocytes in an early murine hair follicle.* (a) *Cartoon showing the distribution of melanocytes in early catagen hair follicle. Loss of differentiated phenotype in some melanotic bulbar melanocytes and apoptotic change in others.* (b) *Transmission electron micrograph of a longitudinally sectioned early catagen murine follicle exhibiting two apoptotic melanotic hair bulb melanocytes. Inset: High power view of pre-melanosomes* (Reproduced with permission from Elsevier and taken from Ref. 13)

self-perpetuating theory would imply a degree of plasticity not seen in most non-malignant cell systems. Moreover, fully differentiated bulbar melanocytes would also need to survive/avoid the extensive apoptosis-driven regression of the hair bulb[70,71] by actively suppressing apoptosis.

Thus, our current view suggests that many of the so-called 'redifferentiating' melanocytes in early anagen correspond to newly-recruited immature melanocytes derived from a melanocyte reservoir[67,72] and are not re-activated from pre-existing hair bulb melanocytes that were melanogenically-active during the previous anagen phase. This is supported by the recent identification of a population of immature DCT+ melanocytes in the hair follicle 'bulge' region – a site of epithelial stem cells. Unlike other follicular melanocytes, this sub-population was not affected by an anti-ckit blocking antibody.[73] Having said that, it is possible that some of the 'new generation' melanogenically-active melanocytes derive from a population of catagen-surviving post-melanogenically active cells. Indeed, low numbers of apparently dendritic melanocytes can be detectable in the retreating epithelial strand of catagen hair follicles undergoing active resorption *via* apoptosis.[65,19] However, these weakly or non-melanogenic cells, lack tyrosinase and TRP-1 expression, and may in fact represent the poorly differentiated melanocytes that co-distribute with pigmented bulbar melanocytes in anagen hair matrix.[23]

Explaining most, if not all, of the aforementioned observed changes during catagen is perhaps the demonstration that the highly melanotic (terminally-differentiated?) hair bulb melanocytes do not survive catagen.[74] Deletion of individual melanotic melanocytes by apoptosis was confirmed using well-described ultrastructural features and TUNEL/TRP-1 co-localisation, and the vast majority of cells attached to the basal lamina of catagen hair bulb are indeed epithelial, rather than melanocytic.

3.6.1.4 Pigment Incontinence During Catagen

Curiously, not all the pigment formed during anagen-associated melanogenesis is actually incorporated into the hair shaft. Why more melanin should be produced than can be incorporated into the hair shaft is unclear but raises the possibility that follicular melanin may serve other purposes in skin homeostasis. The 'excess' pigment appears to be removed, though not immediately, from the hair follicle. Its origin appears to be related to melanocyte degeneration by apoptosis and the resultant pigment-containing apoptotic fragments enter the follicular papilla (Figure 3.6), epithelial strand or connective tissue sheath (CTS) of catagen. Pigment incontinence may also be detected in the epithelial sac of the telogen hair follicle. The precise mechanism(s) of this pigment redistribution is unclear, although it is likely to involve phagocytosis, particularly by macrophages[75,76] and Langerhans cells,[77,18] whose numbers increase during hair follicle regression,[75] or *via* ingestion by the follicular papilla fibroblasts themselves. Indeed, Langerhans cells have been observed to remove pigment from the regressing hair matrix to the follicular papilla *via* direct phagocytosis or *via* Langerhans granule-associated endocytosis of (pre)-melanosomes.[77]

3.6.2 Hormonal Influences and Pigmentation

3.6.2.1 Positive Regulators of Follicular Melanogenesis

Among positive regulators of follicular melanogenesis are the proopiomelano-cortin (POMC)-derived peptides ACTH, α-MSH and β-MSH.[51,60,78–80] These

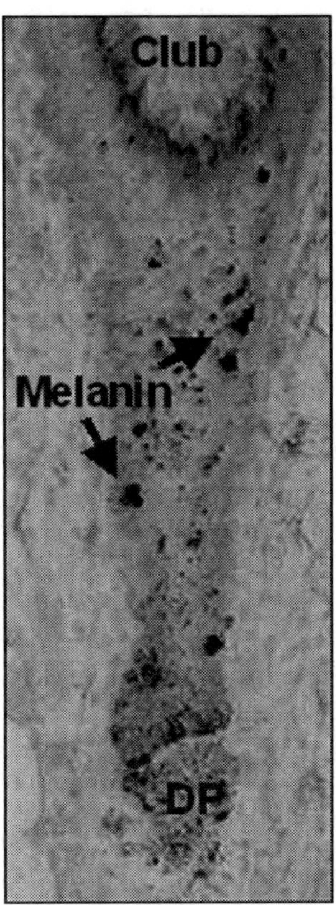

Figure 3.6　*Light micrograph of melanin incontinence in advanced catagen in human scalp hair follicle. Note the presence of melanin granules throughout the trailing epithelial strand below the developing telogen club and within the dermal papilla (DP)*

neuropeptides are also produced locally in anagen skin or can be released locally from nerve endings.[80–83] Anagen-associated melanogenic activity is accompanied by POMC gene and gene product expression,[80–83,62] as well as an increased expression of the MC1 receptor (MC1-R) gene – both of which then decrease during the anagen-catagen transition.[84] Moreover, this accumulation of POMC products is found predominantly in the keratinocytes of the hair follicle rather than in the epidermis – suggesting that the activity of the local POMC/MC1-R system plays an important role in the physiological regulation of anagen-associated hair pigmentation. In this manner, polymorphism in the MC1-R gene has been strongly linked to red hair and fair skin in humans.[85] The binding of α-MSH and ACTH to the MCR-1 receptor induces a signal transduction cascade that results in the activation of adenylate cyclase. Subsequent cAMP production results in increased melanocyte proliferation, melanogenesis and dendrite formation.[86]

We rely heavily on murine studies for relevant information regarding the regulation of human hair pigmentation and so we need to interpret the rodent coat colour literature carefully before extrapolating to humans. For example, studies in guinea pigs have shown that α-MSH increased the proportion of black to gray hairs when administered intramuscularly,[87] while the injection of both α-MSH and potent synthetic analogue (Nle^4, D-Phe^7)-α-MSH into human skin was observed to increase melanogenesis particularly of sun-exposed skin and no effect seen in hair follicles.[88] In support of this, we have recently found that the expression of α-MSH peptide is very low to undetectable in pigmented hair bulb melanocytes in stark contrast to their epidermal counterparts both *in vivo* and *in vitro*.[89]

Recently, we have proposed a role for the β-endorphin/μ-opiate receptor system in the regulation of both epidermal and follicular melanocyte biology (dendricity, proliferation and melanogenesis),[90,91] suggesting that POMC peptides may also modulate melanocyte biology in a non-MC1-R-dependent manner. Moreover, the most proximal hormone of the HPA-axis corticotrophin-releasing hormone (CRH), can also induce phenotypic change in hair follicle melanocytes *in vitro* by upregulating cell dendricity and pigmentation levels,[92] though it is not yet clear if this activity operates *via* direct or indirect mechanisms.

3.6.2.2 Negative Regulators of Follicular Melanogenesis

Agouti signalling protein (ASP) is likely to be an important regulator of hair pigmentation as it can act as an antagonist to α-MSH and an inverse agonist at the MC1-R[93] – so can be classified as both a modulator and an inhibitor of melanogenesis.[52] Transient synthesis of agouti protein switches eumelanogenesis (brown/black) to pheomelanogenesis (red/yellow),[94] a change that can be associated with decreased tyrosinase activity overall. Moreover, the products of Attractin (Atrn) – a type I transmembrane protein – and of Mahogunin (Mgrn1) – a E3 ubiquitin ligase – may also act as modulators of pigmentation as they appear to prevent hair follicle melanocytes from responding to the agouti protein.[95] This system has so far been described in rodents, though a similar system is thought likely to operate in humans.

Several other negative regulators of hair pigmentation include melatonin, glucocorticoids, interferon gamma (INF-γ) and dopaminergic and cholinergic agonists.[53,84,93,94,96,97] Melatonin, albeit at pharmacological doses, inhibits tyrosinase activity in the organ cultures of murine anagen skin,[98] and anagen epidermis and hair follicles express melatonin binding sites.[99]

3.6.3 Nutrition and Hair Pigmentation

Nutritional factors that are crucial for melanin synthesis include the amino acids L-phenylalanine, L-tyrosine and L-Dopa.[100,101] L-Dopa is thought to act not only as a substrate for melanogenesis but also as a positive regulator in the conversion of L-tyrosine to melanin.[101–106]

3.6.4 Transfer and Degradation of Melanin Granules to Hair Shaft Keratinocytes

The cytoplasm of melanocytes in the hair bulb commonly contain a high proportion of fully mature stage IV melanosomes, in contrast to epidermal melanocytes which rarely collect significant numbers of mature melanosomes. In the latter, these maturing melanosomes are swiftly transferred to surrounding epidermal keratinocytes. There is evidence supporting several mechanisms of melanosome transfer including the 'cytophagic' theory (where keratinocytes phagocytose the tips of dendrites containing stage IV melanosomes),[107] the 'discharge' theory (mature melanosomes are released into the intercellular space to be internalised by adjacent keratinocytes), the 'fusion' theory (mature melanosomes pass from melanocyte to keratinocyte *via* fusion of their respective plasma membranes.[108] More recently, filopodia from melanocyte dendrites have been observed to act as conduits for melanosome transfer to keratinocytes,[109] and so melanocyte dendricity is an important phenotypic regulator of melanin transfer to keratinocytes. Notably, myosin Va (encoded by the *dilute* gene, which when mutated can be associated with dilution of hair colour *e.g.* in Griscelli syndrome[110] has also been proposed as the molecular motor involved in dendrite outgrowth and melanosome movement in mammalian melanocytes.[111,112] Moreover, phagocytosis of melanosomes by keratinocytes is mediated *via* the activation of the protease-activated receptor 2 (PAR2) on keratinocytes and inhibitors of PAR2 retard melanosome transfer.[113] Transfer of melanin granules to cortical and medullary keratinocytes of the growing hair shaft is presumed to involve the same mechanism as that in the epidermis, although its ultimate fate, *i.e.* degradation, is strikingly different compared to its status in the epidermis. Melanin granules transferred into hair cortical keratinocytes remain minimally digested, contrasting with epidermis where melanin is almost completely degraded in the differentiating keratinocytes (further discussion in Chapter 2 of this volume). As a consequence of differential processing, eumelanic white individuals may have black hair but very fair skin and blue eyes (*e.g.* in parts of Ireland). The regulation of melanosomes movement and their transfer to keratinocytes or release into the extracellular environment is an important and continuously evolving field of skin biology, which may provide novel information not only on mechanisms of epidermal and hair follicle pigmentation but also on the putative role of melanin granules in the local regulation of cutaneous homeostasis.

3.7 Aging of the Hair Follicle Pigmentary Unit

The hair pigmentary unit goes through significant age-associated changes that range from short fine and usually unpigmented *lanugo* hair at about three months of intra-uterine life, to the white *terminal* hair of old age. In between is childhood *intermediate* hair that after puberty may become, as with scalp hair, more deeply pigmented *terminal* hair. A partial reverse sequence of events is seen with reduction of pigmentation in miniaturising hair follicles in androgenetic alopecia (common male pattern baldness). However, it is in graying/canities where one observes the most dramatic age-related change in hair pigmentation.

3.7.1 Molecular Aspects of Melanocyte Aging

By far the most popular theory for cell aging and perhaps one most easily tested *in vitro* is the 'free radical' theory of aging.[114,115] While the accumulation of oxidative damage is an important determinant in the rate of aging, it is unclear whether it is the primary cause of aging. Here, reactive oxygen species (ROS) damage DNA (both nuclear and mitochondrial), which leads to the accumulation of mutations, induces oxidative stress and consequentially also induces antioxidant mechanisms. It is possible that this antioxidant system becomes impaired with age leading to uncontrolled damage to the melanocyte itself generated from its own melanogenesis-related oxidative stress. A less than effective antioxidant system will lead to an increased sensitivity to peroxidising agents.[116]

Age-associated accumulation of mutations occurs at a higher rate in tissues exposed to high levels of oxidative stress. Melanin synthesis, by its very nature, produces mutagenic intermediates.[117] The extraordinary melanogenic activity of pigmented bulbar melanocytes, which may continue for up to ten years in some hair follicles,[118] is likely to generate large amounts of ROS *via* the oxidation of tyrosine and Dopa to melanin.[119] If not adequately removed, an accumulation of these ROS may generate significant oxidative stress in both the melanocyte itself and also in the highly proliferative anagen hair bulb epithelium. H_2O_2, even in low doses, can induce senescence in cultured fibroblasts.[120] Thus, in these circumstances, melanogenic bulbar melanocytes are perhaps best suited to assume a post-mitotic, terminally differentiated '(pre-)senescence' status to prevent cell transformation.[121]

It has been observed that loss of melanocyte replicative potential is associated with increased melanin content. Millimolar L-tyrosine (melanin precursor) abrogates proliferation in cultured pigmented melanoma cells, with proliferation continuing only in amelanotic cells.[122] Pigmented melanocytes under these conditions assume large, epithelioid, stellate morphologies and have been described as 'pre-senescent' cells. They exhibit increased expression of some cyclin-dependent kinase inhibitors (CDK).[123] On reaching senescence, melanocytes express increased levels of the CDK inhibitors p21 and p16; interestingly the latter is commonly mutated or deleted in familial melanoma.[124] Lost proliferative capacity in hyperpigmented cells appears to result from an inability to activate the MAP kinase pathway necessary for proliferation.[125]

ROS have been implicated directly in hair follicle melanocyte injury. Dilution of hair colour is found in bcl-2-deficient mice[126] by 4–5 weeks of age. Significantly, the anti-apoptotic protein BCL-2 is expressed in sites where ROS are generated,[127,128] *e.g.* mitochondria, endoplasmic reticulum and nuclear membranes where it prevents oxidative damage to cellular constituents including lipid membranes. The lower BCL-2 expression in melanotic hair bulb melanocytes, compared with epidermal melanocytes, may render these cells more susceptible to apoptosis. As the primary cellular target for oxidative damage appears to be the energy-generating mitochondrion, the so-called 'mitochondria theory of aging' has emerged,[129] albeit with significant caveats.[130] Age-related mitochondrial DNA deletions can cause mitochondrial degeneration that, without adequate removal, can compromise cell survival.[131]

Hepatocyte growth factor (HGF) may regulate the survival of melanocytes, as well as melanocyte proliferation and differentiation *in vivo* in a paracrine manner. HGF may even alter melanocyte distribution within the skin *via* down-regulating E-cadherin expression on melanocytes.[132]

3.7.2 Onset and Progression of Hair Graying

Canities appears to be inherited in an autosomal dominant manner; entire kinships may present with marked early graying throughout. The mechanisms of melanocyte loss in both the epidermis and hair follicle are unknown and as yet there has been no convincing explanation for the observed differences in the aging 'state' of these two skin melanocyte sub-populations.

For every decade after 30 years of age the number of pigment-producing melanocytes in exposed/unexposed epidermis decreases by 10–20%.[133] This age-associated loss of dopa-positive epidermal melanocytes occurs body-wide, and is associated with a very gradual reduction in skin colour. By contrast, age-linked loss of colour from hair is dramatic, suggesting that the hair pigmentary unit has a different 'melanogenetic clock'. It is possible that the antioxidant system of both epidermal- and follicular-melanin units becomes variably impaired with age leading to uncontrolled damage to the melanocyte itself from its own melanogenesis-related oxidative stress.

The onset and progression of hair graying correlates closely with chronological aging and occurs to varying degrees in all individuals, regardless of gender or race. Age of onset also appears to be hereditary, occurring usually late in the fourth decade.[134] Thus, the average age for Caucasians is mid 30s, for Asians late 30s and for Africans mid 40s. Hair is said to gray *prematurely* if it occurs before the age of 20 in Whites, before 25 in Asians and before 30 in Africans. The progress of canities is entirely individual: a good rule of thumb is that by 50 years of age, 50% of people have 50% gray hair.[134,135] Clearly, the darker the hair the more noticeable early graying will be. However, graying can be more extensive in dark hair before total whitening is apparent; the reverse is true for blond hair. Graying first appears usually at the temples, and spreads to the vertex and then the remainder of the scalp, affecting the occiput last. Beard and body hair is usually affected later. Graying often follows a wave that spreads slowly from the crown to the occiput.

A characteristic feature of bulbar melanocytes is their extremely high melanin load throughout the entire time that the pigmented hair fibre is forming during anagen – up to ten years in the scalps of some people. This represents a phenomenal synthetic capacity for melanin production (Figure 3.2), whereby a relatively small number of melanocytes can, in a single hair growth cycle, produce sufficient melanin to intensely pigment up to 1.5 m of hair shaft. Moreover, they do this within the context of a melanin-laden cytoplasm, unlike melanogenically-active epidermal melanocytes that retain few fully-mature melanosomes in their cytoplasm at any one time. This 'melanin loading' of bulbar melanocytes is likely to make these cells much more vulnerable than epidermal melanocytes to the toxic elements of melanogenesis.

The synthetic capacity of bulbar melanocytes is greatest during youth when the scalp follicular melanin unit is only a few cycles old. On average therefore, an individual scalp hair follicle will experience approximately 7–15 melanocyte seedings/replacements from the presumptive reservoir in the outer root sheath to the hair bulb in the average 'gray-free' life span of 45 years.[134] Interestingly, repeated plucking of hair from vibrissae follicles leads to the eventual regrowth of gray hair,[136] though the associated tissue injury complicates the interpretation of this finding. Furthermore, the precise rate of canities progression is also complicated by the observation that with advancing age hair follicles remain in the telogen 'resting' phase for longer, suggesting that epithelial stem activation/migration may also become more sluggish with age.

3.7.3 Pathogenesis of Loss of Hair Pigmentation

'Gray hair' has been considered illusory, more an impression of grayness provided by an admixture of fully white and fully pigmented hair. However, canities can indeed affect individual hair follicles with either a gradual loss of pigment over time and over several cycles, a gradual loss of pigment along the same hair shaft (*i.e.* within the growing anagen phase of a *single* hair cycle), or the hair fibre may appear to 'grow in' fully depigmented (Figure 3.7). While few pigment granules are present in truly white hair shafts, melanin granules can be readily detected within the pre-cortex of gray hair follicles. Pigment loss in graying hair follicles is due to a marked reduction in melanogenically active melanocytes in the hair bulb of gray anagen hair follicles (Figure 3.8).[137,138] True gray hairs show a much-reduced, but detectable, dopa reaction as an indicator of tyrosinase activity compared with fully pigmented hair follicles, while white hair bulbs are broadly negative. However, there appears also to be a specific defect of melanosome transfer in graying hair follicles, as keratinocytes may fail to contain any melanin granules despite being in close proximity to melanocytes with a moderate number of melanosomes.[16] Further evidence of some defect in melanocyte/keratinocyte interaction is provided by the observation of significant melanin debris both in the graying hair bulb and sometimes also in the surrounding dermis. This abnormality is due to either defective melanosomal transfer to the cortical keratinocytes and/or melanin incontinence due to melanocyte degeneration. The remaining hair bulb melanocytes in canities-affected anagen hair follicles often appear hypertrophic, although this may reflect a reduction in dendricity rather than an overall increase in cell volume.[139]

Ultrastructural analysis of the human gray hair matrix reveals melanocytes with highly variable levels of melanogenesis.[139] In gray/white hair bulbs, remaining melanocytes contain fewer and smaller melanosomes and less supporting organelles, *e.g.* Golgi apparatus. Interestingly, the remaining melanosomes may be packaged within auto-phagolysosomes suggesting that these melanosomes are defective, perhaps even leaking reactive melanin metabolites. Auto-phagolysosomal degradation of melanosomes is usually followed by the degeneration of the melanocyte itself.[140,141] The involvement of ROS in the histopathology of canities is suggested by the observation that melanocytes in graying and white hair bulbs

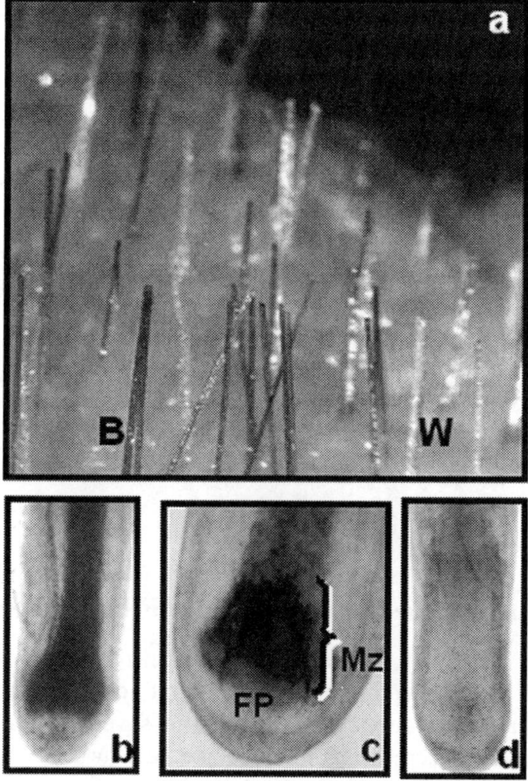

Figure 3.7 (a) *Admixture of hair shafts in scalp with variable pigmentation level; white (W) and black (B). Isolated fully pigmented (b,c) and depigmented (d) intact anagen hair bulbs. Note close association of dermal papilla (DP) with the melanogenic zone (Mz)*

may be vacuolated – a common cellular response to increased oxidative stress.[142] Degenerative change in canities-affected hair bulbs may resemble apoptosis and is reminiscent of melanocyte degeneration in acute alopecia areata where pigmented hair follicles are preferentially targeted by an aberrant immune response.[143] Loss of melanocytes from canities-affected hair bulbs can apparently occur very rapidly. Evidence for this can be found in the pigment incontinence located in the follicular papilla and/or connective tissue sheath of hair follicles that lack any morphological evidence of melanogenesis or melanocytes in their hair bulb. The loss of active melanocytes from the hair bulb of graying and white hair follicles may be associated with a parallel increase in the number of dendritic cells (including Langerhans cells).[144] The relocation of these antigen-presenting phagocytic cells from the upper hair follicle to the lower follicle may be in response to degenerative change within the melanocyte population, with its likely associated antigenic effects.

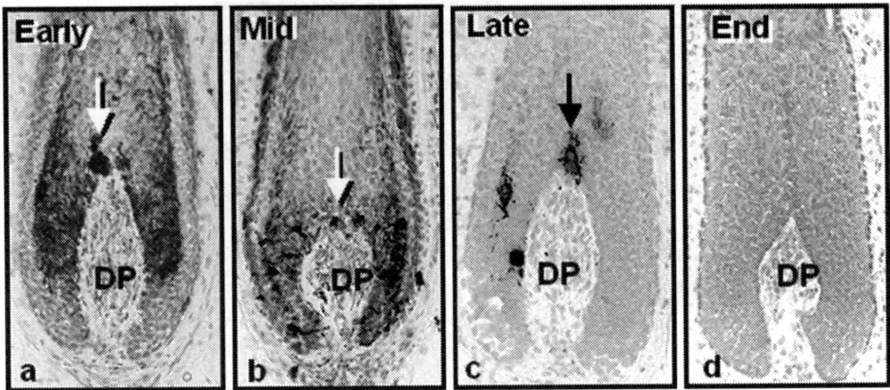

Figure 3.8   *Longitudinal sections of canities-affected human scalp anagen hair follicles. The numbers of melanocytes and level of melanin production (arrows) varies considerably from early* (a), *mid* (b), *late* (c) *and final stages of this condition*

3.7.4 Impact of Pigment Loss on Hair Fibre Structure

Given the close interaction between melanin-transferring melanocytes and hair shaft–forming/melanin-accepting pre-cortical keratinocytes, it is likely that bulbar melanocytes influence keratinocyte behaviour in several ways. Melanin transfer appears to promote a decrease in keratinocyte cell turnover and an increase in keratinocyte terminal differentiation – perhaps by altering intercellular calcium levels. White beard hair has been shown to grow at up to three times the rate of adjacent pigmented hair.[145] In this way melanosomes donated to keratinocytes may act as regulators that control their level of cell differentiation and even metabolic status.[146] Melanocytes may also influence neighbouring keratinocytes *via* the production of various cytokines, growth factors, eicosinoids, adhesion molecules and extracellular matrix.[147] Similarly, the ability of melanins to provide a buffer for calcium is likely to have implications for cell function, given the critical second messenger/cell signalling role for calcium in melanogenesis, melanosome transfer and epithelial cell differentiation.[148] The saturation binding of transition metals (*e.g.* iron, copper *etc.*) to melanin provides yet another effective anti-oxidant defense mechanism for the melanosome-receiving keratinocyte.

Further clinical evidence of melanocyte-keratinocyte interactivity can be seen in the anecdotal impressions that gray hair is coarser, wirier and more unmanageable than pigmented hair. Indeed, gray hair is often unable to hold a 'permanent' or 'temporary set' and may be more resistant to incorporating artificial colour. These observations suggest significant change to the underlying sub-structure of the hair shaft, whereby aging hair follicles may reprogram their matrix keratinocytes to increase production of medullary, rather than cortical, keratinocytes.

3.7.5 Can Canities Serve as a Marker for Disease?

Although controversial, there is some suggestive evidence that graying may be a marker for general health status. Cigarette smoking has been linked with premature gray and even hair loss,[147] although this may rather reflect smoking-related pathology that increases aging of many body systems including pigmentation, or to involve smoke genotoxin-induced apoptosis. Moreover, it has been reported that individuals with premature canities are more likely to develop osteopenia than individuals without canities[149] and that people who grayed before their 20s have lower bone mineral density compared to those who grayed later. Purported associations between early onset of gray hair and cardiovascular disease[150] or studies showing graying of hair, male baldness and facial wrinkling as additional risk factors for myocardial infarction are less clear.

3.7.6 Is Canities Reversible?

Canities-associated pigment loss results from the loss of melanocytes from the melanogenic zone of the hair follicle. Such 'senile' white hair follicles retain, however, amelanotic melanocytes in the outer root sheath. These cells for the most part remain not only dopa-negative and but also negative for most melanocyte-specific markers.[21] While the precise role of outer root sheath amelanotic melanocytes in hair and skin biology is far from clear, these cells appear to be recruitable for repigmentation/repopulation of the epidermis if necessary (*e.g.* in vitiligo).[151] Their failure to contribute to the pigmentation of senile white hair follicles may reflect the lack of a permissive environment for their migration to the melanogenic zone during early anagen. Only melanocytes that have successfully migrated to the hair bulb appear to be susceptible to local pigmentation-inducing influences, *e.g.* anagen-induced secretion of follicular dermal papilla-derived factors. The provision of some of these stimuli *in vitro* can result in the induction of melanogenesis in the amelanotic cells of senile white hair follicles.[16,152] This finding suggests that these cells retain the melanogenic machinery intact and so could be induced to become active again in more permissive *in vivo* micro-environments.

The deficit in canities-affected hair follicles is likely to be multi-factorial. Primary amongst these may be defective migratory stimuli, particularly during the critical stages of the hair cycle when cell-cell and cell-matrix interactions are highly active. Spontaneous scalp hair re-pigmentation has been reported after radiation therapy for cancer[153] or after inflammatory events, *e.g.* erythrodermic eczema and erosive candidiasis of the scalp.[154] Here, it is most probable that the reversal of canities resulted from a radiation/cytokine-induced activation of outer root sheath melanocytes. These clinical observations raise the attractive possibility that these melanocytes may be induced to migrate and differentiate to naturally re-pigment graying hair follicles. Another clinical scenario that provides an insight into both the pathomechanism of canities and possibilities for pigment recovery is the not too uncommon partial spontaneous reversal of canities that occurs during the early stages of canities. Here, melanogenesis in de-activated bulbar melanocytes may re-start during anagen VI of the same hair growth cycle.[18] Study of hair follicles at this point in canities may provide several clues to help us identify the subtle changes in the hair follicle's two melanocyte sub-populations.

3.8 Conclusion

The evolutionary selective pressures for hair pigmentation must have been significant for such an investment of energy resources. Still, its value to modern humans, beyond social and sexual communication, remains enigmatic. The distinct, but open, melanocyte compartments in the epidermis and hair follicle attest to the bi-functionality of melanocytes in the upper hair follicle, *i.e.* these cells aid the re-pigmentation of both epidermis and the new anagen hair bulb. In this way, this follicular melanocyte population provides a very important cell reservoir function, *e.g.* after epidermal injury. Moreover, amelanotic melanocytes in the outer root sheath, by not devoting resources to melanin synthesis, remain plastic and are available to participate in other, as yet poorly defined, inter-cellular processes. Melanotic bulbar melanocytes on the other hand, appear to be largely devoted to producing very large amounts of this melanin. However, the bio-active characteristics of melanin beyond colour are likely to be considerable, especially after transfer of melanin granules to the hair shaft-forming keratinocytes.

We are still only beginning to unravel the mysteries of hair pigmentation in humans. While geneticists continue to study pigmentation-associated gene polymorphism that will help track human migrations throughout pre-history, the diversity of differentiation states exhibited by follicular melanocyte sub-populations, and natural aging of these neural crest-derived cells, continues to offer the biologist new insights into the full potential of this part-time pigmenter, immune cell, homeostasis regulator etc.

3.9 Acknowledgements

Work of the author described in this chapter was supported in part by grants from National Alopecia Areata Foundation, NIH, Proctor & Gamble Ltd. (UK) and Wella Cosmital (Germany).

3.10 References

1. J.M. Wood, K. Jimbow, R.E. Boissy, A. Slominski, P.M. Plonka, J. Slawinski, J. Wortsman and J. Tosk, *Exp. Dermatol.*, 1999, **8**, 153.
2. E. Morgan in *The Ascent of Woman*, Souvenir Press, London, 1985.
3. A. Bertazzo, C. Costa, M. Biasiolo, G. Allegri, G. Cirrincione and G. Presti, *Biol. Trace. Elem. Res.*, 1996, **52**, 37.
4. M. Pagel, W. Bodmer, *Proc. R. Soc. Lond. B. Biol. Sci.*, 2003, **270 (Suppl. 1)**, S117.
5. J.A. Mackintosh, *J. Theor. Biol.*, 2001, **211(2)**, 101.
6. J.L. Rees, *Pigment Cell Res.*, 2000, **13**, 135.

7. M.E. Rawles, *Physiol. Zool.*, 1947, **20**, 248.

8. I.J. Jackson, *Annual Rev. Genet.*, 1994, **28**,189.

9. M. Nakamura, D.J. Tobin, B. Richards-Smith, J.P. Sundberg and R. Paus, *J. Dermatol. Science.*, 2002, **28**, 1.

10. R.A. Fleischman, T. Gallardo and X. Mi, *J. Invest. Dermatol.*, 1996, **107**, 703.

11. K.A. Holbrook, A.M. Vogel, R.A. Underwood and C.A. Foster, *Pigment Cell Res. (Suppl)*, 1988, **1**, 6.

12. H.B. Chase, H. Rauch and V.W. Smith, *Physiol. Zool.*, 1951, **25**, 1.

13. D.J. Tobin and R. Paus, *Exp. Gerontol.*, 2001, **36**, 29.

14. A. Slominski and R. Paus, *J. Invest. Dermatol.*, 1993, **101**, 90S.

15. J.J. Nordlund and J.-P. Ortonne, The normal colour of human skin, in: J.J. Nordlund, R.E. Boissy, V.J. Hearing, R.A. King and J.-P. Ortonne (ed) *The Pigmentary System: Physiology and Pathophysiology*. Oxford University Press: New York, 1998, 475.

16. D.J. Tobin in *Skin, Hair, and Nails – Structure and Function*, B. Forslind and M. Lindberg (ed), Marcel Dekker, New York, 2004, 319.

17. M. Ito, *Arch. Dermatol. Res.*, 1991, **283**, 274.

18. D.J. Tobin and J.A. Cargnello, *Arch. Dermatol.*, 1992, **129**, 789.

19. J.R. O'Sullivan, A. Williams, C. Gummer and D.J. Tobin, *J. Invest. Dermatol.*, 2000, **114**, 861.

20. J.M. Grichnik, W.N. Ali, J.A. Burch, J.D. Byers, C.A. Garcia, R.E. Clark and C.R. Shea, *J. Invest. Dermatol.*, 1996, **106**, 967.

21. T. Horikawa, D.A. Norris, T.W. Johnson, T. Zekman, N. Dunscomb, S.D. Bennion, R.L. Jackson and J.G. Morelli, *J. Invest. Dermatol.*, 1996, **106**, 28.

22. D.J. Tobin and J.C. Bystryn, *Pigment Cell Res.*, 1996, **9**, 304.

23. D.J. Tobin, S.R. Colen and J.C. Bystryn, *J. Invest. Dermatol.*, 1995, **104**, 86.

24. T.B. Fitzpatrick and A.S. Breathnach, *Dermatol. Wochenschr.*, 1963, **147**, 481.

25. M. Hara, M. Yaar, H.R. Byers, D. Goukassian, R.E. Fine, J. Gonsalves and B.A. Gilchrest, *J. Invest. Dermatol.*, 2000, **114**, 438.

26. G. Vancoillie, J. Lambert, A. Mulder, H.K. Koerten, A.M. Mommaas, P. Van Oostveldt and J.M. Naeyaert, *J. Invest. Dermatol.*, 2000, **114**, 421.

27. R.I. Garcia, *Pigment Cell Res.*, 1979, **4**, 299.

28. A.A. Nascimento, R.G. Amaral, J.C. Bizario, R.E. Larson and E.M. Espreafico, *Mol. Biol. Cell.*, 1997, **8**, 1971.

29. Q. Wei, X. Wu and J.A. Hammer, *J. Mus. Res. & Cell Motility.*, 1997, **18**, 517.

30. K. Jimbow, J.S. Park, F. Kato, K. Hirosaki, K. Toyofuku, C. Hua and T. Yamashita, *Pigment Cell Res.*, 2000, **13(4)**, 222.

31. T. Kushimoto, V. Basrur, J. Valencia, J. Matsunaga, W.D. Vieira, V.J. Ferrans, J. Muller, E. Appella and V.J. Hearing, *Proc. Natl. Acad. Sci. USA.*, 2001, **98(19)**, 10698.

32. A. Bomirski, A. Slominski and J. Bigda, *Cancer Metastasis Rev.*, 1988, **7(2)**, 95.

33. K. Jimbow, Y. Miyake, K. Homma, K. Yasuda, Y. Izumi, A. Tsutsumi and S. Ito, *Cancer Res.*, 1984, **44(3)**, 1128.

34. M. Inazu and Y. Mishima, *J. Invest. Dermatol.*, 1993, **100(Suppl. 2)**,172S.

35. L. Oyehaug, E. Plahte, D.I. Vage and S.W. Omholt, *J. Theor. Biol.*, 2002, **215(4)**, 449.
36. A. Napolitano, S. Memoli and G. Prota, *J. Org. Chem.*, 1999, **64**, 3009.
37. A. Slominski, D.J. Tobin, S. Shibahara and J. Wortsman, *Physiol. Rev.*, 2004, **84(4)**, 1155.
38. B. Bilinska, *Spectrochim. Acta A Mol. Biomol. Spectrosc.*, 2001, **57(12)**, 2525.
39. G. Prota, *Fortsch. Cehm. Orga. Natur.*, 1995, **64**, 93.
40. M.S. Blois, A.B. Zahlan and J.E. Maling, *Biophys. J.*, 1964, **156**, 471.
41. B. Commoner, J. Townsend and G.E. Pake, *Nature.*, 1954, **174(4432)**, 689.
42. S. Gidanian and P.J. Farmer, *J. Inorg. Biochem.*, 2002, **89(1–2)**, 54.
43. M.E. Lacy, *J. Theor. Biol.*, 1984, **111(1)**, 201.
44. W. Osak, K. Tkacz, H. Czternastek and J. Slawinski, *Biopolymers.*, 1989, **28**, 1885.
45. S. Ito, *Pigment Cell Res.*, 2003, **16**, 230.
46. G. Prota, Melanins and melanogenesis. Academic, New York 1992.
47. U. Mars and B.S. Larsson, *Pigment Cell Res.*, 1999, **12**, 266.
48. M.H. Slawson, D.G. Wilkins and D.E. Rollins, *J. Anal. Toxicol.*, 1998, **22(6)**, 406.
49. R.C. Sealy, J.S. Hyde, C.C. Felix, I.A. Menon and G. Prota, *Science.*, 1982, **217(4559)**, 545.
50. R.C. Sealy, J.S. Hyde, C.C. Felix, I.A. Menon, G. Prota, H.M. Swartz, S. Persad and H.F. Haberman, *Proc. Natl. Acad. Sci. USA.*, 1982, **79(9)**, 2885.
51. A. Slominski and J. Wortsman, *Endocr. Rev.*, 2000, **21(5)**, 457.
52. V.J. Hearing, *J. Investig. Dermatol. Symp. Proc.*, 1999, **4(1)**, 24.
53. J.M. Carroll, T. Crompton, J.P. Seery and F.M. Watt, *J. Invest. Dermatol.*, 1997, **108(4)**, 412.
54. M. Martinez-Esparza, F. Solano and J.C. Garcia-Borron, *Cell Mol. Biol. (Noisy-le-grand).*, 1999, **45(7)**, 991.
55. A. Slominski and D. Pruski, *Exp. Cell Res.*, 1993, **206(2)**, 189.
56. V.B. Swope, Z. Abdel-Malek, L.M. Kassem and J.J. Nordlund, *J. Invest. Dermatol.*, 1991, **96(2)**, 180.
57. P. Valverde, E. Benedito, F. Solano, S. Oaknin, J.A. Lozano and J.C. Garcia-Borron, *Eur. J. Biochem.*, 1995, **232(1)**, 257.
58. J. Reynolds, *J. Anat.*, 1954, **88**, 45.
59. R. Paus, K.S. Stenn and R.E. Link, *Br. J. Dermatol.*, 1990, **122**, 777.
60. A. Slominski and R. Paus, *J. Invest. Dermatol.*, 1993, **101(Suppl. 1)**, 90S.
61. A.F. Silver and H.B. Chase, *Devl. Biol.*, 1970, **21**, 440.
62. A. Slominski, R. Paus and R. Constantino, *J. Invest. Dermatol.*, 1991, **96**, 172.
63. S. Sugiyama, K. Saga, Y. Morimoto and M. Takahashi, *J. Dermatol.*, 1995, **22**, 396.
64. K.U. Schallreuter, W.D. Beazley, N.A. Hibberts, D.J. Tobin, R. Paus and J.M. Wood, *J. Invest. Dermatol.*, 1998, **111(4)**, 545.
65. S. Commo and B.A. Bernard, *Pigment Cell Res.*, 2000, **13**, 253.
66. K. Jimbow, S.I. Roth, T.B. Fitzpatrick and G. Szabo, *J. Cell Biol.*, 1975, **66**, 666.

67. D.J. Tobin, A. Slominski, V. Botchkarev and R. Paus, *J. Investig. Dermatol. Symp. Proc.*, 1999, **4**, 323.
68. S. Sugiyama and A. Kukita, Melanocyte reservoir in the hair follicles during the hair growth cycle: an electron microscopic study in *Biology and Disease of the Hair*. University of Tokyo Press, Tokyo, 1976, 81.
69. S. Sugiyama, *J. Ultrastructural Res.*, 1979, **67**, 40.
70. D. Weedon and G. Strutton, *Acta Derm. Venerol.*, 1981, **61**, 335.
71. G. Lindner, V.A. Botchkarev, N.V. Botchkareva, G. Ling, C. van der Veen and R. Paus, *Am. J. Pathol.*, 1997, **151**, 1601.
72. J.M. Grichnik, W.N. Ali, J.A. Burch, J.D. Byers, C.A. Garcia, R.E. Clark and C.R. Shea, *J. Invest. Dermatol.*, 1996, **106**, 967.
73. N.V. Botchkareva, M. Khlgatian, B.J. Longley, V.A. Botchkarev and B.A. Gilchrest, *FASEB J.*, 2001, **15**, 645.
74. D.J. Tobin, E. Hagen, V.A. Botchkarev, R. Paus, *J. Invest. Dermatol.*, 1998, **111**, 941.
75. R. Paus, Immunology of the hair follicle. in *The Skin Immune System*, J.D. Bos (ed) CRC Press, Boca Raton, 1997, 377.
76. T. Christoph, S. Muller-Rover, H. Audring, D.J. Tobin, B. Hermes, G. Cotsarelis, R. Ruckert and R. Paus, *Br. J. Dermatol.*, 2000, **142**, 862.
77. D.J. Tobin, *Br. J. Dermatol.*, 1998, **138**, 795.
78. R.D. Cone, D. Lu, S. Koppula, D.I. Vage, H. Klungland, B. Boston, W. Chen, D.N. Orth, C. Pouton and R.A. Kesterson, *Recent Prog. Horm. Res.*, 1996, **51**, 287.
79. A.N. Eberle, *The Melanotropins: Chemistry, Physiology and Mechanism of Action*, Karger, New York, 1988.
80. A. Slominski, J. Wortsman, T. Luger, R. Paus and S. Solomon, *Physiol. Rev.*, 2000, **80(3)**, 979.
81. J. Furkert, U. Klug, A. Slominski, S. Eichmuller, B. Mehlis, U. Kertscher and R. Paus, *Biochim. Biophys. Acta.*, 1997, 1336(2), **315**, 22.
82. J.E. Mazurkiewicz, D. Corliss and A. Slominski, *J. Histochem. Cytochem.*, 2000, **48(7)**, 905.
83. R. Paus, V.A. Botchkarev, N.V. Botchkareva, L. Mecklenburg, T. Luger and A. Slominski, *Ann. N.Y. Acad. Sci.*, 1999, **885**, 350.
84. G. Ermak and A. Slominski, *J. Invest. Dermatol.*, 1997, **108(2)**, 160.
85. P. Valverde, E. Healy, I. Jackson, J.L. Rees and A.J. Thody, *Nat. Genet.*, 1995, **11**, 328.
86. R. Halaban, *Pigment Cell Res.*, 1994, **7(2)**, 89.
87. R.S. Snell, Hormonal control of hair colour in *Pigmentation: Its Genesis and Biologic Control*, V. Riley (ed), Meredith, New York, 1972, 193.
88. N. Levine, S.N. Sheftel, T. Eytan, R.T. Dorr, M.E. Hadley, J.C. Weinrach, G.A. Ertl, K. Toth, D.L. McGee and V.J. Hruby, *J.A.M.A..*, 1991, **266**, 2730.
89. S. Kauser, A.J. Thody, K.U. Schallreuter, C.L. Gummer, D.J. Tobin. *Endocrinology*, 2005, **146(2)**, 523.
90. S. Kauser, K.U. Schallreuter, A.J. Thody, C. Gummer and D.J. Tobin, *J. Invest. Dermatol.*, 2003, **120(6)**, 1073.

91. S. Kauser, A.J. Thody, K.U. Schallreuter, C. Gummer and D.J. Tobin, *J. Invest. Dermatol.*, 2004, **123(1)**, 184.

92. S. Kauser, A. Slominski, E.T. Wei and D.J. Tobin, *Exp. Dermatol.*, 2004, **13(9)**, 582.

93. G. Barsh, *Nat. Med.*, 1999, **5(9)**, 984.

94. W.K. Silvers (ed), *The Coat Colours of Mice: A Model for Mammalian Gene Action and Interaction*, Springer, New York, 1979.

95. L. He, A.G. Eldridge, P.K. Jackson, T.M. Gunn and G.S. Barsh, *Ann. N Y. Acad. Sci.*, 2003, **994**, 288.

96. S.A. Burchill and A.J. Thody, *J. Endocrinol.*, 111, **233**, 1986.

97. A. Logan and B. Weatherhead, *J. Invest. Dermatol.*, 1980, **74(1)**, 47.

98. A. Slominski, N. Chassalevris, J. Mazurkiewicz, M. Maurer, R. Paus, *Exp. Dermatol.*, 1994, **3(1)**, 45.

99. A. Slominski, A. Pisarchik, B. Zbytek, D.J. Tobin, S. Kauser and J. Wortsman, *J. Cell. Physiol.*, 2003, **196(1)**, 144.

100. K.U. Schallreuter and J.M. Wood, *Biochem. Biophys. Res. Commun.*, 1999, **262(2)**, 423.

101. A. Slominski, G. Moellmann, E. Kuklinska, A. Bomirski and J. Pawelek, *J. Cell. Sci.*, 1988, **89(Pt 3)**, 287.

102. A. Slominski and R. Costantino, *Experientia.*, 1991, **47(7)**, 721.

103. A. Slominski and R. Costantino, *Life Sci.*, 1991, **48(21)**, 2075.

104. A. Slominski and T. Friedrich, *Pigment Cell Res.*, 1992, **5(6)**, 396.

105. A. Slominski and R. Paus, *Mol. Cell. Endocrinol.*, 1994, **99(2)**, C7.

106. A. Slominski, G. Moellmann and E. Kuklinska, *Pigment Cell Res.*, 1989, **2(2)**, 109.

107. R.I. Garcia, *Pigment Cell Res.*, 1979, **4**, 299.

108. K. Okazaki, M. Uzuka, F. Morikawa, K. Toda and M. Seiji, *J. Invest. Dermatol.*, 1976, **67(4)**, 541.

109. G. Scott, S. Leopardi, S. Printup and B.C. Madden, *J. Cell. Sci.*, 2002, **115(Pt 7)**, 1441.

110. G. Menasche, C.H. Ho, O. Sanal, J. Feldmann, I. Tezcan, F. Ersoy, A. Houdusse, A. Fischer and G. de Saint Basile, *J. Clin. Invest.*, 2003, **112(3)**, 450.

111. J. Lambert, G. Vancoillie and J.M. Naeyaert, *Cell. Mol. Biol. (Noisy-le-grand).*, 1999, **45(7)**, 905.

112. X. Wu, B. Bowers, Q. Wei, B. Kocher and J.A. Hammer, *J. Cell. Sci.*, 1997, **110(Pt 7)**, 847.

113. M. Seiberg, C. Paine, E. Sharlow, M. Andrade-Gordon Postanzo, M. Eisinger and S. Shapiro, *J. Invest. Dermatol.*, 2000, **115(2)**, 162.

114. D. Harman, *J. Gerontol.*, 1956, **11**, 298.

115. J.M. Gutteridge and B. Halliwell, *Ann. N Y. Acad. Sci.*, 2000, **899**, 136.

116. P. Grammatico, V. Maresca, F. Roccella, M. Roccella, L. Biondo, C. Catricala and M. Picardo, *Exp. Dermatol.*, 1998, **7**, 205.

117. B.N. Ames, M.K. Shigenaga and T.M. Hagen, *Proc. Natl. Acad. Sci. USA.*, 1993, **90**, 7915.

118. A.M. Kligman, *J. Invest. Dermatol.*, 1959, **33**, 307.

119. Z.L. Hegedus, *Toxicology.*, 2000, 14, **145(2–3)**, 85.

120. Q.M. Chen, J.C. Bartholomew, J. Campisi, M. Acosta, J.D. Reagan and B.N. Ames, *Biochem. J.*, 1998, **15:332(Pt 1)**, 43.
121. J. Campisi, *J. Investig. Dermatol. Symp., Proc.*, 1998, **3**, 1.
122. D.C. Bennett, *Cell.*, 1983, **34**, 445.
123. M.M. Haddad, W. Xu, D.J. Schwahn, F. Liao and E.E. Medrano, *Exp. Cell. Res.*, 1999, **253(2)**, 561.
124. C.J. Sherr and J.M. Roberts, *Genes. Dev.*, 1995, **9(10)**, 1149.
125. E.E. Medrano, F. Yang, R. Boissy, J. Farooqui, V. Shah, K. Matsumoto, J.J. Nordlund and H.Y. Park, *Mol. Biol. Cell.*, 1994, **5(4)**, 497.
126. K. Nakayama, I. Negishi, K. Kuida, H. Sawa and D.Y. Loh, *Proc. Natl. Acad. Sci. USA.*, 1994, **91**, 3700.
127. S.J. Korsmeyer, X.M. Yin, Z.N. Oltvai, D.J. Veis-Novack, G.P. Linette, *Biochim. Biophys. Acta.*, 1995, 24, **1271(1)**, 63.
128. M.D. Jacobson, *Trends Biochem. Sci.*, 1996, **21(3)**, 83.
129. A. Kowald, *Exp. Gerontol.*, 1999, **34(5)**, 605.
130. N.K. Fukagawa, *Proc. Soc. Exp. Biol. Med.*, 1999, **222(3)**, 293.
131. B. Kadenbach, C. Munscher, V. Frank, J. Muller-Hocker and J. Napiwotzki, *Mutat. Res.*, 1995, **338(1–6)**,161.
132. G. Li, H. Schaider, K, Satyamoorthy, Y. Hanakawa, K. Hashimoto and M. Herlyn, *Oncogene.*, 2001, **20**, 8125.
133. D.C. Whiteman, P.G. Parsons and A.C. Green, *Arch. Dermatol. Res.*, 1999, **291**, 511.
134. E.V. Keogh, and R.J. Walsh, *Nature.*, 1965, **207**, 877.
135. T. Lloyd, F.L. Garry, E.K. Manders, J.G. Marks Jr., *Br. J. Dermatol.*, 1987, **116(4)**, 485.
136. L. Ibrahim and E.A. Wright, *Br. J. Dermatol.*, 1978, **99**, 371.
137. C.E. Orfanos, H. Ruska, G. Mahrle. Arch Klin, *Exp. Dermatol.*, 1970, **236**, 395.
138. S. Commo, O. Gaillard and B.A. Bernard, *Br. J. Dermatol.*, 150, **435**, 2004.
139. M. Toyoda and M. Morohashi, *Br. J. Dermatol.*, 1998, **139(3)**, 444.
140. I. Weisse, *Ophthalmic. Res.*, 1995, **27(1 Suppl.)**, 154.
141. R. Bowers and D.Q. Chun, Ultrastructural study of senescence of regenerating feather melanocytes in the jungle fowl in *Biological, Molecular and Clinical Aspects of Pigmentation - Pigment Cell*, K. Bagnara and M. Schartl (ed), University of Tokyo Press, Tokyo, 1985, 347.
142. W. Westerhof, D. Njoo and K.E. Menke, Miscellaneous hypomelanoses: disorders characterized by extra-cutaneous loss of pigmentation in *The Pigmentary System: Physiology and Pathophysiology*, J.J. Nordlund, R.E. Boissy, V.J. Hearing, R.A. King, J.-P. Ortonne (ed), Oxford University Press, New York, 1998, 475.
143. D.J. Tobin, D.A. Fenton and M.D. Kendall, *J. Invest. Dermatol.*, 1990, **94**, 803.
144. S. Sato, A. Kukita and K. Jimbow, *Pigment. Cell.*, 1973, **1**, 20.
145. W. Nagl, *Br. J. Dermatol.*, 1995, **132**, 94.
146. A. Slominski, R. Paus and D. Schadendorf, *J. Theor. Biol.*, 1993, **164**, 103.
147. A. Tang, M.S. Eller, M. Hara, M. Yaar, S. Hirohashi and B.A. Gilchrest, *J. Cell. Sci.*, 1994, **107**, 1983.

148. F. D'Agostini, R. Balansky, C. Pesce, P. Fiallo, R.A. Lubet, G.J. Kelloff and S. De Flora, *Toxicol. Lett.*, 2000, **114**, 117.
149. C.J. Rosen, M.F. Holick and P.S. Millard, *J. Clin. Endocrinol. Metab.*, 1994, **79**, 854.
150. I. Eisenstein and J. Edelstein, *Angiology*, 1982, **33(10)**, 652.
151. J. Cui, L.Y. Shen and G.C. Wang, *J. Invest. Dermatol.*, 1991, **97**, 410.
152. D.J. Tobin, R. Paus, *Exp. Gerontol.*, 2001, **36**, 29.
153. M. Shetty, *Br. Med. J.*, 1995, **311**, 1582.
154. J. Verbov, *Br. J. Dermatol.*, 1981, **105**, 595.

Part 2
Application of Hair Biology to Environmental Assessments

Hair in Forensic Toxicology with a Special Focus on Drug-Facilitated Crimes

PASCAL KINTZ and MARION VILLAIN

4.1 Introduction

In the 1960s and 1970s, hair analysis was used to evaluate exposure to toxic heavy metals, such as arsenic, lead or mercury. This was achieved using atomic absorption spectroscopy that allowed detection in the nanogram range. At that time, examination of hair for organic substances, especially drugs, was not possible because analytical methods were not sensitive enough. Examination by means of drugs labelled with radioactive isotopes, however, established that these substances can move from blood to hair and be deposited there. Ten years after these first investigations, it was possible to demonstrate the presence of various organic drugs in hair by means of radioimmunoassay (RIA). In 1979, Baumgartner and colleagues[1] published the first report on the detection of morphine in the hair of heroin abusers using RIA. They found that differences in the concentration of morphine along the hair shaft correlated with the time of drug use. Today, gas chromatography coupled with mass spectrometry (GC/MS) is the method of choice for hair analysis and this technology is routinely used in forensic science to document repetitive drug exposure.

The major practical advantage of hair testing compared to urine or blood testing for drugs is that it has a larger surveillance window (weeks to months, depending on the length of the hair shaft, against 2–4 days for most drugs in blood and urine). For practical purposes, the two tests complement each other. Urinalysis and blood analysis provide short-term information on an individual's drug use, whereas long-term histories are accessible through hair analysis.

Hair does not grow continually, but in cycles, alternating between periods of growth and quiescence (see Chapter 1 elsewhere in this volume). A follicle that is actively producing hair is said to be in the *anagen* phase. Hair (scalp) is produced over 4–8 years at a rate of approximately 0.22–0.52 mm per day or 0.6–1.4 cm per month.[2] After this period, the hair follicle enters a relatively short transition period of about two weeks, known as the *catagen* phase, during which cell division stops and the follicle begins to degenerate. Following the transition phase, the hair follicle enters a resting or quiescent period, known as the *telogen* phase (lasting approximately ten weeks), in which the lower hair shaft forms a club root structure in the hair follicle with very low levels of metabolic activity. Factors such as race, disease states, nutritional deficiencies and age are known to influence both the rate of growth and the length of the quiescent period. On the scalp of an adult, approximately 85% of the hair is in the growing phase, 2–3% in the catagen transition phase and the remaining 12% in the resting stage.

Pubic hair, arm hair and axillary hair have been suggested as an alternative source for drug detection when scalp hair is not available. Various studies have found differences in drug concentrations between pubic or axillary hair and scalp hair. The significant differences in the drug concentrations in these studies were explained by a better blood circulation, a greater number of apocrine glands, a totally different *telogen/anagen* ratio and a different growth rate of the hair (axillary hair 0.40 mm per day, pubic hair 0.30 mm per day).[2]

4.2 Mechanisms of Drug Incorporation into Hair

It is generally proposed that drugs can enter into hair by two processes: adsorption from the external environment and incorporation into the growing hair shaft from blood supplying the hair follicle. Drugs can enter the hair from exposure to chemicals in aerosols, smoke or secretions from sweat and sebaceous glands. Sweat is known to contain drugs present in blood. Because hair is very porous and can increase its weight by up to 18% by absorbing liquids,[3] drugs may be transferred easily into hair *via* sweat. Finally, chemicals present in air (smoke, vapours *etc.*) can be deposited onto the hair.

Drugs appear to be incorporated into the hair shaft by at least three mechanisms: from the blood during hair formation, from sweat and sebum, and from the external environment. This model is more able than a passive model (transfer from blood into the growing cells of the hair follicle) to explain several experimental findings such as:

- drug and metabolite(s) ratios in blood are quite different from those found in hair
- drug and metabolite(s) concentrations in hair differ markedly in individuals receiving the same dose. Evidence for the transfer of the drug *via* sweat and sebum can be supported as drugs and metabolites are present in sweat and sebum at high concentrations and persist in these secretions longer than they do in blood.[3,4]

The exact mechanism by which chemicals are adsorbed onto/absorbed into hair is not known. It has been suggested that passive diffusion may be augmented by drug binding to intracellular components of the hair cells such as the hair pigment melanin. For example, codeine concentrations in hair after oral administration are dependent on melanin content.[5] However, this is probably not the only mechanism since drugs are trapped into the hair of albino animals, which lack melanin. Another mechanism proposed is the binding of drugs with sulphydryl-containing amino acids present in hair. There is an abundance of amino acids such as cystine in hair, which form cross-linking S-S bonds to stabilise the protein fibre network. Drugs diffusing into hair cells could be bound in this way.

Various studies have demonstrated that after the same dosage, black hair incorporates much more drugs than blond hair.[6] This has resulted in discussions about a possible genetic variability of drug deposition in hair and this is still under evaluation.

4.3 Specimen Collection

Sampling and collection procedures for the analysis of drugs in hair have not been standardised. In most published studies, the samples are obtained from random locations on the scalp. Hair is best collected from the area at the back of the head, called the *vertex posterior*. Compared with other areas of the head, this area has less variability in the hair growth rate, the number of hairs in the growing phase is more constant and the hair is less subject to age- and sex-related influences. Hair strands are cut as close as possible to the scalp, and their location on the scalp must be noted. Once collected, hair samples may be stored at ambient temperature in aluminium foil, an envelope or a plastic tube. The sample size taken varies considerably among laboratories and depends on the drug to be analysed and the test methodology. Sample sizes reported in the literature range from a single hair to 200 mg.

4.4 Stability of Drugs in Hair

The presence of opiates was detected in five hair shafts (about 7.5 cm in length) from the English poet John Keats 167 years after his death.[7] It is believed he took laudanum (opium) to control the pain of tuberculosis. The scalps of eight Chilean and Peruvian mummies dating from 2000 BC to 1500 AD tested also positive for benzoylecgonine, the main metabolite of cocaine.[8] All these studies indicate that drug incorporation is very stable in hair and by using chromatographic techniques for example, these studies can help archaeologists to determine that as far back as 2000 BC Indians were already using cocaine. Clearly, organic substances are capable of surviving in hair for thousands of years under favourable conditions (ambient temperature, dry atmosphere).

4.5 Hair Analysis

A large number of procedures involved in hair analysis have been described in the last years. Some of them are summarised in reviews.[9–11]

Hair analysis involves at least five steps:

- decontamination of the hair
- preparation of the hair: pulverisation, segmentation into shorter pieces
- incubation: in methanol, acid, sodium hydroxide, buffer

    – extraction: liquid/liquid, solid phase, solid phase micro-extraction
    – analysis: immunoassay screening or/and chromatography (gas, liquid) coupled
      to mass spectrometry.

The critical issue here in hair analysis is that potential contamination of hair from external sources with drugs of abuse can lead to false-positive results. It is unlikely that anyone would intentionally or accidentally apply anything to his or her hair that would contain a drug of abuse. The most crucial issue facing hair analysis is the avoidance of technical and evidentiary false-positives. Technical false-positives are caused by errors in the collection, processing and analysis of specimens, while evidentiary false-positives are caused by passive exposure to the drug. Approaches for preventing evidentiary false-positives due to external contamination of the hair specimens have been described.[12]

Most but not all laboratories use a wash step; however, there is no consensus or uniformity in the washing procedures. Among the agents used in washing are detergents such as shampoo, surgical scrubbing solutions, surfactants such as 0.1% sodium dodecylsulfate, phosphate buffer, or organic solvents such as acetone, diethyl ether, methanol, ethanol, dichloromethane, hexane or pentane of various volumes for various contact times. Generally, a single washing step is used; although a second identical wash is sometimes performed. If external contamination is found by analysing the wash solution, the washout kinetics of repeated washing can demonstrate that contamination is rapidly removed. According to Baumgartner and Hill,[12] the concentration of drug in the hair after washing should exceed the concentration in the last wash by at least ten times. It has also been proposed that hair should be washed three times with phosphate prior to analysis to remove any possible external contamination and that the total concentration of any drug present in the three phosphate washes taken together should be greater than 3.9 times the concentration in the last wash. Because the decontamination step is limited in time, the bound drugs cannot be totally removed.

Detection of drug metabolite(s) in hair, whose presence could not be explained by chemical hydrolysis or environmental exposure, would unequivocally establish that internal drug exposure had occurred.[13] Cocaethylene and nor-cocaine would appear to meet these criteria, as these metabolites are only formed when cocaine is metabolised. Because these metabolites are not found in illicit cocaine samples, they would not be present in hair as a result of environmental contamination. Thus, the presence of these metabolites in hair could be considered as a marker of cocaine exposure. This procedure can be extended to other drugs. However, there is still a great controversy about the potential risk of external contamination, particularly for crack, cannabis and heroin when smoked, as several authors have demonstrated that it is not possible to fully remove the drugs.[14,15] In conclusion, although it is highly recommended to include a decontamination step, there is no consensus on which procedure performs best and each laboratory must validate its own technique.

An important issue of concern for drug analysis in hair is the change in the drug concentration induced by cosmetic treatment of hair. Hair is continuously subjected to natural factors, such as sunlight, weather, water, pollution *etc.*, which affect and damage the cuticle, but cosmetic hair treatments enhance that damage. Particular attention has been focused on the effects of repeated shampooing, perming, relaxing and dyeing of hair. Repeated shampooing was found to have no significant action on the drug content of hair.[12] After cosmetic treatments, drug concentrations decline dramatically by 50–80% from their original concentration. The products used for cosmetic treatments, like bleaching, permanent waving, dyeing or relaxing, contain strong bases. They will cause hair damage and affect drug content (by loss) or affect directly drug stability.[16]

4.6 Applications of Hair Analysis

By providing information on exposure to drugs over time, hair analysis may be useful in verifying self-reported histories of drug use in any situation in which a history of past rather than recent drug use is desired. In addition, hair analysis may be especially useful when a history of drug use is difficult or impossible to obtain.

Numerous forensic applications have been described in the literature where hair analysis was used to document the case: differentiation between a drug dealer and a drug consumer, chronic poisoning, crime under the influence of a drug, child sedation and abuse, suspicious death, child custody, abuse of drugs in jail, body identification, survey of drug addicts, chemical submission, obtaining a driving licence and doping control.[17–19]

More than 450 articles concerning hair analysis have been published to date reporting applications in forensic toxicology, clinical toxicology, occupational medicine and doping control. The major practical advantage of hair for testing drugs, compared with urine or blood, is its larger detection window, which is weeks to months, depending on the length of hair shaft analysed, against a few days for urine. In practice, detection windows offered by urine and hair testing are complementary: urine analysis provides short-term information of an individual's drug use, whereas long-term histories are accessible through hair analysis. Although there is reasonable agreement that the qualitative results from hair analysis are valid, the interpretation of the results is still under debate owing to unresolved questions such as the influences of external contamination or cosmetic treatment, and possible genetic differences.

More research is required before all of the scientific questions associated with hair drug testing will be satisfied. There is still a lack of consensus among the active investigators on how to interpret the analysis of drugs in hair. Amongst the unanswered questions, five are of critical importance:

- what is the minimal amount of drug detectable in hair after administration?
- what is the relationship between the amount of the drug used and the concentration of the drug or its metabolites in hair?
- what is the influence of hair colour?
- what is the influence of genetic differences in hair testing?
- what is the influence of cosmetic treatments?

Several answers were recently addressed by Wennig[20] and Kintz *et al.*[21] on these specific topics.

4.7 Special Focus on Drug-Facilitated Crimes

The use of a drug to modify a person's behaviour for criminal gain is not a recent phenomenon. However, the sudden increase in reports of drug-facilitated crimes (sexual assaults, robbery *etc.*) has caused alarm in the general public. Drugs involved can be pharmaceuticals, such as benzodiazepines (flunitrazepam, lorazepam *etc.*), hypnotics (zopiclone, zolpidem), sedatives (neuroleptics, some histamine H1-antagonists), anesthetics (gamma hydroxybutyrate or GHB, ketamine), drugs of abuse (cannabis, ecstasy, LSD), or more often ethanol. Most of these substances possess amnesic properties and in sex offences for example, the victims are less able to accurately recall the circumstances under which the offence occurred. As they are generally short-acting, they impair an individual rapidly. Due to their low dosage, except for GHB, a surreptitious administration into beverages such as coffee, soft drinks (cola) or even better alcoholic cocktails is relatively simple.

To perform successful toxicological examinations, the analyst must follow some important rules:

- to obtain as soon as possible the corresponding biological specimens (blood, urine and hair),
- to use sophisticated analytical techniques (LC/MS, headspace/GC/MS, tandem mass spectrometry)
- to take care in the interpretation of the findings.

To address these problems, guidelines for toxicological investigations were recently published in both the United States[22] and France.[23]

Urinalysis for drug use in cases of alleged sexual assault demonstrated in 3303 urine samples that ethanol, either alone or in combination with other drugs, was the most common substance found, followed by cannabis and benzodiazepines.[24] In Paris, the largest study[25] conducted in France revealed that most frequently used drugs were benzodiazepines and related hypnotics. GHB was very seldom found. In our study at Strasbourg,[23] zolpidem appears as the commonest substance, followed by bromazepam.

The narrow window of detection for GHB, six and ten hours in blood and urine, respectively, is an example of the current limitation of these specimens to demonstrate exposure after late sampling.[26] For all compounds involved in drug-facilitated sexual assault (DFSA), the detection times in blood and urine depend mainly on the dose and sensitivity of the method used. Excluding immunoassays and using only combination techniques, *i.e.* using chromatography and specific detection, such as mass spectrometry, substances can usually be found in blood for between six hours and two days and for between twelve hours and five days in urine.[27] Sampling blood or urine is of little value 48 hours after the offence occurred.

To respond to these important limitations that are characteristic of blood and urine, hair was suggested as a valuable specimen.

Several forensic cases, where hair analysis has indicated that drugs were used to sedate the victim quickly in order to commit a crime, are presented in the following pages.

4.7.1 Case 1

Hair strands were obtained from a 19-year-old girl who claimed to have been sexually assaulted after drinking a soft drink spiked with a drug. She had no memory of the crime and went to the police five days after the rape. After contact with the police, this laboratory recommended her to wait for about one month in order to have the corresponding growing hair between the root and the tip. Full-length hair samples (8 cm long) were taken at the surface of the skin from the vertex and stored in plastic tubes at room temperature. Segmentation revealed an increase of GHB concentrations at the corresponding time (Table 4.1) to 2.4 and 2.7 ng/mg confirming exposure, when compared with basal physiological concentrations around 0.7 ng/mg. The rapist, who was arrested several days after the assault, did not challenge this result.

4.7.2 Case 2

A 21-year-old woman was hospitalised for gastric disorders. One night, she was offered by a male nurse a coffee that made her unconscious. When recovering she noticed an assault, but, afraid of the consequences, did not report it to the police immediately. This was done after she left the hospital, six days later. Given this delay, a blood or urine collection was of no value, and we were requested to analyse the victim's hair, sampled 15 days after the alleged offense. Zolpidem was identified in the proximal segment (root to 2 cm) at 4.4 pg/mg (Figure 4.1), while the distal segment (2 to 4.5 cm) remained negative.

4.7.3 Case 3

Blood analysis (collected nine hours after the crime) from a sexually assaulted woman was positive for zolpidem at 390 ng/ml. As the claims of the victim to the police were confused, four weeks later we received an 8 cm hair strand to test for zolpidem. The analysis of four hair segments (4×2 cm) revealed the presence of

Table 4.1 *Gamma hydroxybutyrate (GHB) in hair after segmentation in a case of drug-facilitated sexual assault (DFSA)*

Segment	*GHB* (ng/mg)
0 (root-end)–0.3 cm	1.3
0.3–0.6 cm	0.6
0.6–0.9 cm	0.8
0.9–1.2 cm	2.4
1.2–1.5 cm	2.7
1.5–1.8 cm	0.7
1.8–2.1 cm	0.8
2.1–2.4 cm	0.7
2.4–2.7 cm	0.8
2.7–3.0 cm	0.7

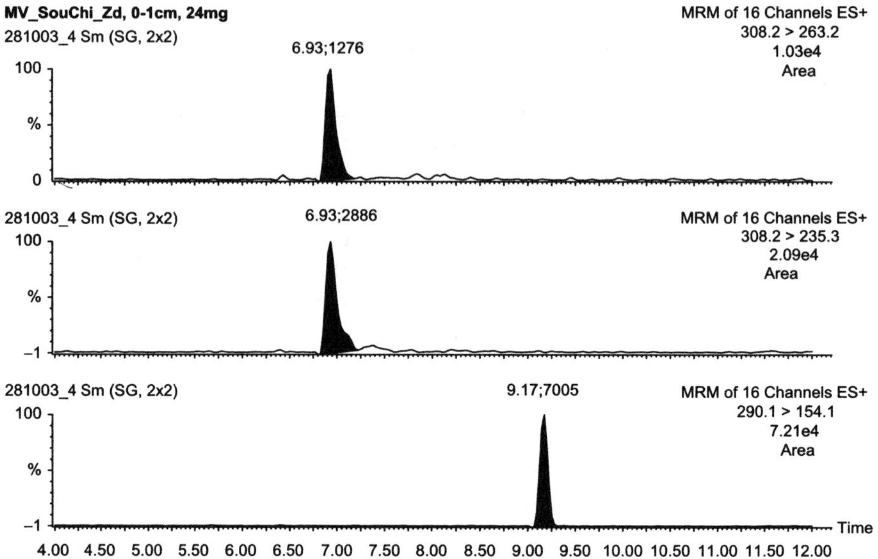

Figure 4.1   *Chromatogram of a hair extract. Top: two daughter ions of zolpidem (at 4.4 pg/mg), Bottom: daughter ion of diazepam-d₅, used as internal standard*

zolpidem at concentrations of 22, 47, 67 and 9 pg/mg from the root to the tip. This demonstrated repetitive exposure to zolpidem before the alleged assault and therefore made the blood result inconclusive.

4.7.4 Case 4

During a party, a 42-year-old man was offered an alcoholic drink. Soon after, he lost all recollection of events and awoke four hours later in a bed with a woman. Terrified, as he was married, he privately requested us to perform some analyses in an attempt to identify the sedative drug. Hair was collected at the laboratory 21 days after the event. 7-Amino-flunitrazepam was identified in the proximal segment (root to 2 cm) at 5.2 pg/mg, while the proximal segment (2 to 4 cm) remained negative. No flunitrazepam, the parent drug, was detected.

4.7.5 Case 5

A 39-year-old woman, experiencing marital difficulties with her husband, felt sleepy for 24 hours after having drunk a coffee, at home. A blood sample, collected 20 hours later, revealed the presence of 51 ng/ml of bromazepam, whereas hair sampled at the same time was bromazepam-free. Another strand of hair was collected one month after the event and the proximal 2 cm long segment was positive for bromazepam at 10.3 pg/mg, the other segments (2–4 and 4–6 cm) remaining negative. These results were in accordance with a single exposure to this drug. The positive analysis of bromazepam in the coffee resulted in an admission by the husband, who did not challenge the biological conclusions.

4.7.6 Case 6

A young woman was victim of a sexual assault in a highway gas station. She declared that the perpetrator forced her to absorb a white quadri-divisible tablet before abusing her. A blood sample, collected 18 hours after the offence, revealed the presence of 151 ng/ml of bromazepam. A strand of hair was collected three weeks after the event and the proximal segment (0–2 cm) was positive for bromazepam at 5.7 pg/mg (Figure 4.2), the consecutive segment (2–4 cm) was positive at 0.9 pg/mg and the last segment remained bromazepam-free.

These results were in accordance with a single exposure to this drug.

4.7.7 Case 7

A 16-year-old girl claimed to have been raped during an afternoon, while sedated. Due to the long delay (six days) between the alleged event and her deposition to the police, there was no value in sampling blood or urine. We were requested to analyse the victim's hair, collected nine weeks after the offense. Benzodiazepines and hypnotics were tested by LC-MS/MS and the first 3 cm segment was positive for zopiclone at a concentration of 4.2 pg/mg (Figure 4.3), the second (3–5 cm) at 1.0 pg/mg, whereas the last segment (5–7 cm) was zopiclone free. This appeared to be consistent with a single exposure to the drug.

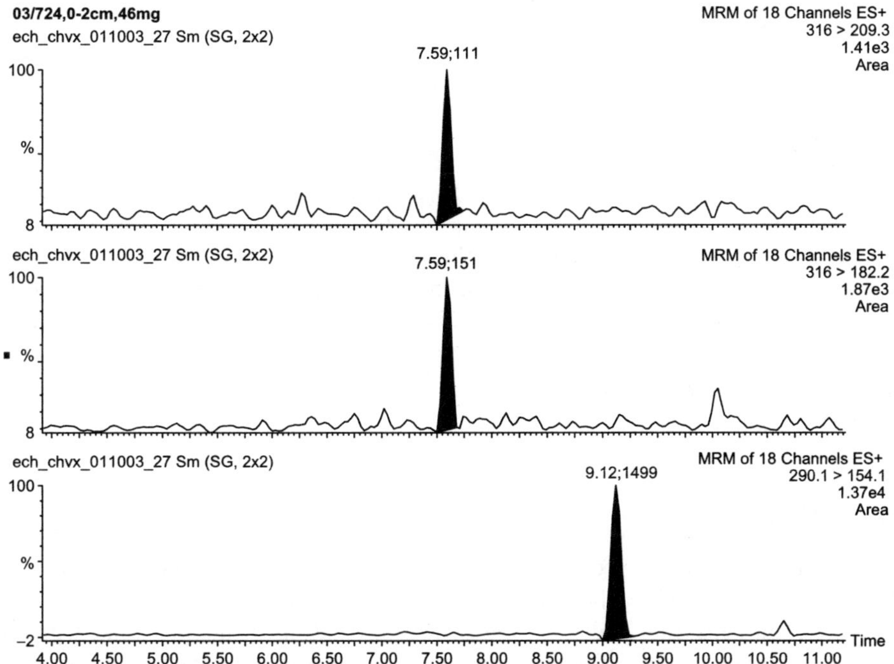

Figure 4.2 *Chromatogram of a hair extract from a victim of an alleged assault. Bromazepam concentration is 5.7 pg/mg. Top: two daughter ions of bromazepam, Bottom: daughter ion of diazepam-d$_5$*

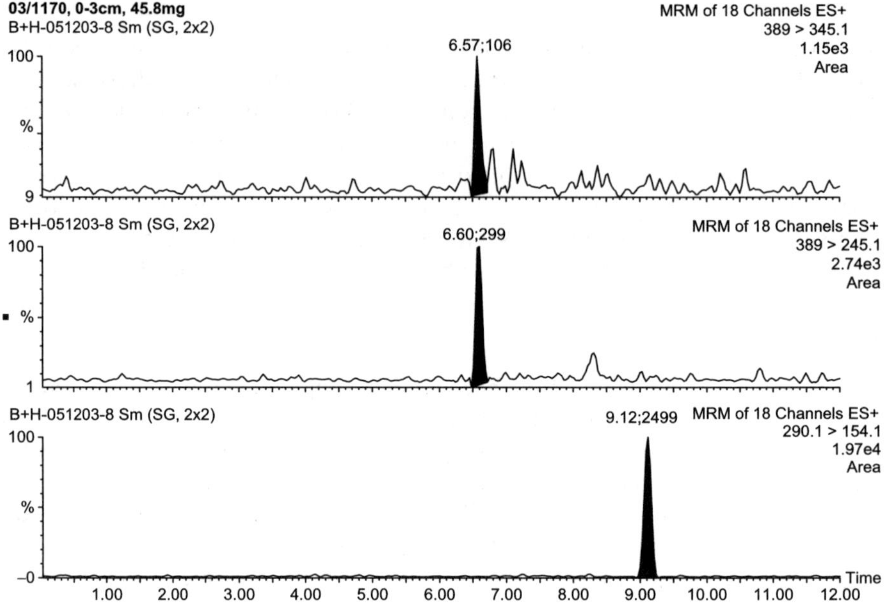

Figure 4.3 *Chromatogram of a hair extract from a victim of a alleged assault. Zopiclone concentration is 4.2 pg/mg. Top: two daughter ions of zopiclone, Bottom: daughter ion of diazepam-d$_5$*

4.7.8 Case 8

A 42-year-old man was offered a drink by a relative during a party. Several hours later, he noticed that his money was missing, but he had no recollection of events during the previous period. He went to the police, but no specimen was collected at that time. After several similar cases in the same region of France, the judge in charge of the case requested a hair analysis. 7-Amino-flunitrazepam, the major metabolite of flunitrazepam and its marker in hair, was detected (Figure 4.4) in the corresponding segment of hair at 31.7 pg/mg, while the distal segment was negative.

4.7.9 Case 9

A 21-year-old woman was confined illegally for 12 days and continuously raped by three men. To the police, she claimed that she had no recollection of the event due to incoherent behaviour and excessive sedation. Analysis of the proximal segment (root to 3 cm) of a strand of her hair demonstrated exposure to clonazepam, an anti-epileptic drug with sedative and amnesic properties. Analysis revealed the identification of 7-aminoclonazepam, its marker, at 135 pg/mg, while the distal segment (3–6 cm) remained negative (Figure 4.5).

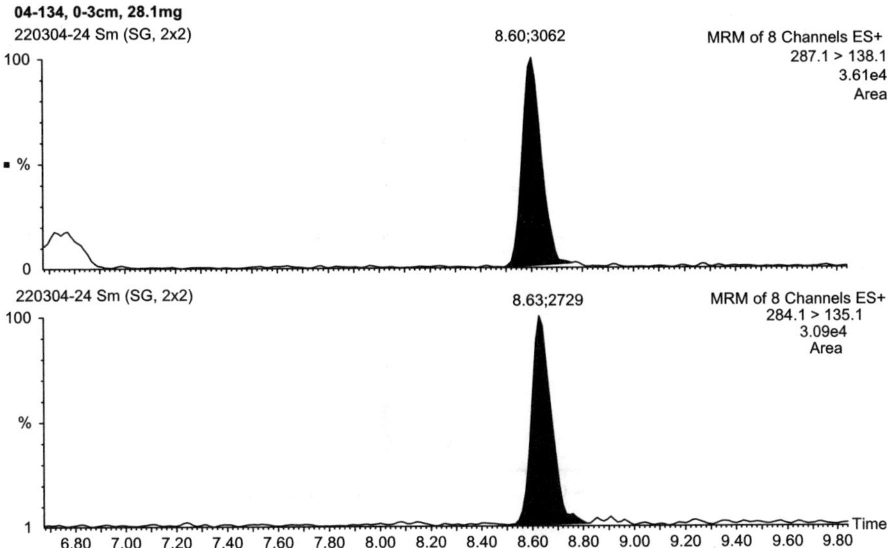

Figure 4.4 *Chromatogram of a hair extract from a victim of a robber. 7-Aminofluni-trazepam concentration is 31.7 pg/mg. Top daughter ion of 7-aminofluni-trazepam-d₃, Bottom: daughter ion of 7-aminoflunitrazepam*

4.7.10 Case 10

A 19-year-old woman went to the police to declare a rape after having a drink that may have been laced with ecstasy (MDMA). At the medico-legal unit of the hospital, a urine sample was collected (about ten hours after the rape) that revealed the presence of MDMA and its metabolite MDA at 1852 and 241 ng/ml, respectively, confirming her previous declarations. To the judge in charge of the case, she claimed that she never took ecstasy and directly gave the name of the rapist, who was rapidly arrested and send to jail. As the circumstances were unclear, the judge requested a hair analysis that demonstrated the simultaneous presence of various stimulants, with the following concentrations: 21.3, 31.6 and 6.7 ng/mg for MDMA, MDEA and MDA, respectively. These results were inconsistent with the claim of being drug-free. During a later confrontation with the judge, she admitted that it was a false notification, that no rape occurred and that it was a revenge on the alleged rapist.

4.8 Discussion

Despite late sampling or even lack of collection of traditional biological fluids, such as blood and/or urine, results of hair testing allow us to document the use of hypnotics in drug-facilitated sexual assault (DFSA).

The literature in this area is growing very rapidly. Frison *et al.*[28] detected thiopental (150 to 300 pg/mg) and its metabolite pentobarbital (200 to 400 pg/mg)

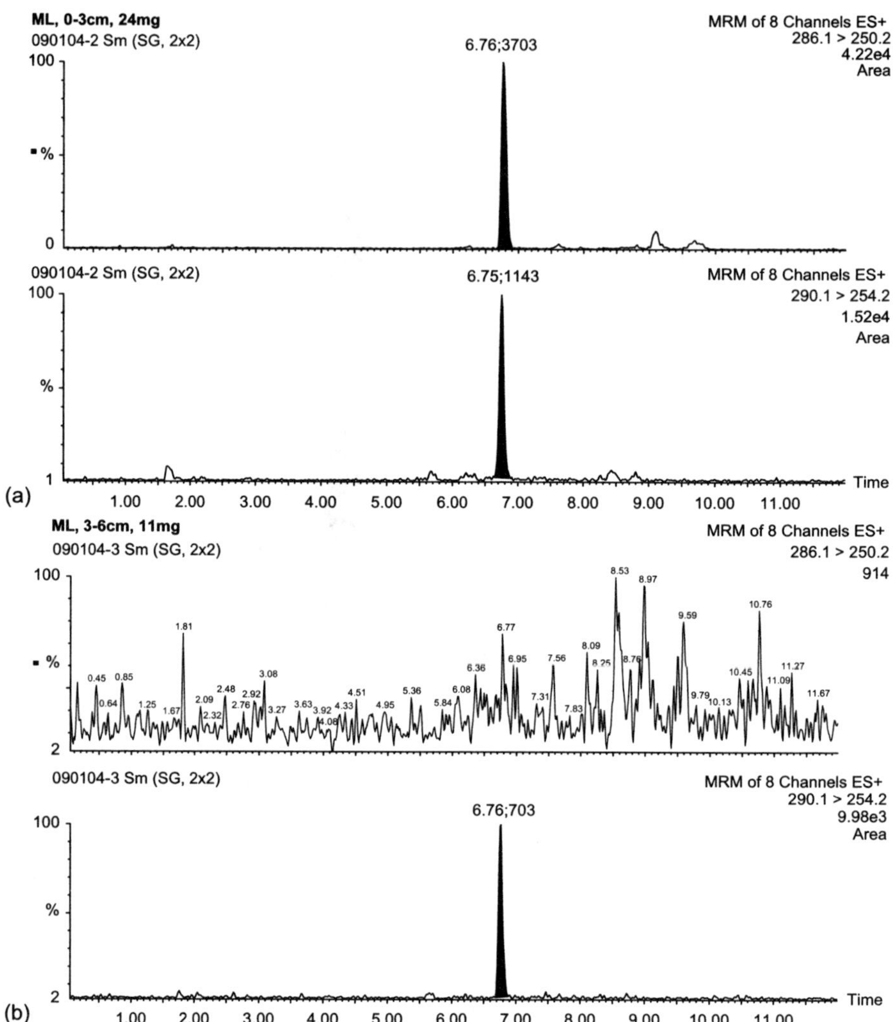

Figure 4.5 *Chromatogram of a hair extract from a victim of an alleged assault. A: Proximal segment, positive for 7-aminoclonazepam at 135 pg/mg; Top: daughter ion of 7-aminoclonazepam, Bottom: daughter ion of 7-aminoclonazepam–d₄. B: Distal segment, negative for 7-aminoclonazepam. The peaks are the background noise of the machine, see the very low abundance. This pattern is classic from a chromatogram without a drug peak*

in three different proximal segments, corresponding to the time of the assault, while distal segments remained negative. In two separate cases, Pépin *et al.*[29] detected 7-aminoflunitrazepam (19 pg/mg) and zopiclone (13 pg/mg). The same authors[30] demonstrated GHB exposure by comparing the concentrations along the hair shaft. Basal GHB concentrations were about 0.7 ng/mg, in comparison with a 5.3 ng/mg concentration in the segment corresponding to the time of the assault.

From our own data and data from the literature, it is obvious that the target concentrations in hair after a single exposure are in the range of a few pg mg^{-1}. To obtain the required ultra-low limits of detection together with suitable mass spectrometry information, tandem mass spectrometry appears to be a prerequisite. The selectivity and sensitivity of these methods have increased extraordinarily to almost completely suppressing the noise level. In comparison with the concentrations that are measured with drugs of abuse, such as heroin or cocaine, in cases of DFSA the concentrations are at least 1000 times lower.

As is the case with other applications (survey of addicts, doping control, driving license regranting *etc.*) hair testing is a valuable approach to increase the window of drug detection. Embarrassment associated with urine collection, particularly after sexual assault, can be greatly mitigated through hair analysis. It is always possible to obtain a fresh, identical hair sample if there is any trouble during analysis, *e.g.* a specimen mix-up or a breach in the chain of custody. This makes hair analysis essentially fail-safe, in contrast to blood or urine analysis, since an identical blood or urine specimen cannot be obtained at a later date. The comparison between urine and hair is given in Table 4.2.

The discrimination between a single exposure and long-term use can be documented by multi-sectional analysis. With the concept of absence of migration along the hair shaft, a single spot of exposure must be present in the segment corresponding to the period of the alleged event, using a growth rate for hair of 1 cm per month. As this growth rate can vary from 0.7 to 1.4 cm per month, the length of the hair section must be calculated accordingly. A delay of three to four weeks between the offence and hair collection and analysis of 2 cm sections is considered satisfactory in order to include the spot of exposurein the hair shaft sample. The hair must be cut as close as possible to the scalp. Particular care is also required to ensure that the individual's hair strands retain the positions they originally had beside one another. Our typical collection procedure is presented in Figure 4.6.

The unique possibility to demonstrate a single drug exposure through hair analysis has some additional interests. In case of late crime declaration, positive hair findings are of paramount importance for a victim, in order to start, under

Table 4.2 *Comparison between urine and hair for testing drugs of abuse and pharmaceuticals*

Parameters	Urine	Hair
Drugs	all, except hormones	all, except hormones
Major compound	metabolites	parent drug
Detection period	2–5 days,	weeks, months
Type of measure	incremental	cumulative
Screening	yes	no
Invasiveness	high	low
Storage	$-20\,^{\circ}$C	ambient temperature
Risk of false negative	high	low
Risk of false positive	low	undetermined
Risk of contamination	high	low
Control material	yes	needed

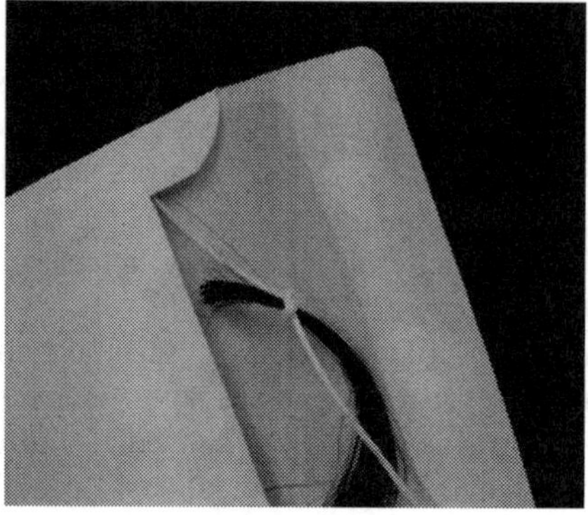

When

3 to 5 weeks after the alleged event

How much

four strands of hair of about 100 hairs

Where

in vertex posterior

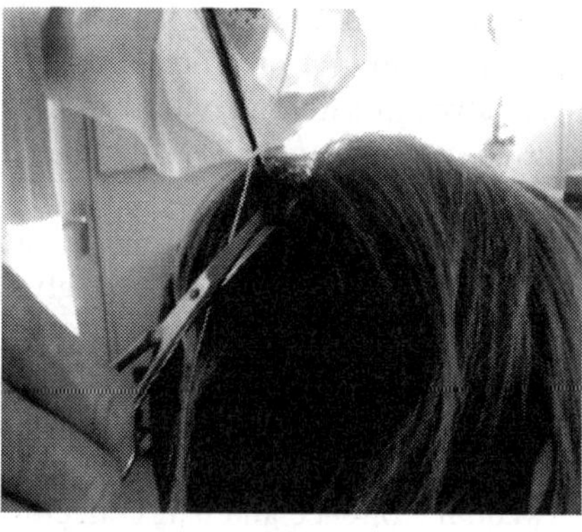

How

- root and tip ends must be distinguished, using a string 1 cm from the root
- hair must be cut by scissors as close as possible from the scalp

Do not pull out !
Do not use adhesive !

→ Storage in an envelop at ambient temperature

To be asked to the victim

- Use of pharmaceuticals before the alleged event?
- Use of pharmaceuticals after the alleged event?
- If yes, which compounds, at which dosage?
- Cosmetic treatments (bleaching, colouration ...) since the alleged event?

Figure 4.6 *Protocol for hair collection in case of drug-facilitated crimes*

suitable conditions, a psychological follow-up. It can also help in the discrimination of false reports of assault, for example in the case of revenge. These cases are often sensitive with little other forensic evidence. Tedious interpretations, *e.g.* in cases of concomitant intake of hypnotics as a therapy for sleeping disorders, are avoided when investigations are done using hair instead of urine. Finally, in the absence of hair testing it is always possible for the advocate of the defendant, to claim during the criminal trial that the detected drug was ingested by the victim him/herself and had no connection with the crime.

4.9 References

1. A.M. Baumgartner, P.F. Jones, W. Baumgartner and C.T. Black, *J. Nuclear Med.*, 1979, **20**, 748.
2. M. Saitoh, M. Uzaka, M. Sakamoto and T. Kobori in *Advances in Biology of Skin*, W. Montagna and R.L. Dobson (ed), Pergamon Press, Oxford, 1969, 183.
3. E. Cone, *Ther. Drug Monit.*, 1996, **18**, 438.
4. G.L. Henderson, *Forensic Sci. Int.*, 1993, **63**, 19.
5. R. Kronstrand, S. Förstberg-Peterson, B. Kagedal, J. Ahlner and G. Larson, *Clin. Chem.*, 1999, **45**, 1485.
6. G.L. Henderson, M.R. Harkey and C. Zhou, *J. Anal. Toxicol.*, 1998, **22**, 156.
7. W.A. Baumgartner, V. Hill and W. Blahd, *J. Forensic Sci.*, 1989, **34**, 1433.
8. L.W. Cartmell, A. Aufderhide and C. Weems, *J. Okla. State Med. Assoc.*, 1991, **84**, 11.
9. H. Sachs and P. Kintz, *J. Chromatogr. B*, 1998, **713**, 147.
10. M.R. Moeller, *J. Chromatogr.*, 1992, **580**, 125.
11. M. Uhl, *Forensic Sci. Int.*, 1997, **84**, 281.
12. W.A. Baumgartner and V.A. Hill in *Recent Developments in Therapeutic Drug Monitoring and Clinical Toxicology*, I. Sunshine (ed), Marcel Dekker, New York, 1992, 577.
13. E.J. Cone, D. Yousefnejad, W.D. Darwin and T. Maguire, *J. Anal. Toxicol.*, 1991, **15**, 250.
14. D.L. Blank and D.A. Kidwell, *Forensic Sci. Int.*, 1995, **70**, 13.
15. D.A. Kidwell and D.L. Blank, in *Drug Testing in Hair*, P. Kintz (ed), CRC Press, Boca Raton, 1996, 17.
16. V. Cirimele, P. Kintz and P. Mangin, *J. Anal. Toxicol.*, 1995, **19**, 331.
17. P. Kintz, *Toxicol. Letters*, 1998, **102–103**, 109.
18. H. Sachs, in *Drug Testing in Hair*, P. Kintz (ed), CRC Press, Boca Raton, 1996, 211.
19. M.R. Moeller, P. Fey and H. Sachs, *Forensic Sci. Int.*, 1993, **63**, 43.
20. R. Wennig, *Forensic Sci. Int.*, 2000, **107**, 5.
21. P. Kintz, V. Cirimele and B. Ludes, *Forensic Sci. Int.*, 2000, **107**, 325.
22. M. LeBeau, W. Andollo, W.L. Hearn *et al.*, *J. Forensic Sci.*, 1999, **44**, 227.
23. P. Kintz, V. Cirimele, M. Villain and B. Ludes, *Ann. Toxicol. Anal.*, 2002, **14**, 361.
24. I. Hindmarch, M. Elsohly, J. Gambles and S. Salamone, *J. Clin. Forensic Med.*, 2001, **8**, 197.
25. F. Questel, G. Lagier, D. Fompeydie *et al.*, *Ann. Toxicol. Anal.*, 2002, **14**, 371.
26. P. Kintz, V. Cirimele, C. Jamey and B. Ludes, *Ann. Toxicol. Anal*, 2002, **14**, 129.
27. A. Verstraete, *Ann. Toxicol. Anal.*, 2002, **14**, 390.
28. G. Frison, D. Favretto, L. Tedeschi and S.D. Ferrara, *Forensic Sci. Int.*, 2003, **133**, 171.
29. G. Pépin, M. Chèze, G Duffort and F. Vayssette, *Ann. Toxicol. Anal.*, 2002, **14**, 395.
30. G. Pépin, Y. Gaillard, M. Chèze and J.P. Goullé, *J. Med. Leg. Droit. Méd.* 2003, **46**, 93.

Hair and Human Identification

BRUCE A. BENNER, JR. and BARBARA C. LEVIN

5.1 Introduction

Human beings are unique in many ways – externally and internally – and with society's increasing emphasis on security, techniques and tools that can be used to identify individuals definitively have become part of our daily lives. Law enforcement personnel are particularly interested in identifying alleged perpetrators and distinguishing them from victims and have used both established (*e.g.* fingerprints) and creative (*e.g.* pet hair) comparisons for associating crime scene evidence with samples of known origin. Employers are continually evaluating the best way to ensure that only authorised personnel gain access to sensitive company locations. In addition to crime and work-place security challenges, identification is also needed for a uniquely personal and private concern – establishing blood relationships between adults and children. Over the last several years, there have been numerous examples of identification by deoxyribonucleic acid (DNA) sequencing techniques that have justified reuniting a child with his/her family as well as the release of an incarcerated yet innocent individual.

One of the external characteristics that can make humans at least somewhat unique (depending upon the colour and style) is our hair, essentially a non-living tissue that we are continually growing and losing. Hair is one of the most common forms of forensic trace evidence, which explains why crime scene investigators are extremely interested in developing methods using hair for identifying and distinguishing people. Microscopic inspection of hair can suggest its origin (body location, ethnicity) and is typically the first step in forensic hair analysis. Tremendous progress in nuclear DNA (nDNA) and mitochondrial DNA (mtDNA) sequence analysis during the last 18 years has enabled the characterisation of both minute amounts of DNA and degraded DNA. Included in these advancements would be Kary Mullis's development of the polymerase chain reaction (PCR) whereby a specific segment of DNA from a minute amount of cells (*e.g.* a drop of blood, saliva, a small number of skin cells, a segment from a single hair) is amplified virtually without limits[1] and then can be sequenced for identification purposes. Both nDNA and mtDNA analyses have been applied to hair samples.

Plucked hairs usually have live follicle cells that can be used for nDNA sequence analysis, whereas cut hair or hair lost naturally during the telogen phase (quiescent) will likely not have live cells associated with the sample. For cut or shed hair, mtDNA sequencing methods can be used to associate or distinguish hair samples from different individuals.

Given that hair is often retrieved from crime scenes and that it can be readily obtained in a noninvasive manner from alleged perpetrators and victims, it is a sample often used for forensic identification. This chapter will review the basic characteristics of hair, its composition and the current state-of-the-art analytical methods using hair for human identification. These methods include technologies based on physical (microscopy) and chemical (nDNA, mtDNA, internal and surface organic analysis) attributes.

5.2 Background

5.2.1 General Composition and Characteristics of Hair

Mammals have hair growing from their skin, or more specifically from structures in the skin called follicles. This characteristic of mammals is one that distinguishes them from other animals. The general function of hair is to protect the mammal from extreme weather conditions and environmental risks.[2] Human hair is composed of proteins ($\approx$65% to 90%), water ($\approx$15% to 35%), and lipids ($\approx$1% to 9%).[3] Specific proteins called keratins associate through disulfide bonds to form fibre-like structures, and this cross-linking engenders the remarkable stability of hair and its resistance to degradation, both chemical and biological.[4] This stability of hair makes it a valuable forensic sample, as it remains intact at the crime scene long after blood and body fluids have experienced significant, if not complete degradation. The majority of forensic investigations of hair use terminal hair, hair from the head and pubic regions.[4]

The anatomy of a hair and follicle (Figure 5.1) includes an outer layer of scale-like structures that together form the cuticle which covers the cortex and which typically comprises the thickest portion of the hair's cross-section. The medulla, a channel-like structure, constitutes the inner core of the hair shaft and runs most of the length of the hair. The medulla of human hair can look dark under transmitted light because this structure is often filled with air, which provides an insulating layer and thus temperature control for the person. The medulla can be classified as continuous, discontinuous and fragmented,[5] yet all three descriptions have been used for the medulla of hair samples from the same individual. Aspects of these structures of hair along with more obvious characteristics such as colour, thickness and curliness can all be used by an experienced forensic microscopist to link hairs retrieved from a crime scene with samples collected from specific individuals. The key is having a number of hairs from both the scene and from identified sources in order to determine the natural intra-sample variation and thus increase the probability of associating known facts with those collected at the crime scene.

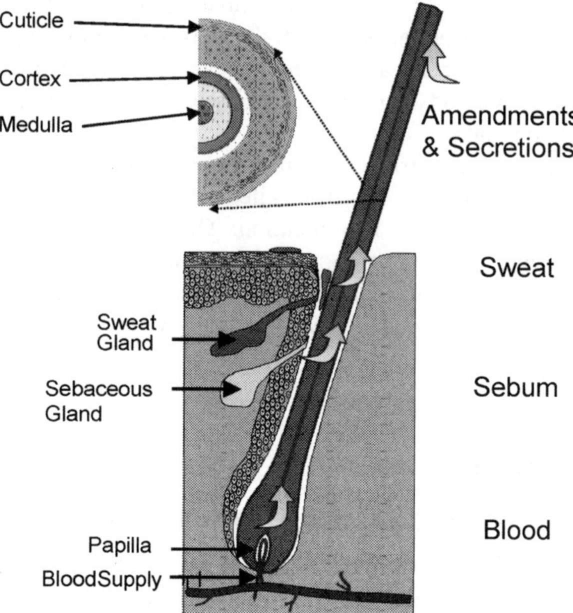

Figure 5.1 *Anatomy of a hair and follicle*

5.2.2 Growth Phases of Hair

Though earlier chapters of this text have more thoroughly detailed treatments of hair growth (see Chapters 1–3 in this volume), a brief description of the hair growth process is appropriate here to emphasise some unique and visibly distinctive features of hair in those different phases. The life of a hair follicle (and thus the hair itself) has four principle phases: the anagen phase during which the follicle is growing hair; the catagen phase when the lower two-thirds of the anagen hair follicle undergoes an apoptosis-driven tissue regression such that hair growth slows to a complete stop; the telogen phase where the hair remains anchored in the follicle by its club root until exogen, a separate and active phase characterised by the loosening of the hair fibre such that it becomes loosely associated with the scalp and can be either shed or easily removed from the follicle by pulling. In a healthy scalp, 80% to 90% of the follicles are in the anagen phase, 10% to 18% are in the telogen phase, and ≈2% of the follicles are in the catagen phase. Active hair growth (anagen phase) is estimated to last for 1000 days with a subsequent resting period (telogen phase) of another 100 days, suggesting that a specific hair follicle could grow and firmly retain a strand of hair for about three years.[4]

Considering an average growth rate of 1 cm per month, hair could grow as long as 36 cm (14 in). Most of us have encountered individuals with scalp or facial hair longer than 36 cm, so other factors must stabilise facial and scalp hair in these extreme cases.

One aid in distinguishing the anagen from the catagen and telogen phases is the position of the start of the medulla with respect to the root bulb. The medulla is not generated during the catagen/telogen phase so one would observe that the medulla and root bulb are distinctly separated in catagen hair and blending into one another in anagen hair.[6] Even so, there is no actual sharp break between the anagen and catagen phases of hair growth so they are often indistinguishable.

One important characteristic of the hair follicle is that it is well nourished by the skin's vascular system,[2,3] such that blood components, including nutrients, medications, illicit drugs and environmental pollutants are incorporated into the internal parts (cortex and medulla) of the hair during the active growth anagen phase. If these components are unique or unusual with respect to the population as a whole, they can be used for distinguishing individuals (see below). Along with components that are incorporated into the core (cortex/medulla) of the hair from the vascular system (*e.g.* blood-borne nutrients, drugs *etc.*), the external surfaces of hair are bathed with the secretions of sebaceous and sweat glands present in close proximity to the hair follicle. These natural lipid materials that coat the hair are well known[2,7] and include fatty alcohols, fatty acids, squalene, cholesterol and large fatty acid esters. As opposed to integral lipids incorporated into the hair from both sebaceous gland and bloodstream involvement,[8,9] the high density of sebaceous glands in the scalp[7] strongly suggests that surface lipids of the hair are mainly from secretions from those glands. This surface-deposited material can accumulate and represent a significant mass fraction of the hair, as evidenced by the 'greasy' appearance of unwashed hair.[2] Methods used for characterising these surface components of hair will be discussed below.

5.2.3 Hair as Forensic Evidence: Interpretation and Legal Considerations

Circumstantial evidence is often used to associate an alleged perpetrator and/or victim with a crime scene or one another. Unlike circumstantial evidence, direct evidence is not inferred but is typically offered as a statement by an individual who personally witnessed the crime. Circumstantial evidence includes fingerprints, all sources of DNA, hair, fibres, glass, footprints, ammunition (discharged or loaded), weapons or virtually anything obtained at the crime scene that investigators believe was left by the victim and/or alleged perpetrator.[10] All forensic evidence is therefore circumstantial evidence and for it to be used in court, it, along with the analytical methods used in characterising the evidence and the individual who performed the analyses, must be approved for admission into court. Furthermore, the forensic method used in a criminal investigation must have been developed using sound scientific procedures and published in the peer-reviewed scientific literature. In this manner, the forensic method is deemed as generally accepted in the scientific community and its application to a specific piece of forensic evidence can then be admissible in court.[10,11] The individual presenting the forensic evidence must also be accepted and admitted into the court as an 'expert witness', as per the Federal Rule of Evidence Rule 702 in the US.[12]

The admissibility of forensic biological evidence has been well established. For example, hair, breath, blood, fingerprints *etc.* are considered 'non-testimonial' means of identification, so the use of such evidence does not violate an individual's US Constitutional Fifth Amendment rights against coerced self-incrimination. Specifically, the ruling of Schmerber vs. California[13] stated that the use of bodily evidence from a crime suspect does not impair his Fifth Amendment protection against self-incrimination. There have been challenges to such evidence under the Fourth Amendment rules against unreasonable searches and seizures, and though it is possible for a suspect to refuse to 'donate' such a sample, the suspect can be compelled to provide one by a court order. Alleged perpetrators can voluntarily donate hair samples and may prefer to allow hair sampling because it is less invasive than blood or urine donation.

5.3 Forensic Analysis of Hair for Identification

5.3.1 Microscopy

Ever since the development of the stereomicroscope (*ca.* 1671 by Cherubin d'Orleans[14]), analysts have investigated features of human and animal hair in three dimensions and at modest magnifications ($2\times$ to $3\times$). Modern stereomicroscopes offer a wide range of magnifications ($2\times$ to $540\times$).[14] Using a stereomicroscope, a forensic investigator can observe gross characteristics of hair including colour, curliness, the condition of the roots (if present) and tips, and if any extraneous material (*e.g.* dried blood, vermin, fungi, dandruff, soil, gunshot residue *etc.*) are associated with the hair samples.[4] Instruments of higher magnification, such as the compound or polarising microscopes ($>50\times$) can highlight more detailed features of the hair samples, including pigment distribution, medulla continuity, scale structure and other less evident characteristics. Adding solvent such as water to the hair (wet-mounting) can improve the visual detection of tiny internal features such as air pockets in the cortex (cortical fusi, Figure 5.2) and structures in the medulla that may be indicative of animals other than humans.

For example, one way of distinguishing human from non-human hair is the petal-like scales covering the outside of the shaft of animal hairs (fur of a mink or a cat),

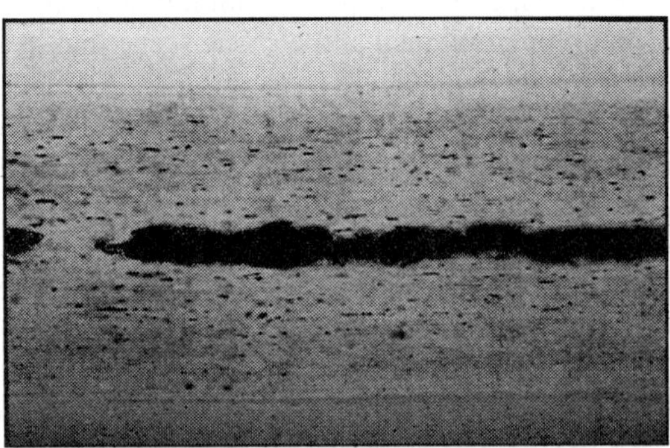

Figure 5.2 *Photomicrograph of a human hair showing cortical fusi*

that are not observed on human hair.[5] (See below for other characteristics of animal hair that are useful in discriminating it from human hair.)

One practical treatise on the forensic microscopy of hair for crime laboratory personnel[5] describes many characteristics that could be used in distinguishing/associating evidential hair from/with those collected from known individuals. These include features that can suggest ethnicity, body area determination (scalp, pubic region, extremities), degree of medulla fragmentation, scale structure (for determining if the hair is from a human or animal source), vermin infestation and root condition (useful in determining if the hair was forcibly removed). Hair from the three ethnic groups – Caucasian, Negroid and Mongoloid – can have significant morphological differences observable under a microscope. Mongoloid hair generally has an exceptionally thick cuticle, Negroid hair mostly has a particularly dense packing of pigment granules and Caucasian hair generally have shafts with moderate though consistent thicknesses and an even distribution of pigment.

Deedrick[15] provided an updated guide to the forensic microanalysis of hair including some information of the utility of DNA sequence analysis of hair evidence. One important point discussed by both Hicks[5] and Deedrick is that it is not possible to identify definitively the age or gender of the individual who donated/lost the hair based solely on microscopic analysis, though some inferences from noted treatment, length and coarseness can be helpful. Tanabe *et al.*[16] described a technique using cellophane tape to fix a single hair to a slide for microscopic analysis followed by sectioning of the medulla and generation of blood type information by immunohistochemistry using the biotin-antibiotin method. Choudry and co-workers[17] treated hair samples with several reagents and observed that a 12 hour treatment with mercaptoacetic acid, a disulfide reducing agent, exposed the intertwined fibre structure of the hair strand. Electron microscopy of these treated hair samples illuminated differences between samples from different sources suggesting that the technique could be used for individualising hair samples. Another group,[18] also using electron microscopy, observed microfibres in the medulla of human hair not present in the medullas of hair from other mammals.

Inspection of the roots of hairs collected at a crime scene can aid the investigators, particularly if the hairs are in the active growth or anagen phase.[6,19] Petraco and co-workers[6] discussed morphological differences in hair roots as related to the three stages of growth as well as the manner in which the hair was removed. As an example, hair in the anagen (active growth) phase will not be easily shed and if removed during a struggle (as opposed to being purposely plucked), the root will appear to be angled with respect to the hair shaft. Also, the anagen hair will have attached follicular cells that offer an opportunity for genetic testing (see below) not technically possible in 1988 when Petraco *et al.* published their work.[6] Tafaro[19] discussed two homicide cases that illustrated the usefulness of documented post-mortem changes in hair roots in associating hair obtained at crime scenes with those sampled during autopsies. Post-mortem banding, air spaces that appear dark when viewed by a light microscope, occurs less than 1 mm from the end of the root, and should be observable in hair obtained from a decedent at the scene as well as after an autopsy. In addition, hair plucked from a deceased subject will have a fragmented or 'bushy' root, offering another way of distinguishing the victim from the alleged perpetrator. However, Linch and Prahlow[20] found in a study of 22 cases with reliable time-of-death information that post-mortem banding was observed in only four cases. Two other post-mortem characteristics, the hard keratin root points and the 'bushy' root appearances, were observed in the majority of the samples from the decedents.

Bisbing and Wolner[21] designed and completed an ambitious study of the microscopic characterisation of hairs from 17 pairs of twins and one set of identical triplets. Through the laborious microscopic inspection of multiple hairs from the twins, all the hairs were distinguishable and associated with their true source, although the authors cautioned that a single human hair 'can never be associated with one person to the exclusion of all others'. Therein lays the importance of inspecting multiple hairs from the same individual. Another important and related point emphasised by Bisbing and Wolner is that while multiple hairs from the same individual may display wide variations in characteristics, the breadth of these variations may be 'unique' to an individual in a sample set and aid the investigator in associating the samples.

A number of groups have discussed the controversial estimation of probabilities with respect to forensic microscopic hair comparisons.[22–26] The efforts of Gaudette and co-workers[22–24] were admirable because the use of realistic probability statements can have a significant impact on a criminal case, as noted with respect to human identification based on DNA sequences (see below). However, other groups[25,26] had concerns of possible experimental bias and the statistical treatment of the data. Hoffmann[27] perhaps put the debate in perspective by noting that the significance of the results with all microscopic examinations of hair is heavily dependent on the experience and the expertise of the analyst (hair examiner), and stated the need for a standardised method for evaluating evidential hair. All the workers involved in this debate have emphasised the need to be cautious in the interpretation phase when associating/distinguishing evidential hair with samples of known origins, so as not to falsely incriminate a suspect.

5.3.2 DNA Sequencing

5.3.2.1 Nuclear DNA

The recent passing of Francis H.C. Crick, co-discoverer with James D. Watson of the double helical structure and mode of replication of deoxyribonucleic acid (DNA), offers an opportunity to reflect on the tremendous progress that has been achieved in the health and forensic fields through study of this biomolecule. Since Watson's and Crick's remarkable announcement concerning DNA in 1953,[28] researchers have developed many methods to probe for the genetic information programmed by the sequence of four nitrogen bases that couple the helices. Two copies of DNA are present in the nucleus of every cell in our body (except red blood and sex cells). These DNA, often referred to as nuclear DNA or nDNA, are distributed among two identical sets of 23 chromosomes. The Human Genome Project (HGP), an enormously ambitious international effort, set out to map and sequence all the genes (the genome) in humans. Multiple contributions summarising preliminary results and related aspects of the HGP were published in 2001[29] with the full sequence presented in 2004.[30] The final report reduced the size of the human genome from 40,000–60,000 genes to 20,000–25,000 genes. Certainly, the benefits of the HGP to humanity will be appreciated well into the future, as this knowledge sheds insight on what proteins are expressed by the $\approx$20,000 genes in the human body.

General results from the HGP and earlier genetic research suggest that a small minority of the human genome – $\approx$3% – are actual genes that instruct cells to make specific proteins, while the majority of the DNA sequence is referred to as non-coding,[31,32] and whose functions are not known. Interestingly, from a strictly genotype standpoint, human beings are remarkably similar, with $>$99.7% of our DNA sequences being identical leaving only 0.3% of the genome ($\approx$1 million nucleotides) useable for distinguishing individuals. These 1 million sequences, including many repetitive regions, can yield an astronomically high probability of distinguishing/associating individuals (see below).

In the mid 1980s, Prof. Alex Jeffreys and his co-workers found that much of the non-coding regions of DNA contained sequences that were repeated many times, often referred to as variable number of tandem repeats (VNTR), and the number of the repeats in these VNTRs were unique to the individual. They used restriction fragment length polymorphism (RFLP), an enzyme-based technique, to clip the regions surrounding the VNTRs, added a radioactive tag and a probe strand of DNA to bind with specific VNTRs, and separated the mixture of different base-pair (bp) units (typically 10–100 bp) by gel electrophoresis.[33] Jeffreys and his group applied their sequencing techniques to two rape-homicides in Narborough, England in which a few thousand blood samples volunteered by residents of the town were analysed for a 33 bp VNTR fragment repeated over 20 times near a gene that Jeffreys was fortuitously studying.[34] Results of the blood samples were compared with semen samples obtained at the crime scenes and the perpetrator, whose DNA in his reluctantly provided blood sample matched the evidential DNA, was subsequently arrested, admitted guilt for both crimes and is presently serving a life sentence. These cases were the first high-profile use of VNTR fragments for aiding the solution of a crime and it encouraged both the application and refinements of DNA sequencing as applied to human identification for law enforcement, paternity testing, and even human evolutionary study. The National Institute of Standards and Technology (NIST) has developed a Standard Reference Material (SRM 2390) for VNTR testing.[35] This SRM is intended for:

- standardisation of forensic and paternity quality assurance procedures for restriction fragment length polymorphisms (RFLP) testing using HaeIII restriction enzymes
- instructional law enforcement or non-clinical research purposes.

Butler[31] wrote a comprehensive treatment of short tandem repeat (STR) units and defined them as repeated units of 2–6 bp. Tetranucleotide STRs (four nucleotides in the repeated unit) are currently the most frequently applied to human identification because they are readily amplified by the polymerase chain reaction (PCR) without the possible copying infidelity of larger VNTRs. The STRs are also more applicable to sequencing degraded DNA because, being shorter in length, these STRs are more likely to be found intact even in highly degraded DNA samples. When different STRs on multiple chromosomes are examined, the discriminatory power increases such that the statistical probability that one has correctly identified the individual is very high. For example, the greater the number of STRs that are examined, the greater the probability that two individuals will be distinguishable (or the greater the improbability of a random match), with probability estimates sometimes suggesting that the suspect can be associated with a crime scene to the exclusion of all other people living on Earth. For more information on STRs, check the informative web site developed at NIST.[36]

A number of groups[37,38] have reported the analysis of nDNA from hair samples. Huguchi and co-workers[37] detailed the results of nDNA analysis of single hairs and showed that DNA provided discriminating data in addition to that from the morphological analysis of hair.

5.3.2.2 Mitochondrial DNA

Application of DNA sequencing to hair depends, of course, on the type of DNA that is associated with the sample. If the hair sample has a viable root (live cells), nDNA techniques may be employed using PCR to generate STR profiles for comparison of crime scene samples with those obtained from victims and suspected perpetrators. Even with all the advances in techniques using nDNA for human identification (STRs, PCR, *etc.*), telogen hair (hair in the resting phase with minimal live root cells) has so little associated nDNA that even the most sensitive techniques cannot provide sufficient nDNA for sequencing.[39] Fortunately, nature provides an alternative source of genetic information – a genome that resides in mitochondria (the cell's energy producers). Hundreds and even thousands of mitochondria (each with 2–5 copies of mitochondrial DNA (mtDNA)) are found in a cell, compared with only two copies of nDNA (distributed among the 46 chromosomes). Human mtDNA is a double-stranded closed circular loop of DNA containing approximately 16,569 bp (Figure 5.3). Anderson *et al.*[40] published the sequence of mtDNA in 1981. Earlier, Hutchison and co-workers[41] had proposed that mammalian mtDNA is inherited from the maternal line, with evidence from the comparison of the mtDNA of equine hybrids and their respective parents. MtDNA encodes for 2 ribosomal RNAs, 22 transfer RNAs and 13 polypeptides involved in oxidative phosphorylation. The remaining sequence is a non-coding region referred to as either the control region, displacement or D-loop and includes the regions most used by the forensic community for human identification.[31,40]

Perhaps of greatest concern in the use of mtDNA results for human identification is the possibility of heteroplasmy in samples – the presence of a single base pair difference in the mtDNA in different tissues in a single individual or when comparing sequences in maternal relatives. Wilson *et al.*[42] were perhaps the first to raise this as a potential issue in the use of control region mtDNA in human identification. Examination of blood, buccal swabs, and hair root and hair shaft

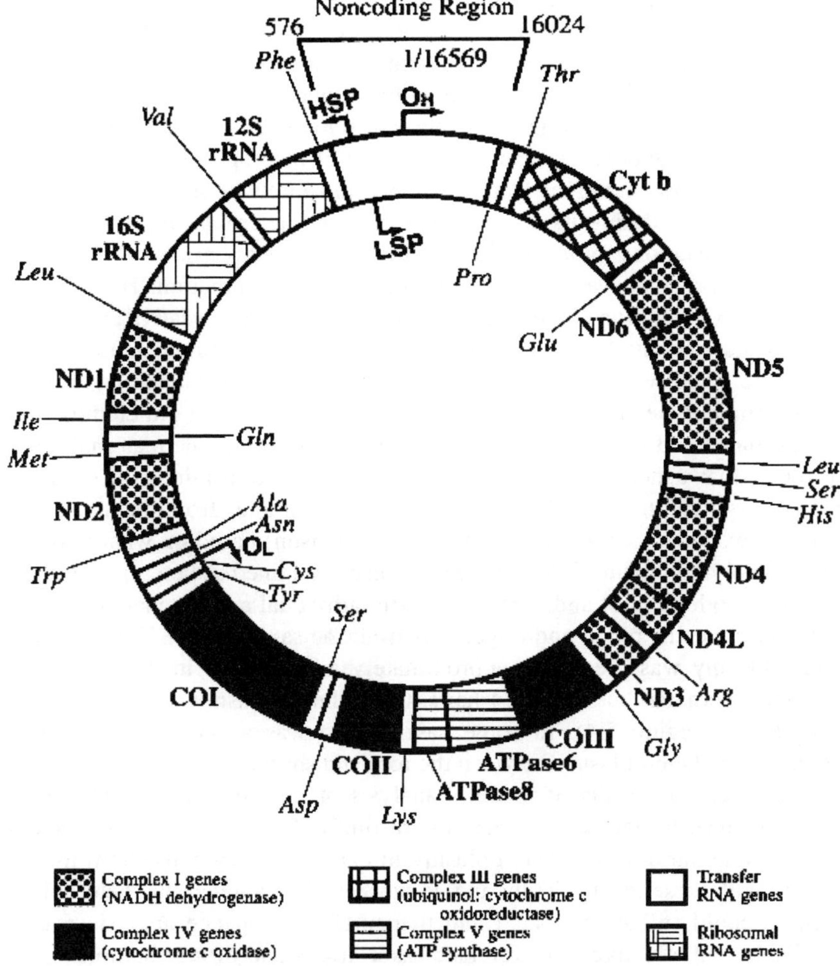

Figure 5.3 *The Human Mitochondrial Genome. Definitions of abbreviations: Phe – phenylalanine; Thr – threonine; Pro – proline; Glu – glutamic acid; Leu – leucine; Ser – serine; His – histidine; Arg – arginine; Gly – glycine; Lys – lysine; Asp – aspartic acid; Tyr – tyrosine; Trp – tryptophan; Cys – cystine; Asn – Asparagine; Ala – alanine; Met – methionine; Gln – glutamine; Ile – isoleucine; Val – valine*

samples from a mother, her son and her daughter revealed the presence of a cytosine/thymine (C/T) heteroplasmy in all of the samples. However, the predominance of C over T or the T over C differed depending on the tissue examined. They cautioned that the detection of only one of the heteroplasmic base pairs in one forensic sample and the other heteroplasmy in another sample from the suspect/victim could be misinterpreted as two individuals, when, in fact, they are from the same person. This error could cause an exclusion of evidential samples. Similarly, Parsons and co-workers[43] performed a more comprehensive study of mtDNA sequences from 134 independent families and 327 generational events, observing a sequence substitution rate of 1 per 33 generations or greater than 20 times the frequency predicted from phylogenic (family tree) analysis. Grzybowski[44] observed high frequencies of heteroplasmy in the HVI region of hair roots from 35 people, suggesting that heteroplasmy in mtDNA is more the rule than the exception. He hypothesised that the high rate of heteroplasmy in the hair roots might be due to how mutations from the maternal source are incorporated into the individual hair root during its development. Though heteroplasmy in the mtDNA can make associations more difficult, Grzybowski emphasised that in some situations it can actually aid in association and/or discrimination. Huhne *et al.*[45] refined mtDNA extraction procedures for hair shafts, blood and saliva and, contrary to Grzybowski's results,[44] matched the mtDNA sequences to the donors regardless of the tissue type, with no observed heteroplasmy. Sekiguchi and co-workers[46] performed mtDNA sequence analysis on a variety of tissue types, including buccal cells, hair, blood and fingernails from multiple generations of one maternal lineage. They found that a heteroplasmy was not consistently transmitted to subsequent generations and recommended analysis of multiple samples from the same source so as not to exclude a sample (in a forensic comparison) based on what may be an infrequent polymorphism. Sekiguchi *et al.* also examined a heteroplasmy in 24 single hairs (both the root and shaft) from one individual and compared the results with the buccal cells, blood and fingernails from the same person. They found that the heteroplasmy was present at approximately a 50% level in the buccal cells, blood and fingernails, but in the 24 hairs, the % heteroplasmy ranged from about 10% to 100% (Figure 5.4). In some cases, there was a statistically significant difference in the heteroplasmy between the root and shaft of the same hair. A recent paper[47] detailed the results of mtDNA analysis of multiple hairs from the same person performed by ten laboratories. Interestingly, for a number of single hairs, they also found variations in heteroplasmy in extracts of segments from the same hair. Even with these results, Tully and co-workers[47] concluded that heteroplasmic differences could still be managed and that mtDNA sequence analysis could be used reliably and effectively for forensic associations and discriminations.

MtDNA has the advantage of allowing samples from maternal relatives to confirm the identity of suspects or unknown crime victims, abducted children and unknown soldiers whose remains are recovered after many years and has been used to identify the World Trade Center victims.[48,49] Recent advances in sequencing mtDNA enable genetic characterisations of small human hair samples with or without root sheath cells,[45,50–57] as well as domestic dog hair.[58–60] Wilson and co-workers[51] described a process where hair is ground and extracted after which mtDNA is amplified by PCR and specific noncoding segments are analysed by capillary electrophoresis and compared with the mtDNA isolated from the blood of the same individual. Another study[52] found that the ability to sequence mtDNA was significantly better for scalp hair (75% success rate) than for either pubic (66% success rate) or axillary hair (52% success rate), though the actual sequence was the same in the respective hair types. Linch *et al.*[53] discussed fundamental aspects of hair growth and the incorporation of mitochondria (source of mtDNA) probably from the melanocytes. They also remarked that the bottleneck theory of mtDNA

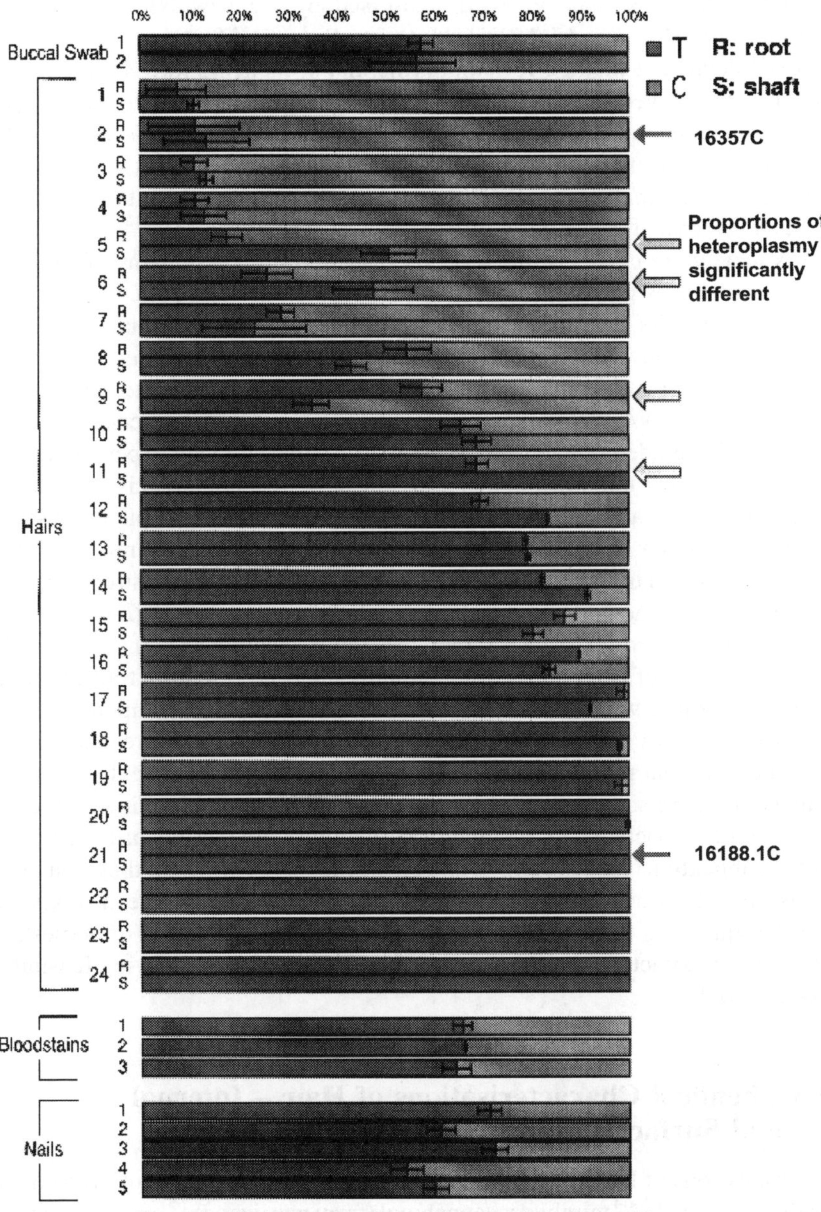

Figure 5.4 *Heteroplasmy in 24 hairs from one individual*

segregation could cause a heteroplasmic site found in the maternal mtDNA to become homoplasmic in the mtDNA of the offspring. It is also possible that current DNA sequencing techniques are not sensitive enough to detect low-frequency heteroplasmies. This issue is being addressed by the new NIST SRM 2394 – *Heteroplasmic Mitochondrial DNA Mutation Detection Standard.* Allen and co-workers[54] detailed a method for amplifying and sequencing mtDNA for shed hair (telogen hair) as well as saliva stains and saliva on stamps, all frequent types of evidence encountered by criminologists. They reported a success rate of greater than 90% for their method for these types of evidence and applied the technique to a number of criminal cases. Perhaps the most impressive results of their work was their estimated sensitivity of as little as 33 fg (33×10^{-15} g) of mtDNA (approximately ten copies of the mtDNA). The questioned remains of 'wild west' outlaw Jesse James played prominently in a report by Stone *et al.*[56] in which two teeth and two hairs retrieved from a burial site, nearly 100 years after James's death, yielded consistent mtDNA sequences that were indistinguishable from samples donated by a person in James' maternal lineage. This study demonstrated the utility of mtDNA techniques in providing useful DNA sequence information from old and degraded remains and supports the belief of many experts that we will no longer have any unknown soldiers or war dead. To support this resolve, every person entering the armed forces now has to provide a blood sample to be stored in a repository. Houck and Budowle[55] discussed the uses of both microscopic and mtDNA techniques in forensic hair comparisons, emphasising the strengths of these two independent methods in providing valuable information from evidentiary samples. Another comparison of two dissimilar techniques[57] involved mtDNA sequence analysis of hairs taken from ten individuals representing four family units that was compared with the measurement of surface organic components of hairs from the same set of samples. The mtDNA results of this study were consistent with the major tenet that mtDNA sequences from different families are distinguishable, but those from the same maternal lineage are indistinguishable. Finally, the use of dog pet hair to associate a victim and/or perpetrator with a crime has been reported in three publications by Savolainen and co-workers[58–60] in which they contend that there is sufficient variation in the control region of mtDNA of domestic dogs for the mtDNA sequencing to be a useful forensic technique. The nDNA of domestic cats has also been extracted, amplified and sequenced,[61] likewise showing feasibility as a forensic tool.

5.3.3 Chemical Characterisations of Hair – Internal and Surface Components

Most discussions of hair analysis typically involve measurements of drugs of abuse including cocaine,[62–64] tetrahydrocannabinol (from marijuana),[65] amphetamines,[66,67] and anabolic steroids.[68] Sachs[69] reviewed the measurement of toxins and illicit drugs in hair, emphasising that the mechanisms for incorporation of blood-borne chemicals into the hair is not understood. In another review, Tagliano *et al.*[70] summarised the state-of-the-art methodologies of hair analysis and, as did

Sachs,[69] stated the importance of understanding how drugs accumulate in hair. In particular, the authors stressed the understanding of the time record of drug use implied by their detection in hair. Considering the lack of knowledge of the relationship between blood components and species incorporated into the hair from the bloodstream over time, some workers present spirited arguments that hair analysis for therapeutic and illicit drugs for proof of use was of little or no value.[71,72] Still, the measurement of a unique drug (either illicit or therapeutic) in hair samples could help associate a person with a crime scene, especially if corroborated with another tissue sample (*e.g.* blood).

Measurement of the surface components of hair for forensic identification has been a much smaller field of study compared with those of microscopic evaluations, DNA analyses (nDNA and mtDNA) and drugs in hair. The natural lipid materials that coat the hair are well known[3,7] and include fatty alcohols, fatty acids, squalene, cholesterol, and long-chain fatty acid esters. As opposed to integral lipids incorporated into the hair from both sebaceous gland and blood,[8,9] the high density of sebaceous glands in the scalp[7] strongly suggests that surface lipids of the hair are mainly from secretions from those glands.

An early study by Nicolaides[7] found that individual differences in enzyme concentrations, pH, body temperature and other parameters could significantly affect the relative concentrations of fatty acids and related compounds, and therefore yield a unique chemical signature for each person. One could also envision chemical changes to these surface lipids by microbial action, air oxidation and even hydrolysis by the solvent used for extraction.[73]

Bernier *et al.*[74] described the chemical characterisations of components generated from within the skin with the ultimate purpose of identifying natural mosquito attractants. The majority of the compounds identified were saturated and unsaturated fatty acids and fatty acid esters. The workers did not report the measurement of fatty alcohols, squalene or cholesterol, prominent components detected in hair samples studied at NIST (see below).[57] This is perhaps due to the heavier influence of perspiration rather than sebum on the skin emanations reported by Bernier and co-workers[74] who sampled from the palms of individuals, a body part with few (if any) sebaceous glands.

Buchanan *et al.*[75] described the characterisation of skin lipids extracted with isopropanol from the fingers of 50 volunteers. They determined that samples collected from children had higher levels of volatile free fatty acids than samples from adults. Adult samples had higher levels of less volatile fatty acid esters than those from children, possibly explaining the experiences of forensic investigators who have found that fingerprints from children do not remain on a surface as long as do those from adults. In related work, Asano and co-workers[76] attempted to use chemical components in fingerprints for the purpose of gleaning personal information such as age, gender, and routine activities to help law enforcement in focusing investigations on more likely perpetrators. Gender discrimination was not possible by fingerprint components, but Asano *et al.*[76] concluded that given the diversity of the compounds that make up fingerprints, and sebum in general, they believed that some of these components could be used to distinguish individuals.

Auwärter *et al.*[77] found significantly higher levels of fatty acid ethyl esters (FAEE) in the hair from heavy drinkers than in non-drinkers.

One of the earliest reports of the measurement of surface compounds in hair was by Fujita and co-workers,[78] who measured residues of hair care products on hair and found that of the 102 brands investigated, residues of 66 brands were measurable on the hair 5–20 days after treatment, suggesting that these species could be used in identifying individuals. Andrasko and Stocklassa[79] reported a liquid chromatographic (LC) method for characterising the methanol/water extracts of hair. Differences in the profiles from individuals were believed to be due to the different shampoos used by the subjects, although no mention was made of natural components from sebum coating the hair. A somewhat different approach to the surface analysis of small samples of hair (100 μg to 1 mg) involved on-line supercritical fluid extraction – gas chromatography/mass spectrometry (SFE-GC/MS)[57] of hair samples from 20 individuals (see Figure 5.5 for schematic of apparatus).

The method offered a number of benefits including greater sensitivity than liquid extraction methods because the entire extractable mass is transferred to the analytical system, compared with only a few % transferred in a conventional liquid extraction/injection. Another benefit of the on-line technique is higher recoveries of volatile species, components that would be lost during a multi-step liquid extraction and concentration. Results from the study suggested, in general, that the relative levels of different surface components measured for an individual's hair was consistent yet sufficiently different from that of another person to enable differentiation. In addition, mtDNA was examined in a subset of the hair samples, representing samples from ten individuals and four families, and their sequences were clearly grouped by maternal lineages. Within one of these lineages was a mother and her two sons, who though indistinguishable based on their mtDNA

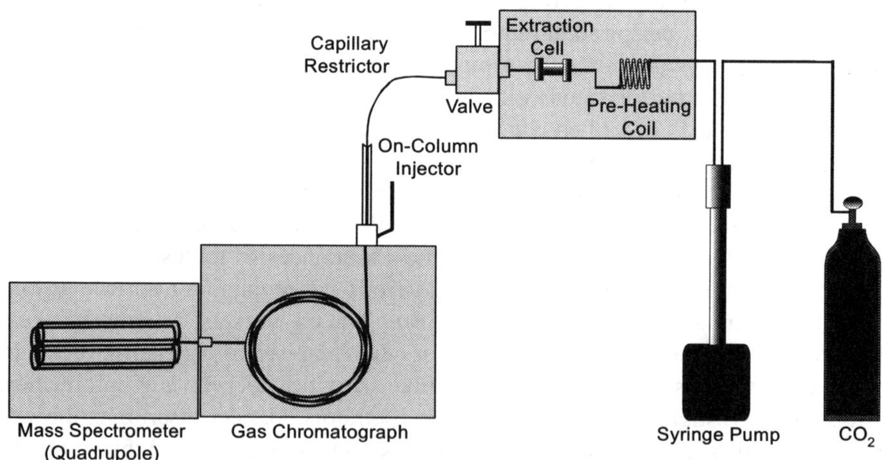

Figure 5.5 *On-line supercritical fluid extraction-gas chromatography/mass spectrometry (SFE-GC/MS) apparatus for the analysis of surface organic components of hair*

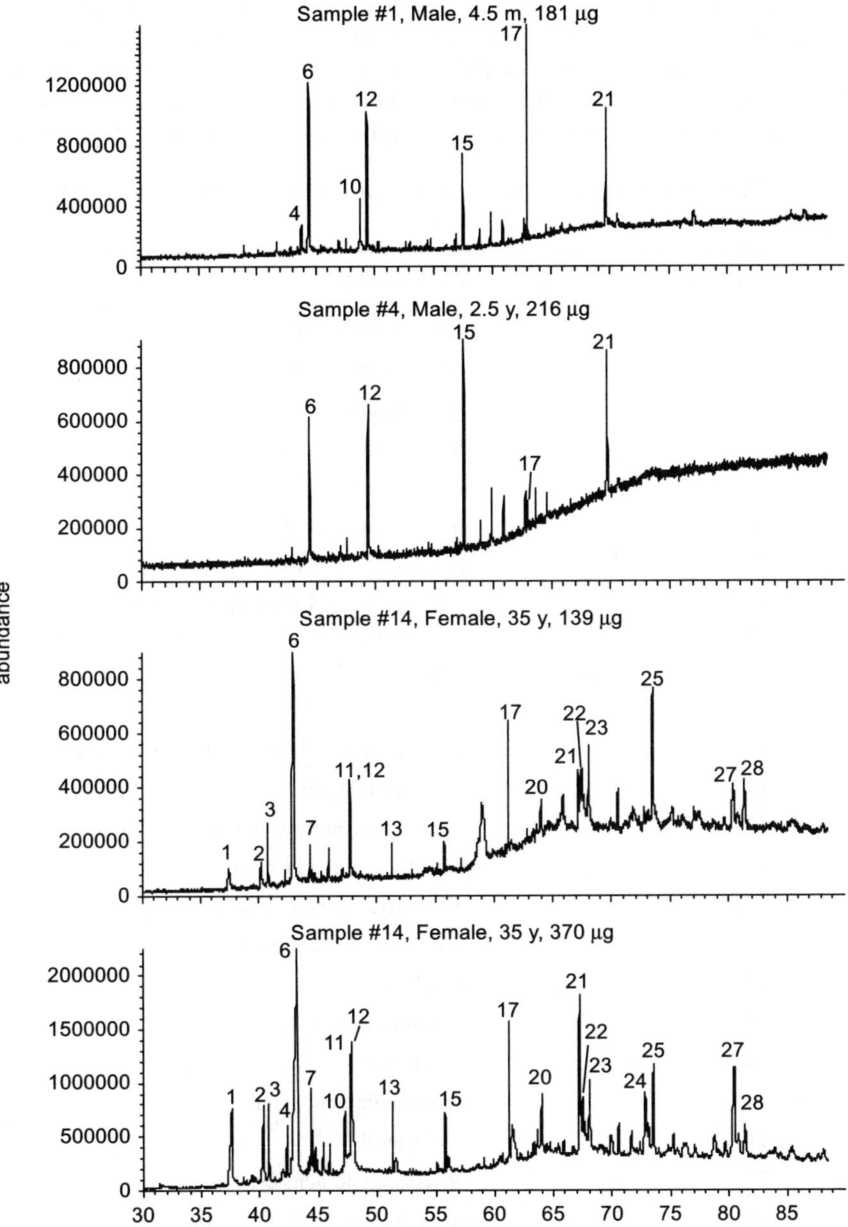

Figure 5.6 *SFE-GC/MS of hair from maternal relations*

sequences, their SFE-GC/MS hair analyses yielded significantly different and distinguishable results (Figure 5.6 and Table 5.1 for peak identifications). In addition to the chemical species deposited naturally by the sebaceous and sweat glands, other components extracted from the surface of hair in this work included phthalates, possibly from residues of shampoo stored in plastic bottles, sunscreen

Table 5.1 *Identification of components from on-line SFE-GC/MS of human scalp hair*

Peak No.	Identification
1.	tetradecanoic acid
2.	pentadecanoic acid
3.	pentadecanol
4.	hexadecenoic acid
5.	hexadecanol
6.	hexadecanoic acid
7.	hexadecanoic acid ester
8.	2-hydroxy-3,3,5-trimethylcyclohexyl-benzoic acid ester (homosalate)
9.	heptadecanol
10.	octadecenoic acid
11.	2-ethylhexylmethoxycinnamate/salicylate
12.	octadecanoic acid
13.	2-ethylhexymethoxycinnamate/salicylate
14.	hexadecanoic acid ester
15.	diisooctyl- or bis(2-ethylhexyl) phthalate
16.	octadecanoic acid ester
17.	squalene
18.	N,N-dimethyl-1-dodecanamine
19.	octadecane
20.	dodecanoic acid ester
21.	cholesterol
22.	hexadecanoic acid ester
23.	tetradecanoic acid, C_{16} ester
24.	hexadecenoic acid, C_{18} or C_{20} ester
25.	hexadecanoic acid, C_{16} ester
26.	octadecanoic acid, C_{16} ester
27.	hexadecenoic acid, C_{18} ester
28.	hexadecanoic acid, C_{18} ester

compounds whose source could be either a sunscreen and/or hair treatment product, and an amine antioxidant (tentatively identified as N,N-dimethyl-1-dodecanamine), also likely from the formulation of a hair care product. It is quite possible that a number of the components present in sebum are also ingredients in hair care products used by both adults and children. Goodpaster *et al.*[80,81] performed related work in evaluating off-line extraction techniques of small hair samples using more aggressive solvents than the supercritical carbon dioxide used for the on-line SFE-GC/MS. They found that while using large-volume injection techniques improved the sensitivity of the off-line liquid extraction methods, the on-line SFE-GC/MS proved to be the most sensitive for extracting and measuring surface organic components from hair samples as small as 40 µg, certainly in the lower mass range of forensic hair samples. These surface organic analytical techniques do not destroy the hair structure and may, theoretically, allow subsequent mtDNA sequencing of the same hair sample, thus providing evidence for human identification from two independent techniques.

5.4 Summary

Since hair is among the most common forensic evidence collected at crime scenes, the techniques used for associating/excluding evidential hair samples from hair of known origins are extremely important tools for law enforcement. Well-established microscopic techniques when used by experienced and knowledgeable analysts can contribute valuable data to criminal investigations. Developments in sequencing DNA from hair, particularly mtDNA, have advanced to the point that ng quantities of DNA can be amplified and sequenced in hours as opposed to weeks. These genetic techniques can often provide compelling evidence for correlation of a human suspect with hair from a crime scene, as well as the statistical probability of the accuracy of that conclusion. These types of results can heavily influence a jury. Finally, organic analysis of associated components of hair, either incurred (drugs) or deposited on the surface (naturally or topically) can contribute complementary data to the investigation. It is possible that future work will demonstrate the advantage of using all three of these forensic techniques in sequence on the same hair sample (*i.e.* microscopy followed by surface chemical analysis followed by mtDNA extraction, amplification and sequencing), thus expediting the criminal investigation and increasing the probability that the correct person or persons are charged and convicted of the crime.

5.5 Acknowledgements

We thank Dr. John V. Goodpaster of the Bureau of Alcohol, Tobacco, Firearms and Explosives (ATF) for his help in writing this chapter, including his thorough review. We credit John P. Pendley, ATF, for his help in illuminating the legal status of hair and Susan M. Ballou (NIST) for reviewing the chapter and for providing valuable forensics texts. Finally, we thank Max Houck (West Virginia University) for the photomicrograph of the hair (Figure 5.2).

5.6 Disclaimer

Mention of trade names or commercial products does not constitute endorsement or recommendation for use by the National Institute of Standards and Technology (NIST), nor does it imply that the materials or equipment identified are necessarily the best available for the purposes.

5.7 References

1. K.B. Mullis, *Ann. Biol. Clin.*, 1990, **48**, 579.
2. H. Harding and G. Rogers in *Forensic examination of human hair*, L.R. Robinson (ed), Taylor and Francis, Philadelphia, PA, 1999, 1–62.
3. M.R. Harkey, *Forensic Sci. Int.*, 1993, **63**, 9.
4. R.E. Bisbing in *Forensic Science Handbook*, R. Saferstein (ed), Prentice-Hall, Inc. Englewood Cliffs, NJ, 1982, 184–211.
5. J.W. Hicks, *Microscopy of Hairs: A Practical Guide and Manual*, Federal Bureau of Investigation, Washington, DC, 1977, 1–41.
6. N. Petraco, N. Fraas, F. Callery and P. Deforest, *J. Forensic Sci.*, 1988, **33**, 68.
7. N. Nicolaides, *Science*, 1974, **186**, 19.
8. P.W. Wertz and D.T. Downing, *Lipids*, 1988, **23**, 878.
9. P.W. Wertz and D.T. Downing, *Comp. Biochem. Physiol.*, 1989, **92B**, 759.
10. T.F. Kiely in *Forensic Science: Introduction to Scientific and Investigative Techniques*, S.H. James and J.J. Nordby (ed), 1st edn, CRC Press, Washington DC, 2003, Chapter 31.
11. G.J. Beggs, *Federal Standard for Expert Testimony Reliability Before Daubert*, Lectric Law Library, http://www.lectlaw.com/files/exp08.htm, 1998, 1–7.
12. Federal Rule Of Evidence 702 (USA), 1975.
13. Schmerber vs. California, U.S. Supreme Court, 1966, **767[384]**.
14. P.E. Nothnagle, W. Chambers and M.W. Davidson, *Introduction to Stereomicroscopy*, http://www.microscopyu.com/articles/stereomicroscopy/stereo-intro.html, 2004.
15. D. Deedrick, *Forensic Sci. Commun.*, 2000, **2**, 1.
16. R. Tanabe, I. Ishiyama and Y. Itakura, *J. Forensic Sci.*, 1988, **33**, 767.
17. M.Y. Choudhry, C.R. Kingston, L. Kobilinsky and P. Deforest, *J. Forensic Sci.*, 1982, **28**, 293.
18. J.L. Clement, R. Hagege, A.L. Pareux, J. Connet and G. Gastaldi, *J. Forensic Sci.*, 1981, **26**, 447.
19. J. Tafaro, *J. Forensic Sci.*, 2000, **45**, 495.
20. C.A. Linch and J.A. Prahlow, *J. Forensic Sci.*, 2001, **46**, 15.
21. R.E. Bisbing and M.F. Wolner, *J. Forensic Sci.*, 1984, **29**, 780.
22. B.D. Gaudette and E.S. Keeping, *J. Forensic Sci.*, 1974, **19**, 599.
23. B. Gaudette, *J. Forensic Sci.*, 1976, **21**, 514.
24. B. Gaudette, *J. Forensic Sci.*, 1978, **23**, 758.
25. P. Barnett and T. Ohira, *J. Forensic Sci.*, 1982, **27**, 272.
26. C.G.G. Aitken and J. Robertson, *J. Forensic Sci.*, 1987, **32**, 684.
27. K. Hoffmann, *J. Forensic Sci.*, 1991, **36**, 1053.

28. J.D. Watson and F.H.C. Crick, *Nature,* 1953, **171**, 737.

29. International Human Genome Sequencing Consortium, *Nature,* 2001, **409**, 860.

30. International Human Genome Sequencing Consortium, *Nature,* 2004, **431**, 931.

31. J.M. Butler, *Forensic DNA Typing: Biology and Technology Behind STR Markers,* Academic Press, New York, 2001.

32. G.T. Duncan, M.L. Tracey and E. Stauffer in *Forensic Science: An Introduction to Scientific and Investigative Techniques,* S.H. James and J.J. Nordby (ed), CRC Press, Washington, DC, 2003, 221–250.

33. A.J. Jeffreys, V. Wilson and S.L. Thein, *Nature,* 1985, **314**, 67.

34. A.J. Jeffreys, V. Wilson, and S.L. Thein, *Nature,* 1985, **316**, 76.

35. B.C. Levin, H. Cheng, M.C. Kline, J.W. Redman and K.L. Richie, *Fres. J. Anal. Chem.,* 2001, **370**, 213.

36. *Short Tandem Repeats (STRs).* National Institute of Standards and Technology (NIST), http://www.cstl.nist.gov/div831/strbase/, 2004.

37. R. Higuchi, C.H. von Beroldingen, G.F. Sensabaugh and H.A. Erlich, *Nature,* 1988, **332**, 543.

38. R. Uchihi, K. Tamaki, T. Kojima, T. Yamamoto and Y. Katsumata, *J. Forensic Sci.,* 1992, **37**, 853.

39. J.A. Dizinno, M.R. Wilson, B. Budowle and J. Robertson (ed), *Forensic Examination of Hair,* Taylor & Francis: Philadelphia, PA, 1999; Chapter 3.

40. S. Anderson, A.T. Bankier, B.G. Barrell, M.H.L. deBrujin, J. Drouin, I.C. Eperon, D.P. Nierlich, B.A. Roe, F. Sanger, P.H. Schreier, A.J.H. Smith, R. Staden and I.G. Young, *Nature,* 1981, **290**, 457.

41. C.A. Hutchison III, J.E. Newbold, S.S. Potter and M.H. Edgell, *Nature,* 1974, **251**, 536.

42. M.R. Wilson, D. Polanskey, J. Replogle, J. Dizinno and B. Budowle, *Human Genetics,* 1997, **100**, 167.

43. T.J. Parsons, D.S. Muniec, K. Sullivan, N. Woodyatt, R. Alliston-Greiner, M.R. Wilson, D.L. Berry, K.A. Holland, V.W. Weedn, P. Gill and M.M. Holland, *Nat. Gen.,* 1997, **15**, 363.

44. T. Grzybowski, *Electrophoresis,* 2000, **21**, 548.

45. J. Huhne, H. Pfeiffer, K. Waterkamp and B. Brinkmann, *Int. J. Legal Med.,* 1999, **112**, 172.

46. K. Sekiguchi, K. Kasai and B. Levin, *Mitochondrion,* 2003, **2**, 401.

47. G. Tully, S.M. Barritt, K. Bender, E. Brignon, C. Capelli, N. Dimo-Simonin, C. Eichmann, C.M. Ernst, C. Lambert, M.V. Lareu, B. Ludes, B. Mevag, W. Parson, H. Pfeiffer, A. Salas, P.M. Schneider and E. Staalstrom, *Forensic Sci. Int.,* 2004, **140**, 1.

48. M. Lesney, *Today's Chemist at Work,* 2002, **11**, 33.

49. K. Miller, *The Scientist,* 2002, **16**, 40.

50. R. Hopgood, K.M. Sullivan and P. Gill, *Biotechniques,* 1992, **13**, 82.

51. M.R. Wilson, D. Polanskey, J. Butler, J.A. Dizinno, J. Replogle and B. Budowle, *Biotechniques,* 1995, **18**, 662.

52. H. Pfeiffer, J. Huhne, C. Ortmann, K. Waterkamp and B. Brinkmann, *Int. J. Legal Med.,* 1999, **112**, 287.

53. C.A. Linch, D.A. Whiting and M.M. Holland, *J. Forensic Sci.,* 2001, **46**, 844.

54. M. Allen, A.-S. Engstrom, S. Meyers, O. Handt, T. Saldeen, A. von Haeseler, S. Paabo and U. Gyllensten, *J. Forensic Sci.*, 1998, **43**, 453.

55. M.M. Houck and B. Budowle, *J. Forensic Sci.*, 2002, **47**, 964.

56. A.C. Stone, J.E. Starrs and M. Stoneking, *J. Forensic Sci.*, 2001, **46**, 173.

57. B.A. Benner, Jr., J.V. Goodpaster, J.A. DeGrasse, L.A. Tully and B.C. Levin, *J. Forensic Sci.*, 2003, **48**, 554.

58. P. Savolainen, B. Rosén, A. Holmberg, T. Leitner, M. Uhlén and J. Lundeberg, *J. Forensic Sci.*, 1997, **42**, 593.

59. P. Savolainen and J. Lundeberg, *J. Forensic Sci.*, 1999, **44**, 77.

60. P. Savolainen, L. Arvestad and J. Lundeberg, *J. Forensic Sci.*, 2000, **45**, 990.

61. M.A. Menotti-Raymond, V.A. David, J.C. Stephens, L.A. Lyons and S.J. O'Brien, *J. Forensic Sci.*, 1997, **42**, 1039.

62. W. Baumgartner, V. Hill and W. Blahd, *J. Forensic Sci.*, 1989, **34**, 1433.

63. J.F. Morrison, S.N. Chesler, W.J. Yoo and C.M. Selavka, *Anal. Chem.*, 1998, **70**, 163.

64. M.J. Welch, L.T. Sniegoski and S. Tai, *Anal. Bioanal. Chem.*, 2003, **376**, 1205.

65. M.R. Moeller, P. Fey and H. Sachs, *Forensic Sci. Int.*, 1993, **63**, 43.

66. J. Liu, K. Hara, S. Kashimura, M. Kashiwagi and M. Kageura, *J. Chromatogr. B,* 2001, **758**, 95.

67. O. Suzuki, H. Hattori and M. Osano, *J. Forensic Sci.*, 1984, **29**, 611.

68. V. Dumestre-Toulet, V. Cirimele, B. Ludes, S. Gromb and P. Kintz, *J. Forensic Sci.*, 2002, **47**, 211.

69. H. Sachs, *J. Forensic Sci.*, 1997, **84**, 7.

70. F. Tagliaro, Z. DeBattisti, G. Lubli, C. Neri, G. Manetto and M. Marigo, *J. Chromatogr. B,* 1997, **689**, 261.

71. L. Potsch, *Int. J. Legal Med.*, 1996, **108**, 285.

72. A. Tracqui, P. Kintz and P. Mangin, *Forensic Sci. Int.*, 1995, **70**, 183.

73. K.V. Curry and S. Golding, *J. Soc. Cosmet. Chem.*, 1971, **22**, 681.

74. U.R. Bernier, M.M. Booth and R.A. Yost, *Anal. Chem.*, 1999, **71**, 1.

75. M.V. Buchanan, K. Asano and A. Bohanon, *Proceedings of the International Society for Optical Engineers (SPIE),* 1997, **2941**, 89.

76. K.G. Asano, C.K. Bayne, K.M. Horsman and M.V. Buchanan, *J. Forensic Sci.*, 2002, **47**, 805.

77. V. Auwärter, F. Sporkert, S. Hartwig, F. Pragst, H. Vater and A. Diefenbacher, *Clin. Chem.*, 2001, **47**, 2114.

78. Y. Fujita, M. Nakayama, K. Kanbara, N. Nakayama, N. Mitsuo, H. Matsumoto and T. Satoh, *Eisei Kagaku (Japanese Journal of Toxicology and Environmental Health),* 1989, **35**, 37.

79. J. Andrasko and B. Stocklassa, *J. Forensic Sci.*, 1990, **35**, 569.

80. J.V. Goodpaster, B.C. Drumheller and B.A. Benner Jr., *J. Forensic Sci.*, 2003, **48**, 299.

81. J.V. Goodpaster, J.J. Bishop and B.A. Benner Jr., *J. Sep. Sci.*, 2003, **26**, 137.

CHAPTER 6

Hair and Metal Toxicity

STEFANOS N. KALES and DAVID C. CHRISTIANI

6.1 Overview

Hair is most relevant to metals toxicology as a biological medium or biopsy material for analysis. Hair can be non-invasively obtained, easily stored and transported, and later analysed for the presence of certain metals.[1–3] Hair analysis is most informative when the metal of interest is a xenobiotic and the exposure route is ingestion. In these cases, it is most likely that hair analysis reflects an internal dose of the metal and not external contamination of the hair. Determination of hair mercury to estimate dietary methylmercury exposure is the best example.[4] A large body of epidemiological evidence correlates hair mercury concentrations to blood mercury levels and both of these to fish consumption in a dose-response fashion. In addition, hair is one of several biomarkers used in epidemiological studies of arsenicosis and arsenic-contaminated drinking water. Further, hair analysis can be used clinically and forensically to document thallium poisoning, an intoxication that also results in pathological changes in the hair.

Hair's utility as a biomarker is considered to be limited in most occupational environments where exposures are airborne. In these situations, hair is subject to exogenous contamination by the toxicant of interest. Therefore, distinguishing metal content internally distributed and excreted into the hair after absorption from metals that externally contaminate the hair surface is difficult, if not impossible.[2,4] Likewise, hair has limited usefulness when the metal of interest is both a potential occupational exposure and an essential, dietary trace element. In these cases, the metal may naturally be present in hair in varying amounts.

Timing is also important. Due to the hair shaft being non-vascular and to its growth rate, hair metal content is unlikely to accurately reflect very recent exposures (hours to several days). In addition, hair is also unlikely to document exposures that have occurred more than one year before the time of analysis.[4] In individuals with hair of sufficient length, however, segmental hair analysis may provide information regarding exposure over time.[4–6]

For each major metal of occupational and environmental interest, this chapter will:

  a. briefly summarise the toxicology and kinetics in relation to hair
  b. present the relative advantages and disadvantages of hair compared to other biomarkers of exposure
  c. present the situations where hair analysis may be indicated to monitor and/or document human exposure.

Further, we will discuss the misuse of 'commercial' hair tests for panels of metals and minerals whose results are promoted as indicators of health, nutritional status and metal toxicity.

Finally, we will present an overview of methodological issues when using hair as a biomarker for metals.

6.2 Mercury

6.2.1 Toxicology

Mercury is a xenobiotic used in chloralkali production, some switches, fluorescent light bulbs and certain batteries. Elemental mercury (dental amalgams, thermometers and gas meters), methylmercury (fish) and ethylmercury (thiomerosal in vaccines) are the forms and potential exposures most relevant to the general population[7,8] (See Table 6.1). For most people, background mercury exposure and variability in blood mercury result primarily from the consumption of fish containing methylmercury, which bioaccumulates in seafood worldwide.[7-10]

Acute exposures to elemental mercury can produce acute lung injury, while chronic exposures may produce renal dysfunction and neuropathy. Frank poisoning results in additional neurological disturbances including tremor, behavioural changes ('erethism'), gingivitis[11,12] and progression to delirium/hallucinations.[13]

Clinical methylmercury intoxication is characterised by ataxia, tremor, constriction of visual fields, and cortical and cerebral atrophy[12,13] and is generally associated with hair levels greater than 50 parts per million (ppm) (>200 μg Hg L^{-1} in blood).[14] There is considerable debate regarding the 'safe' dose of

Table 6.1 *Major chemical forms of mercury associated with the general population exposure*

Form	Absorption/ source of exposure	Major target organs	Major excretory pathway	Approximate whole body half-life	Usual biomarkers
Hg^0	Inhalation of vapour	Kidney, CNS, PNS	Urine and faeces	60 days	Urine mercury
Methyl Hg	Ingestion of seafood	CNS	faeces	40–70 days	Blood and hair mercury
Ethyl Hg	Parenteral from vaccines	CNS, Kidney	faeces	7–20 days	Blood mercury

Adapted from Clarkson *et al.* NEJM 2003[8]

methylmercury from fish in the diet, particularly for children and women of childbearing age.[10,15–17]

Thimerosal or thiomersal contains ethylmercury and has been used safely as a vaccine preservative since the 1930s. Ethylmercury has neurological effects similar to methylmercury, but is considerably less toxic, with higher blood concentrations required to cause poisoning.[18] Although adverse effects have not been documented, recently, thimerosal has been almost completely removed from US licensed vaccines to decrease childhood exposure as a precaution, but continues to be used in other countries.[8]

6.2.2 Kinetics and Relation to Hair

The different chemical forms of mercury influence the absorption, distribution, toxicological manifestations, excretion and useful methods of biological monitoring (Table 6.1). Metallic (elemental) mercury, methylmercury and ethylmercury all cross the blood-brain barrier. Elemental mercury also accumulates in the kidneys and renal excretion is important in its elimination from the body.[8,19]

Methylmercury penetrates into the central nervous system to a greater extent than ethylmercury due to the latter's larger size and faster decomposition, which may explain, in part, why ethylmercury is less neurotoxic than methylmercury.[18] In addition, ethylmercury is converted more rapidly to inorganic mercury and has a considerably shorter half-life than methylmercury.[8] A recent toxicokinetic study demonstrated that after immunisation with thiomerosal-containing vaccines, infants rapidly eliminated ethylmercury from the blood through the faeces (estimated half-life of seven days) with urine mercury undetectable in most samples.[20]

Methylmercury is distributed widely and concentrated in the blood. While methylmercury excretion is primarily faecal, it is also excreted into the hair where it accumulates and reaches concentrations ranging from 140 to 370 times that of blood.[21] The concentration in new hair is directly proportional to blood methylmercury concentration.[22]

6.2.3 Hair vs. Other Biomarkers

For elemental and inorganic mercury, toxicity thresholds and laboratory evaluation are based primarily on urinary excretion, its main route of elimination. A 24-hour urine collection is the best indicator of recent or chronic exposure to elemental or inorganic mercury.[19,23,24] Blood mercury can also detect intense, acute exposures to inorganic and elemental mercury.[23] Hair is not well-suited for monitoring exposure to elemental mercury vapour. First, it is limited by external contamination of the hair.[25] Second, most mercury in hair is in the form of methylmercury and reflects dietary exposure from seafood. Although one study found that average levels of mercury in scalp and pubic hair were significantly higher in dentists occupationally exposed to elemental mercury than controls,[26] the observed hair concentrations may still have reflected fish consumption.[27]

Blood mercury is the best test for current methylmercury exposure.[23] Because urinary excretion of methylmercury is low, an elevated blood mercury concentration (with undetectable urine mercury) or a high ratio of blood to urine mercury supports exposure to an organic form of mercury. Investigators have used hair mercury measurements as a biomarker in populations with traditionally high fish or marine mammal consumption in attempts to elucidate benchmark doses for various effects.[10,28–30] In methylmercury exposure, the blood mercury concentration is directly proportional to the mercury concentration in new hair, and because methylmercury reaches much higher concentrations in hair, this tissue is more easily analysed. Methylmercury accounts for about 80% of total mercury in hair among those who eat fish,[10] but may account for less when methylmercury exposures are quite low.[31] The average ratio of hair mercury to blood mercury is about 250:1,[21] which means that a hair methylmercury value in ppm can be multiplied by four to find the corresponding estimated blood concentration in μg L^{-1}. Due to hair's growth rate, when measured from the scalp, each cm of hair reflects the average blood concentration for the last month.[22] Therefore, hair measurements provide the advantage of an integrated measure of exposure.[10] Blood concentrations, on the other hand, may fluctuate to a greater extent when fish consumption is intermittent and blood levels are not steady state.

Of major importance to studies of methylmercury's potential effects on the developing central nervous system, investigators have demonstrated that maternal hair and blood levels of mercury correlate well with the concentrations in umbilical cord blood, infant blood and infant brain tissue.[32–34] Grandjean *et al.*[33] found that the hair mercury in Faroese mothers who consumed a diet high in pilot whale was significantly correlated with umbilical cord blood taken immediately after delivery. Cernichiari *et al.*[34] examined the brains of 27 deceased Seychellois infants with matching infant blood samples, and matching maternal hair and blood. Maternal hair mercury was highly correlated (correlation coefficients 0.6–0.8) to concentrations in six infant brain regions and to maternal blood (correlation coefficient 0.82).

Table 6.2 summarises hair mercury values found in selected studies and advisory guidelines and relates them to blood mercury concentrations and exposure settings. Both background blood and hair mercury concentrations are determined by the frequency and type of fish consumption. The highest levels are seen in those consuming fish frequently, especially when predator fish high in mercury, such as swordfish and shark, are consumed regularly. In the US recent data demonstrate that average blood mercury concentrations are quite low,[35,36] which correspond to hair levels of 0.23 to 0.5 ppm.[31,35] Among frequent fish consumers in the US, however, it is not uncommon to observe higher blood mercury concentrations (10 to $>$50 μg L^{-1}),[7,49–51] corresponding to the hair levels of approximately 2–12 ppm, as observed by Airey[37] (See Table 6.2).

Neurotoxicity thresholds vary depending on the outcome in question. In addition, there are likely qualitative differences in exposure to methylmercury from fish, as opposed to methylmercury fungicides. Studies among populations that engage in high fish-consumption have found exposures that overlap with the lowest observed level for pre-natal effects of 10 ppm in hair determined during the Iraqi epidemic of poisoning due to fungicide-contaminated bread. Nonetheless, they have not observed cases of clinical congenital poisoning.[29,48]

Table 6.2 *Hair Mercury values in selected major studies and advisory guidelines; relation to blood mercury and exposure settings*

Exposure setting/location	Hair mercury concentration in ppm	Blood mercury concentration in µg/L	Mercury concentration in fish in ppm	Comments	Reference
Background					
Illinois, US	1.0+/−2.4 men 0.9+/−0.6 women	N/A	N/A	987 hospital employees and family members	Sky-Peck, 1990[3]
New Jersey, US	0.5 (87% <1.0)	1.0(95% <5)	N/A	189 pregnant women who averaged 1.6 fish meals a week.	Stern *et al.* 2001[35]
US	N/A	1.0(95% <8)	N/A	1079 women from an NHANES sample, 73% reported <3 fish meals in the 30 days prior to sampling	Schober *et al.* 2003[36]
US	0.23(0.05 to 2.3)	N/A	N/A	1276 samples from 356 women reporting no seafood consumption. Hair concentrations reflect methylmercury, not total mercury	Smith *et al.* 1997[31]
International: from 32 locations in 13 countries	2.3+/−2.8 (total)	N/A	N/A	559 samples	Airey, 1983[37]
	1.9+/−1.7			– Total mean excluding daily fish consumers	
	1.4+/−1.3			– Ate fish once a month or less	

Table 6.2 (*Continued*)

Exposure setting/location	Hair mercury concentration in ppm	Blood mercury concentration in µg/L	Mercury concentration in fish in ppm	Comments	Reference
	1.9+/−1.5			− Ate fish twice a month	
	2.5+/−2.2			− Ate fish every week	
	11.6+/−6.6			− Ate fish every day	
Health Advisory Guidelines					
US FDA	<50	<200	<1.0 (action limit)	Endpoint is development of overt neurological disease in adults based on review of 'Swedish expert group'.[38]	US FDA[14]
Canada	<10		<0.5 (guideline)	Endpoint is developmental toxicity. Safety factor of five was applied to adult threshold. Fish mercury guideline excludes large predator species, which should be limited to less than once per month	Health Canada[39]
WHO	<5	<20		Safety factor of ten applied to threshold for overt neurological disease in adults to derive limit	WHO, 1990[21]

US EPA	<1.2	<5.8		Endpoint is developmental neurotoxicity. Safety factor of ten applied to Faroe Islands benchmark dose level (58 ppb in blood/12 ppm in hair) for Boston naming test to derive limit	US EPA[10]

Studies of populations with high rates of fish-consumption

Seychelles			0 to 1.6 (11 species)	36 mother-baby pairs and 40 fishermen. Estimated that Seychellois consume yearly 80–100 kg of fish per person	Matthews 1983[40]
			0.4 to 4.4 (dogtooth tuna)		
	12.0+/−6.6 (4.1 to 32)			-Mothers	
	15.2+/−11.5 (2.1 to 48)			-Babies	
	26.3+/−14.5 (5.5 to 68)	As high as 240		-Fishermen	
Seychelles		16	0.3(98%<0.7) (>20 species)	779 mother-infant pairs. Stated that the average Seychellois consumes 12 fish meals per week. Mean blood concentration based on correlation study of maternal hair and blood and infant brain tissue n = 27 (ref)	Cerniciari et al., 1995[22]
					Cerniciari et al., 1995[34]

Table 6.2 (*Continued*)

Exposure setting/location	Hair mercury concentration in ppm	Blood mercury concentration in µg/L	Mercury concentration in fish in ppm	Comments	Reference
					Myers *et al.*, 2003[41]
	6.9+/−4.5 (0.5 to 26.7)			-Mothers	
	6.3+/−3.3 (0.9 to 25.8)			-Children at 66 months of age	
Faroe Islands	4.3 (2.6 to 7.7 (25–75% range))	23 (13 to 41 (25–75% range))	3.3 (pilot whale)	Cohort of 1022 births where the diet periodically is high in pilot whale. About 50% of the mercury in pilot whale is methylmercury. A survey found that adults consume a daily average of 72 g of fish, 19 g of whale. Hair and blood data are for maternal hair at parturition and umbilical cord blood	Grandjean *et al.*, 1992[29] and 1997[33]
			0.07 (cod)		
	(15% of values > 10)				
New Zealand	8.3 (mean)	N/A	2.2 (shark consumed as 'fish and chips')	Values are for 73 'high' mercury mothers identified from among 935 heavy fish consumers from >10,000 mothers screened	Kjellstrom *et al.*, 1986[42]

	6–10 (78% of sample)				Mitchell *et al.*, 1982[43]
	> 10 (22% of sample)				
Quebec, Canada	>/ = 6.0 (28% of sample)	N/A	N/A	3,599 persons from nine indigenous Cree communities tested in 1993–94. In 1988, 14% had hair mercury levels 15.0 ppm or greater	Dumont *et al.*, 1998[30]
	>/ = 15.0 (3% of sample)				
Brazilian Amazon	11.8+/−8.0 (0.5 to 50)	N/A	0.53 (0.03 to 1.5)	Three riverside communities with mercury contamination from gold mining. N = 220, 327 & 321. Fish mercury concentrations shown are for carnivorous species	Santos *et al.*, 2000[44]
	19.9+/−12.0 (0.1 to 94)		0.49 (0.09 to 1.6)		
	4.3+/−1.9 (0.4 to 12)		0.12 (0.03 to 0.4)		
Brazilian Amazon	4.3+/−2.2 (0.4 to 12)	N/A	0.12 (0.03 to 0.4)	Four riverside communities not affected by gold mining. N = 321, 316, 499 & 214. Fish mercury concentrations shown are for carnivorous species	Santos *et al.*, 2002[45]
	4.0+/−2.1 (0.4 to 12)		0.3 (0.05 to 0.5)		
	5.4+/−3.1 (0.4 to 17)		0.04 (0.01 to 0.5)		

Table 6.2 *(Continued)*

Exposure setting/location	Hair mercury concentration in ppm	Blood mercury concentration in μg/L	Mercury concentration in fish in ppm	Comments	Reference
	8.6+/−6.3 (0.6 to 46)		0.3 (0.01 to 2.5)		
Poisoning Epidemics					
Minimata disease Victims (Minimata and Nigata, Japan)	>/= 50 (threshold for adult disease)	>/= 200 (threshold for adult disease)	9–24 (average), some samples up to 40 ppm, values are for fish	Industrial releases of methymercury into Minimata Bay and Agano River. Over 230 persons poisoned and over 50 deaths	FDA[14]
Iraqi epidemic 1971–1972	1 to >700	0 to 5000 (range)	9.1 (4.8 to 14.6) values are for methylmercury in ppm in wheat flour	Poisoning epidemic due to ingestion of bread baked with methylmercury treated grain. 6530 poisoning victims admitted to hospitals and 459 deaths	Bakir et al., 1973[46]
	10–20 (threshold for fetal effects)	200 to 400 (threshold for adult disease)			Cox et al., 1989[47]
	50 to 100 (threshold for adult disease)	>3000 (23% lethality)			Clarkson and Strain 2003[48]

N/A = not available

Only limited data are available regarding blood mercury concentrations after immunisation with thimerosal-containing vaccines[20,52] or after ethylmercury poisoning.[18] However, these studies do not provide information on the utility of hair mercury in ethylmercury exposures.

6.2.4 Indications for Hair Analysis

Properly handled hair samples are a recognised biomarker for methylmercury exposure in epidemiological studies using standardised protocols. This technique is well-established and provides several advantages over blood mercury as discussed above. In the clinical setting, however, problems may occur due to external contamination, use of hair samples far from the scalp, improper specimen handling and the use of 'commercial' hair tests.[53,54] Carefully collected samples analysed by experienced research laboratories may provide additional corroboration of exposure or longitudinal information in some cases.[23] Hair analysis is not indicated to assess potential toxicity due to elemental mercury from dental amalgams, and hair results should not be used to justify unproven interventions such as amalgam filling removal and chelation. (See below on commercial hair tests.)

6.3 Arsenic

6.3.1 Toxicology

The most common exposure to arsenic in the general population is from naturally occurring organoarsenates, primarily arsenobetaine and arsenocholine. These are concentrated in various types of seafood, including fish and shellfish, but are innocuous. Health effects from arsenic are largely from tri- and pentavalent inorganic arsenic species with occupational exposures from byproducts of mining, smelting and coal-burning and its use in various pesticides.[55] Worldwide, the most important exposures occur in discrete geographic areas with high levels of inorganic arsenic in groundwater.[56–59] Chronic arsenic-related health effects are endemic in these populations.

Large acute exposures generally result from ingestions and produce an inflammatory gastroenteritis with altered vascular permeability that can produce hypovolemia and shock. Chronic arsenic exposure causes a wide variety of skin lesions including melanosis, keratoses and palmar and solar hyperkeratoses, as well as non-melanoma skin cancers.[56,59] It is also associated with peripheral vascular or 'blackfoot' disease, neuropathy, hematopoietic disturbances and lung cancer.[55,56,58,59]

6.3.2 Kinetics and Relation to Hair

Arsenic is absorbed primarily through lungs and the gut.[19] It is distributed widely and rapidly cleared from the blood. Acutely, the highest tissue concentrations are found in the liver and kidney.[60,61] Chronically, arsenic accumulates in the skin, hair and nails due to their high content of sulfhydryl-rich keratin.[19,60,61] Other than some accumulation in the lungs, arsenic is not significantly stored in internal organs. Inorganic arsenic species are detoxified in the body by methylation.[59] Excretion is predominantly renal, and 60–80% is excreted in methylated forms.[59,61] As the dose increases, the amount excreted as inorganic forms increases as well.[61]

6.3.3 Hair vs. Other Biomarkers

Because of its rapid clearance from the blood, blood testing is not utilised to assess occupational or chronic exposures to arsenic. A 24 hour urine collection is the method of choice for diagnosis of intoxication and monitoring occupational or environmental exposures. Urine concentrations can be deceptively elevated, however, by the consumption of seafood rich in organoarsenates within the last 24–72 hours.[19] For this reason, subjects should abstain from seafood for an adequate interval prior to collection and/or both total and inorganic arsenic species should be determined. If seafood consumption is responsible for an elevated total arsenic concentration, the resulting inorganic arsenic will be negligible.

Urine testing may also be of limited value more than 96 hours after exposure has ceased because most internal arsenic will have been eliminated.[61] Hair testing, therefore, has two distinct advantages. First, because arsenic accumulates in hair, hair analysis may identify arsenic exposure for longer periods. Second, hair binds inorganic arsenic, but seafood-derived organic forms of arsenic do not to accumulate in hair.[19,61–64] On the other hand, hair has the disadvantage of airborne contamination in occupational settings and can also bind exogenous arsenic from bathing in contaminated water.[61] Harrington *et al.*[65] found mean hair arsenic levels of 5.7 ppm in subjects bathing in arsenic-contaminated water, but who drank bottled water that was not contaminated. Unfortunately, no method is capable of removing all external arsenic contamination and none can distinguish externally deposited arsenic from that derived from ingestion.[19,61,66,67]

In general, background concentrations of arsenic in hair are usually well below 1ppm,[3,61,68] while health effects may be seen at concentrations exceeding 1ppm.[69] The dose-response relationship between hair arsenic concentrations and the degree of exposure or poisoning is approximate and may be misleading in individual cases.[61] There are several likely explanations, including the relative contribution of external contamination to total concentrations, individual differences in metabolism and hair deposition, and different analytical techniques especially with regard to washing. In settings where arsenic exposure is through inhalation or drinking contaminated water, hair arsenic concentrations reflect a combination of internal dose (personal exposure) and environmental contamination (airborne deposition or washing in contaminated water).[69] Table 6.3 compares Hindmarsh's[61] suggested guidelines for the 'approximate' interpretation of arsenic hair levels with data from selected studies of background concentrations, as well as various exposure settings. Toenails and fingernails can also be analysed for arsenic with similar advantages (chronic accumulation of inorganic arsenic) and disadvantages related to external contamination as observed with hair analysis. In epidemiological studies both hair and nail are often collected along with urine and environmental samples.[56–58,69]

Table 6.3 *Hair arsenic levels in selected studies in relation to urine arsenic and drinking water arsenic and Hindmarsh's[61] suggested guidelines for interpretation*

Exposure setting	Hair arsenic concentration in ppm	Urine arsenic concentration in $\mu g/L^{-1}$ or $\mu g\ g^{-1}\ Cr$	Drinking water arsenic concentration in $\mu g/L^{-1}$	Comments	Reference
'Normal'	<1	N/A	N/A	Suggested guideline	Hindmarsh, 2002[61]
Background – generally unexposed persons	<1.1	N/A	N/A	271 adults in US without occupational exposure	DiPietro et al., 1989[68]
	<1	N/A	N/A	987 Illinois hospital employees and family members	Sky-Peck, 1990[3]
	<1	N/A	N/A	20 unexposed adults	Bencko, 1995[70]
Background – polluted areas	3.1	N/A	N/A	40 adults in an 'exposed' area	Bencko, 1995[70]
	3.2+/−2.4	25+/−25	N/A	23 boys residing in a 'heavily polluted area'	Bencko, 1995[70]
	8.0+/−8.2	71+/−37	N/A	38 volunteers exposed to airborne arsenic through home coal burning, 29% prevalence of arsenic-related skin lesions	Shraim et al., 2003[71]
Endemic areas with high concentrations of Arsenic in Groundwater	3.4+/−3.3	280+/−410	37% > 50, 11% >/= 300	4386 persons from arsenic-affected areas in Bangladesh	Rahman et al., 2001[56]

Table 6.3 (*Continued*)

Exposure setting	Hair arsenic concentration in ppm	Urine arsenic concentration in $\mu g/L^{-1}$ or $\mu g\ g^{-1}\ Cr$	Drinking water arsenic concentration in $\mu g/L^{-1}$	Comments	Reference
	1.5+/−1.6	180+/−2.7	25% > 50, 3% >/= 300	7135 persons from arsenic-affected areas in West Bengal, India	Rahman *et al.*, 2001[51]
	5.5+/−2.0	N/A	40	31 persons from rural areas of Victoria, Australia	Hinwood *et al.*, 2003[69]
	2.8 (0.25–12.4)	799 (24–3696)	57% > 50, 20% > 300	59 hair samples from Bihar, India, 63% prevalence of arsenic-typical neuropathy	Chakraboti *et al.*, 2003[72]
	5.6+/−0.4	140+/−9	212+/−15	59 persons with arsenic-related skin lesions from West Bengal, India	Mahata *et al.*, 2003[73]
	4.5+/−3.7	N/A	248+/−59	47 persons from West Bengal, India drinking contaminated water for 3–10 years	Mandal *et al.*, 2003[64]
Chronic poisoning	>/= 10	N/A	N/A	Suggested guideline, but may be less than ten	Hindmarsh, 2002[61]
	2.6 to 7.5	23 to 277	N/A	Three patients poisoned with arsenic-containing medications, all three with arsenic-related skin lesions and one with cytopenia and polyneuropathy	Chakraboti *et al.*, 2003[74]

Lethal	Usually >45	N/A	N/A	Suggested guideline, can be lower, higher levels have also been reported in survivors	Hindmarsh, 2002[61]
	100	N/A	N/A	Patient died after ingesting 250 ml of copper acetoarsenite	Hindmarsh, 2002[61]
External contamination	Up to thousands	N/A	N/A	Suggested guideline	Hindmarsh, 2002[61]
	29 to 452	N/A	N/A	Five workers from a gold smelter	Hindmarsh *et al.*, 1999[75]
	36 (0.3 to 151)	N/A	N/A	39 antimony smelters	Bencko, 1995[70]

N/A = not available

6.3.4 Indications for Hair Analysis

Hair arsenic can be a useful indicator of chronic arsenic poisoning in clinical and forensic settings, but its presence should only be considered as circumstantial evidence in most cases, unless external contamination can be ruled out.[61] In individual patients, hair levels must be correlated with clinical findings of arsenic toxicity, urine measurements and possible exposure sources. Arsenic in hair has also been a valuable epidemiological tool in studies of individuals drinking arsenic-contaminated water when used along with urine, nail and water arsenic measurements.

6.4 Lead

6.4.1 Toxicology

Lead is a xenobiotic. Exposures may occur during mining, smelting and refining. It is the most widely utilised non-ferrous metal and used to make certain batteries, pigments, ammunition and solders.[76] Average population exposures in countries that have eliminated leaded gasoline have dropped sharply in recent years.[76,77] The presence of lead paint in older homes remains, however, an important source of exposure to children and construction workers.[76,78]

Lead is toxic to the haemal, renal, reproductive, cardiovascular and peripheral and central nervous systems.[76,78] Symptoms of lead toxicity in adults, if present, are varied and non-specific, including abdominal pain, fatigue, headache, arthralgias and myalgias. Central neurologic dysfunction depends on chronicity and severity and may range from subtle neurocognitive deficits to encephalopathy. Epidemiologically, increased lead body burdens otherwise not associated with disease in adults are associated with increases in blood pressure.[79–81]

6.4.2 Kinetics and Relation to Hair

Absorption is highest through inhalation, but can also occur through the gastrointestinal tract, and is increased by iron and calcium deficiencies.[19,76] Once absorbed, lead is distributed widely throughout the body, where it exists in three compartments or pools.[76] The half-life in blood and other rapid exchange tissues is on the order of weeks, while the in other soft tissues the half-life is measured in months.[19] Bone is the major endogenous storage site of lead, with a half-life of 5–15 years.[19,76] Excretion is primarily renal, but small amounts are also excreted through the bile, hair, nails and sweat.[19]

6.4.3 Hair vs. Other Biomarkers

Measurement of lead in whole blood is the best marker of current and recent exposures and correlates well with acute and sub-chronic toxicities.[76] Bone lead, on the other hand, provides the best estimate of body burden and can be determined non-invasively by X-ray fluorimetry (K-XRF).[82–84]

Background lead concentrations in hair have been described as high as 12+/−15 ppm,[3] but others have found much lower levels with the 90th percentile <10 ppm[68] and a geometric mean of 2.4 ppm.[85] Among highly exposed workers, hair lead content is significantly correlated with blood lead, but may represent a mixture of internal dose and external contamination.[19,86] As a screening test for lead poisoning in childhood, hair lead measurement had poor sensitivity for identifying children with blood lead $\geqslant 10$ μg dL^{-1}.[87] Niculescu *et al.*[88] found a strong correlation ($r = 0.72$, $p < 0.001$) between blood lead and hair lead in a group where 84% of blood lead values were $\geqslant 40$ μg dL^{-1}, but a much weaker correlation ($r = 0.03$, $p < 0.05$) where 84% of blood lead values were < 40 μg dL^{-1}. Foo *et al.*[89] found a significant correlation between hair lead and blood lead ($r = 0.85$, $p < 0.0001$) among occupationally exposed persons with a geometric mean blood lead of 34 μg dL^{-1} and range of 3–77 μg dL^{-1}.

6.4.4 Indications for Hair Analysis

Hair analysis for lead has limited epidemiological utility due to external contamination and inadequate sensitivity and specificity at lower levels of exposure, now known to be health concerns. It is not indicated for clinical diagnosis and not indicated for childhood or workplace screening programs. It may have limited applications in ecological and exposure assessment studies as a proxy marker of environmental contamination.[90] Anecdotally, segmental hair analysis was useful in corroborating the timeline of lead ingestion in a case of criminal poisoning.[5]

6.5 Cadmium

6.5.1 Toxicology

Cadmium is a xenobiotic that occurs naturally with zinc and lead. Occupational uses include electroplating, cadmium alloy and battery production, welding solders and cadmium pigments.[91] Tobacco smoking is the most important source of non-occupational exposure to cadmium.[19,91,92] Large acute inhalational exposures can produce acute lung injury, while chronic exposures are nephrotoxic and epidemiologically linked to emphysema and bone demineralization.[91]

6.5.2 Kinetics and Relation to Hair

Similar to lead, cadmium is absorbed well *via* inhalation and to a lesser extent through the gut.[19,91] Absorption is increased by calcium, iron and zinc deficiencies. Cadmium is distributed throughout body tissues, with roughly 50% of the body burden found in kidneys and about 15% in liver. Significant accumulation occurs in both anatomic sites with a biological half-life of 10–30 years in kidney and 5–10 years in liver.[19,91] Smokers have on average twice the body burden of non-smokers.[19,93] Excretion is primarily renal, with lesser amounts excreted in the faeces, saliva, hair and nails.[19]

6.5.3 Hair vs. Other Biomarkers

A number of parameters are useful for assessing different aspects of cadmium exposure and toxicity. Both blood and urine cadmium are considered to reflect a mixture of recent exposure and body burden, which depend on the intensity, duration and continuity of the exposure.[19] Thresholds for nephrotoxicity are best established for urine cadmium.[94–96] In addition, β-2-microglobulin can be quantified in urine and is a marker of early renal toxicity. The concentration of β-2-microglobulin may correlate well with the degree of nephrotoxicity.[97] Finally, XRF can be used to estimate the cadmium body burden by directly measuring the kidney or liver cadmium content.[19]

Background cadmium concentrations in hair are usually less than 0.5–0.8 ppm.[68,85,92,98] Experience with the use of hair cadmium as a biomarker is limited and its utility is severely hampered by the inability to distinguish external contamination from endogenous deposition.[19] It has been used in some parts of Europe as one of several indicators of environmental exposure/contamination.[92,93,98–100] In general, higher cadmium levels in hair are observed in more polluted areas. In addition, men and boys accumulate more cadmium in hair than women and girls, presumably due to spending a greater time outdoors on average.

Several lines of evidence suggest that hair cadmium does not reflect internal dose or body burden. First, cigarette smoking, the major exposure for most adults, has only a minimal effect on hair cadmium concentrations, but is a major determinant of blood and urine cadmium.[92] Second, Erzen and Kragelj[100] found the correlation between blood and hair cadmium concentrations in military recruits to be non-significant. Third, Hac *et al.*[93] compared the cadmium concentrations in renal cortex and hair in 65 paired samples from autopsies and found no correlation between renal and hair cadmium levels. They concluded that hair is not a good indicator of personal exposure to cadmium. Several other animal and human studies have also failed to find significant correlations between cadmium concentrations in hair and liver, kidney or lung.[2]

6.5.4 Indications for Hair Analysis

At present, there are no clinical indications for the measurement of cadmium in hair or as a biological marker of internal cadmium dose or body burden. It may have a role as a proxy marker of human interaction with environmental contamination.

6.6 Manganese

6.6.1 Toxicology

Manganese is an essential trace element that serves as a cofactor for several enzymes.[101,102] It also has a number of industrial uses including manganese alloys, dry cell batteries, paints, fertilisers and in the gasoline additive, methylcyclopentadienyl manganese tricarbonyl (MMT).[101,102] Occupational exposures also occur in mining and smelting. Manganese compounds are respiratory tract irritants.[103]

Chronic inhalation exposure is associated with neuropsychiatric disturbances, with advanced neurological disease closely resembling Parkinsonism.[101,102]

6.6.2 Kinetics and Relation to Hair

Occupational exposure occurs primarily from inhalation, but some gastrointestinal absorption also occurs. Manganese may share certain absorptive and metabolic pathways with iron.[102,104] The major body storage site is the liver, but accumulation also occurs in the brain and kidney.[19] Excretion is primarily faecal through the bile,[102] but small amounts are also excreted in the urine, hair, sweat and nails.[19]

6.6.3 Hair vs. Other Biomarkers

In general, bio-monitoring has proven difficult for manganese. While average manganese concentrations in biological media among exposed groups are higher than in non-exposed groups, significant overlap may occur, and correlation with individual exposures and the severity of toxicity is poor.[19,101] There is also little dose-response consistency among manganese concentrations in blood, urine and hair, especially for individuals.[101] Blood manganese may reflect recent exposure, but due to rapid elimination, there is little correlation with past exposure.[102] On an aggregate basis, urine manganese has shown some correlation with air exposures,[19] but did not discriminate between exposed workers and non-exposed referents in one recent study.[105]

Table 6.4 summarises hair manganese levels across various studies and their relation to concentrations in other biological media and exposure setting. Background hair levels are usually reported as less than 0.3 ppm, however, differences in analytical technique or unaccounted for external contamination may yield higher results. Sky-Peck[3] found mean manganese hair concentrations in healthy volunteers to be slightly above 1 ppm using energy dispersion XRF, while Dipietro *et al.*[68] and Paschal *et al.*[85] used inductively coupled argon plasma emission spectroscopy (ICAP). Higher background levels in hair were also seen among the 'unexposed' referents in the Bader *et al.*[105] and Boojar and Goodarzi[103] studies, perhaps due to a small degree of unaccounted for external contamination.

Hair monitoring generally demonstrates higher values among exposed groups than unexposed groups, but it is unclear to what extent this reflects internal dose vs. external contamination.[103,105,106] Loranger and Zayed[106] compared garage workers exposed to higher ambient levels of manganese through the gasoline additive MMT to other workers with lower ambient exposure. The garage workers had a significantly higher mean hair manganese, 660 ppb, compared to the controls, 390 ppb, but their estimated total manganese intakes were similar when dietary sources were accounted for. When one contrasts the blood and hair levels of the poisoning cases reported by Woolf *et al.*[104] and Ono *et al.*[107] with the heavily exposed workers in the Bader *et al.*[105] and Boojar and Goodarzi[103] studies, clearly, a significant portion of the hair manganese observed in the battery workers and miners results from external contamination. Further, the hair results shown in Table 6.4

Table 6.4 *Hair manganese levels in selected reports in relation to urine and blood manganese concentrations and exposure settings*

Exposure setting	Hair manganese concentration in ppm	Urine manganese concentration in $\mu g/L^{-1}$ or $\mu g/g^{-1}$ Cr	Blood manganese concentration in $\mu g/L^{-1}$	Comments	Reference
Background	0.22 (90% <1.1)	N/A	N/A	271 adults in US without occupational exposure	DiPietro *et al.*, 1989[68]
	0.23 (mean)	N/A	N/A	322 adults in US (extension of DiPietro *et al.* study)	Paschal *et al.*, 1989[85]
	1.1+/−0.6 men	N/A	N/A	987 Illinois hospital employees and family members	Sky-Peck, 1990[3]
	1.2+/−1.3 women				
MMT exposure	0.66 (0.12–2.47)	N/A	7.6 (2.8–14.5)	37 garage mechanic ('exposed'). Garage workers had greater exposures to ambient manganese through MMT in gasoline exhaust, but total estimated manganese exposure including diet similar to control workers.	Loranger and Zayed, 1995[106]
	0.39 (0.06–1.19)	N/A	6.7 (0.2–13.1)	27 other blue collar workers ('less exposed'), controls	
Dry cell battery manufacture	8.2 (0.9–27.7)	0.49 (0.1–2.2)	13.8 (6.1–23.3)	39 highly exposed workers, selected from the area with the highest ambient manganese concentrations	Bader *et al.*, 1999[105]
	5.2 (0.5–17.2)	0.33 (0.1–1.3)	11.7 (3.2–23.0)	22 moderately exposed workers, selected from the area with intermediate ambient manganese concentrations	

	4.6 (0.4–29.8)	0.26 (0.1–1.8)	10.7 (3.9–25.8)	39 low exposure workers, selected from the area with the lowest ambient manganese concentrations	
	2.2 (0.4–6.2)	0.39 (0.1–1.2)	7.5 (2.6–15.1)	17 unexposed referents	
Manganese mining				145 manganese miners tested after 4 to 7 years of exposure	Boojar and Goodarzi, 2002[103]
	23+/−4	14.8+/−3.3	18.6+/−3.1	Smokers	
	20+/−4	13.2+/−3.2	16.7+/−3.5	non-smokers	
	1.4+/−0.4	1.4+/−0.3	1.5+/−0.4	65 matched controls smokers	
	1.6+/−0.5	1.5+/−0.4	1.6+/−0.3	non-smokers	
Poisoning cases	3.1 (nl <0.26)	8.5 (nl <1.1)	38 (nl <14)	10-year-old boy exposed to drinking water with elevated manganese, psychometric testing suggested impaired memory	Woolf et al., 2002[104]
	1.4 (nl <0.1)	<1.0	43 (nl <25)	17-year-old male welder with involuntary myoclonic movements	Ono et al., 2002[107]

N/A = not available; MMT = methycyclopentadienyl manganese tricarbonyl

for Bader *et al.*[105] are for the most proximal, first centimeter of axillary hair. Further analyses of the second and third centimeters of axillary hair demonstrated progressively higher manganese content, suggesting increasing contamination over time.

6.6.4 Indications for Hair Analysis

No single biological medium definitively and reliably reflects individual manganese exposure or toxicity. Therefore, hair manganese analyses should be considered as an additional source of potential exposure information in both individual cases and group studies, along with blood and urine manganese, and when possible environmental manganese determinations. The potential for external contamination must also be considered. Given the wide variation in reported background hair concentrations, control specimens should be obtained and analysed using the same methods. Hair manganese is more likely to be informative regarding biological dose when the exposure is through ingestion rather than ambient air. In the latter case, the hair may be a surrogate measure of environmental manganese contamination.

6.7 Thallium

6.7.1 Toxicology

Thallium salts are toxic, and thallium sulfate has been used as a rodenticide. While a 1970s ban on its use in the US has reduced accidental poisonings, it continues to be used in some countries as a rodenticide, as well as being implicated in homicide attempts.[108–110] Anecdotal occupational exposure to thallium has also been described in the manufacture of a special glass.[111]

Characteristically, the ingestion of thallium salts produces gastrointestinal symptoms, followed by the onset of a painful ascending neuropathy.[108,109] The neuropathy can be progressive with associated weakness. Acute alopecia begins after about a week, and can progress to complete baldness over several weeks.[109] Characteristically, the alopecia is painless, hair can be pulled out in clumps with little effort, and the inner one-third of eyebrows are spared.[108,112–114] In addition to diffuse alopecia, thallium intoxication can cause blackening of the hair roots, which is seen when these are examined microscopically.[108,112,114] Hoffman believes, however, that untrained observers may have difficulty in noting this diagnostic feature.[108,109]

6.7.2 Kinetics and Relation to Hair

Thallium is readily absorbed by inhalation, gastrointestinal absorption and dermally.[109,110] Subsequently, thallium is widely distributed throughout the body and concentrated in the kidneys.[109,110] Excretion is predominantly renal and faecal, but thallium is also excreted into the saliva, nails and hair.[19,115] The elimination half-life for thallium has been reported to be anywhere from 2–30 days.[19,108,115]

Thallium does not have an internal anatomic reservoir, but binds sulfhydryl groups with high affinity like those in keratin, which likely explains its concentration in hair and nails.[108,115]

6.7.3 Hair vs. Other Biomarkers

Thallium can be determined in blood, urine, hair and nails. Because most information relating to thallium concentrations in biological media comes from poisoning cases, little is known about background concentrations or the quantitative relationship between exposures, internal doses and the relative concentrations among different biological media.[19] Acutely, the measurement of thallium from a 24-hour urine collection is the diagnostic test of choice.[108,109] Both urine and blood thallium concentrations are usually elevated in poisoning cases and are commonly measured in these cases.

Because thallium concentrates in the hair, elevated hair thallium concentrations can be used to help confirm the diagnosis of poisoning.[108,109,114,116] Table 6.5 summarises the results of hair analysis in selected reports from the literature. While hair levels in this table are presented in ppm to maintain consistency with the remainder of the chapter, thallium content in hair is usually described in units 1000-fold lower (ppb or ng g^{-1}). Limited data describe background concentrations of thallium in biological media. Available information suggests that these levels should be very low and less than 20 ppb in hair.[111,114,118] Among poisoning victims, hair concentrations range from 48 to 35,000 ppb with most between 150 to 1500 ppb. In Rusyniak *et al.*'s[114] case series, there was a good correlation between 24-hour urine thallium excretion and the thallium concentrations in the hair taken simultaneously (Figure 6.1). With regard to timing, elevated thallium concentrations in hair have been found as early as 2–3 weeks after ingestion in poisonings[6,114,116] and as late as 13 months after the cessation of their occupational exposures.[111] Thallium also concentrates in nails and they have been analysed in poisoning cases,[108,116] but there are insufficient data available to determine the relative advantages of hair vs. nails.

6.7.4 Indications for Hair Analysis

Analysis of thallium content in hair should be considered as complementary to urine and blood testing in confirming the diagnosis of acute thallium poisoning, possibly as early as two weeks after the onset of symptoms. Unlike urine samples, hair is unlikely to demonstrate elevated concentrations immediately after a single ingestion, however, given the relatively short elimination half-life of thallium, hair analysis may be especially useful in cases where several months have elapsed since the cessation of exposure. Analytic methods must be capable of quantifying concentrations in the ppb range.

6.8 Commercial Hair Tests and Their Potential Misuses

Commercial hair analyses are promoted as a means of determining a patient's nutritional status and exposure to toxic heavy metals. Although hair analysis for

Table 6.5 *Hair thallium levels in selected published reports in relation to urine and blood thallium concentrations and exposure settings*

Exposure setting	Hair thallium Concentration in ppm	Urine thallium concentration in $\mu g/L^{-1}$ or μg per 24 hours	Blood thallium concentration in $\mu g/L^{-1}$	Comments	Reference
Background	<0.020	<10	N/A	Source of background values not specified	Rusyniak *et al.*, 2002[114]
	N/A	<5	<2	Source cited as Mulkey and Oehme, 1993[117]	Mercurio and Hoffman, 2002[108]
	<0.012	N/A	N/A	Unexposed control subject's hair analysed. Thallium content below detection limit of 0.012 ppm	Hirata *et al.*, 1998[111]
	<0.016	N/A	N/A	Source cited as Schoer, 1984	Yoshinaga *et al.*, 1993[6]
Poisoning cases	0.252 (0 to 1.324)	1127 (0 to 5885)	N/A	Mean values for 12 victims of a combined thallium and arsenic poisoning taken after about 16 days of symptoms. Suspected to be a criminal act 73% incidence of neuropathy overall and present in all six patients with hair thallium >0.100 ppm. 5/6 patients with hair thallium >0.100 ppm had alopecia; 0/6 patients with hair thallium <0.100 ppm had alopecia	Rusyniak *et al.*, 2002[114]
	0.53	532	275	21-year-old with alopecia, severe ascending neuropathy and coma. Biological media tested for thallium about one month after she was presumably poisoned for a second time	Mercurio and Hoffman 2002[108]

	35	N/A	N/A	38-year-old who eventually died of thallium poisoning. Hair sampled three weeks after hospitalisation	Yoshinaga *et al.* 1993[6]
	1.46	1022	400	Three victims of a thallium poisoning presumably criminal. Timing of hair sampling about three weeks after onset of symptoms	McCormick and McKinney, 1983[114]
	10.69	3838	240		
	12.69	N/A	N/A		
Manufacture of a special glass containing thallium carbonate	0.020	N/A	N/A	Worker 1: handled thallium-containing materials for four years and presented with alopecia and polyneuropathy. He was tested 32 months after exposure cessation	Hirata *et al.*, 1998[111]
	0.576			Worker 2: replaced worker 1 in previous job handling thallium-containing materials. He was tested 13 months after exposure cessation and also complained of alopecia	

N/A = not available

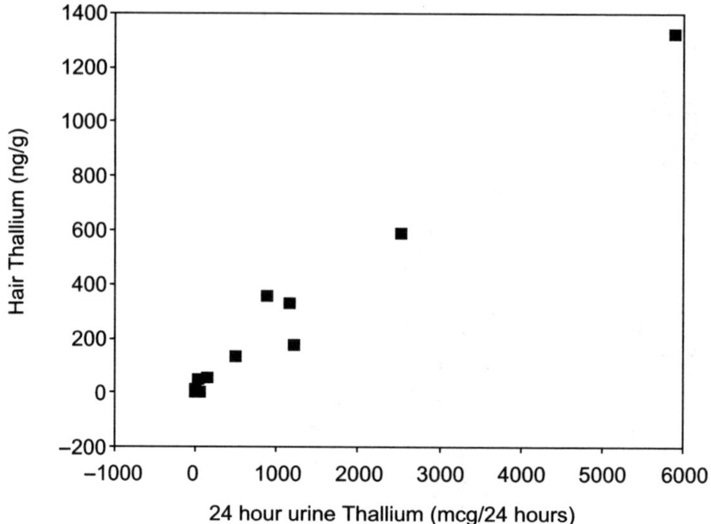

Figure 6.1 *Hair thallium values for 12 victims of a poisoning plotted against 24-hour urine thallium. Both hair and urine were collected after about 16 days of symptoms. Data are from Rusyniak et al.[114]*

heavy metals has a number of valid applications, as described above, commercial analyses that simultaneously determine large numbers of metals and minerals have long been misused. In 1974, Lazar[119] wrote: 'Trace-metal hair analysis, as advertised by several commercial laboratories, is done often on hair samples supplied by clients, often by mail, usually of uncertain origin and without adequate history of previous environmental or personal exposures... Such analysis does provide numerical results, but these results are of very little true value and unfortunately are frequently used to support questionable diagnoses or recommendations'.

Subsequent to Lazar's editorial, commercial hair analyses of this type have been studied, and criticised and discouraged by various investigators and committees.[2,4,7,53,54,120–124] Unfortunately, such unconventional practices continue today, with patients sending in hair samples to commercial laboratories directly or through a variety of 'alternative' practitioners. Seidel *et al.*[53] calculated that nine US laboratories were performing an annual average of 225,000 hair analyses at a cost of almost $10 million.

In our experience, such commercial testing is typically targeted at patients with unexplained somatic symptoms, as well as others with idiopathic diagnoses such as autism and amyotrophic lateral sclerosis. Characteristically, these analyses are not element specific, nor based on exposure history, but test from 15–39 different minerals and metals.[7,53,122] Statistically, this multiple testing increases the probability of 'abnormal' results based on chance alone.[7,123] The results are often used to justify questionable therapies such as extensive vitamin and other supplementation regimens, mercury amalgam filling removal and chelation.[7,53,120–122]

Reliability studies of commercial hair analyses using duplicate hair samples have demonstrated discordant reference range intervals, poor result reproducibility and divergent nutritional recommendations from different laboratories analysing identical samples.[53,120] In 1985, Barrett[120] sent duplicate samples from two healthy volunteers to thirteen different commercial laboratories. He found considerable divergence in results for identical samples sent to different laboratories as well as those sent to the same laboratory. Seidel *et al.*[53] recently updated the Barrett investigation by sending a split sample of scalp hair to six laboratories that perform 90% of the commercial hair analyses in the US. Reported values for 12 minerals varied more than 10-fold on the split sample. Also, statistically significant extreme values for the same element compared to other laboratories' results were found for 14 of 31 minerals analysed by more than three laboratories.

Additionally, case reports of patients labelled as having heavy metal toxicity on the basis of commercial hair tests, which subsequently received second opinions, describe inaccurate diagnoses based on hair testing.[2,7,122,125] Such patients often experienced needless anxiety based on the inaccurate diagnoses. They also usually lacked significant exposure histories, and when conventional blood and urine tests were performed had normal concentrations of the metals in question.

A summary report of the US Agency for Toxic Substances and Disease Registry (ATSDR) panel on hair analysis encapsulated the experts' view on commercial hair analyses with the following statement: 'Universally, the panelists expressed concern about the misuse of hair analysis to justify and support unnecessary and unethical medical therapy'.[4] We agree that in clinical situations, hair testing for heavy metals should be targeted to specific elements based on the patient's occupational and environmental history and/or exposure assessments, should be used only as an adjunct to more reliable biomarkers and should utilise experienced reference or research laboratories.

6.9 Important Methodological Issues in Hair Metal Analysis

There are a number of unresolved methodological issues that limit the clinical use of hair analysis and must be taken into account when hair is used for forensic and/or epidemiological purposes. The ATSDR-convened panel,[4] has discussed and reviewed these in depth and the reader is directed to the full meeting report for additional information and bibliography.[126] While the panel recommended the standardisation of hair sampling protocols, a definitive consensus was not reached. Logically, the ideal methodology may differ depending on the metal of interest.

6.9.1 Sampling

Hair from various scalp locations may grow at variable rates, while hair segments closer to or more distant from the scalp will reflect different chronological periods of exposure. In addition, although hair growth rate is often assumed to be 1.1 cm per month for the purpose of timing exposures, the average growth rate may range from 0.6 to 1.5 cm per month with additional variation among individual subjects.[10] Therefore, one must decide where on the scalp to sample from and how far from the scalp to include in the analysis sample. While a universally agreed upon anatomical location does not exist, consideration should be given to sampling from a reproducible point on the skull.[126] The amount of hair collected should exceed the minimum required by the laboratory for a valid analysis, while not being cosmetically unacceptable to the subject. To avoid sample contamination, cutting tools and sample storage vessels should not be made of or treated with the metals of interest for the analysis to be performed.[126]

6.9.2 Cosmetic/Hygiene Products

Personal care products may contain metals that can be adsorbed onto hair. For example, Grecian formula®, a hair dye, has been reported to contain lead.[3] Other hair treatments such as permanents and hair waving may alter the content of copper, arsenic and mercury.[3,127] Dandruff treatments containing selenium sulfide could potentially alter hair mercury concentrations.[85] Bleaching of the hair alters melanin content and therefore, potentially affects metal binding.[128] The effects of different shampoos and the frequency of hair washing on hair metal content have not been thoroughly studied,[126] but because different sample washing procedures have been noted to affect results (see below), it follows that shampooing habits could potentially affect metal recovery for certain elements.

6.9.3 Sample Washing

Decisions regarding the necessity of washing and the washing method should be substance specific.[126] When the metal of interest can only reach the hair internally (*e.g.* dietary MeHg) and is not an external contaminant, washing is not necessary.[126,129] Washing protocols become important when the metal of interest has an external source, but there is no universal agreement as to an ideal method.[2,70,130–132] Many groups use the acetone washing procedure recommended by the International Atomic Energy Agency.[133] One should recognise several facts about washing hair samples prior to analysis. First, different washing methods utilising different solutions and/or different durations of washing may produce significant variation in results.[67,68,132] Second, washing only incompletely removes external contamination, but may also remove some internally deposited metals.[2,126]

6.9.4 Sample Preparation

Hair proteins bind metals. Therefore, in order to be analysed for metal content, hair must be digested into a form in which metals can be ionised for measurement.[54] Digestion methods differ among laboratories and may be an additional source of variation.[134]

6.9.5 Analytical Methods

No less than six analytical methods exist for measuring metal content in hair.[126] Atomic absorption spectrophotometry (AAS) is applicable to several metals including lead,[5,88–90] mercury,[44,45] arsenic,[58,69,72] cadmium[92,93,98] and manganese.[103,105] Cold vapour atomic absorption (CVAA) is preferred for MeHg analysis.[126] CVAA has been widely applied in epidemiological investigations.[22,29,34,41] Using the 'Magos' reagents, it can distinguish between total and inorganic mercury concentrations (allowing methylmercury to be determined as the difference).[22] Inductively coupled argon plasma spectrophotometry techniques (ICP-OES, ICP-MS, ICP-AES) can be used to simultaneously determine multiple elements and are commonly used.[6,68,71,111,126]

6.9.6 Targeted Analyses

The ATSDR panel did reach general consensus that analytic approaches should be targeted to preferably a single metal and a specific exposure context.[126]

6.9.7 Interpretation

Sampling location, hair growth rate, sample washing and preparation, the potential confounding of cosmetic and/or hygiene products, as well as, for most metals, the lack of universally accepted reference ranges (or adverse effect levels) greatly affect the ability to make accurate estimations of individual patients' metal exposures based on hair analysis. In epidemiological investigations, these limitations may be surmountable by stringently standardising all procedures for all participants and samples. In this way, investigators will be able to mitigate individual variation and compare relative exposures within groups of study participants. Ideally, internal validation studies that correlate hair metal content to appropriate blood and/or urine biomarkers of the same metal should also be done on a subset of the participants. This technique makes results comparable in a quantitative fashion with other studies and levels of metal exposure associated with documented biological effects.

6.10 References

1. T. Suzuki in *Biological Monitoring of Toxic Metals*, T. W. Clarkson, L. Friberg, G. F. Nordberg, and P. R. Sager (ed), Plenum Press, New York, NY, 1988, 623.
2. A. Taylor, *Ann. Clin. Biochem.*, 1986, **23**, 364.
3. H.H. Sky-Peck, *Clin. Physiol. Biochem.*, 1990, **8**, 70.
4. D. Harkins and A. Susten, *Environ. Health Persp.*, 2003, **111**, 576.
5. P. Grandjean, *Human Toxicol.*, 1984, **3**, 223.
6. J. Yoshinaga, Y. Shibata and M. Morita, *Clin. Chem.*, 1993, **39**, 1650.
7. S.N. Kales and R.H. Goldman, *J. Occup. Environ. Med.*, 2002, **44**, 143.
8. T. Clarkson, L. Magos and G. Myers, *N. Engl. J. Med.*, 2003, **349**, 1731.
9. S. Langworth, C.G. Elinder, C.J. Göthe and O. Vesterberg, *Int. Arch. Occup. Environ. Health*, 1991, **63**, 161.
10. National Research Council, *Toxicological Effects of Methylmercury*, National Academy Press, Washington DC, 2000.

11. H.L. Evans in *Environmental and Occupational Medicine*, W.N. Rom (ed), Lippincott-Raven, Philadelphia, Pennsylvania, 1998, 997.
12. D. Campbell, M. Gonzales and J.B. Sullivan in *Hazardous Materials Toxicology*, J.B. Sullivan and G.R. Kreiger (ed), Williams & Wilkins, Baltimore, Maryland, 1992, 824.
13. D.K. Parkinson in *Environmental and Occupational Medicine*, W.N. Rom (ed), Little, Brown, Boston, Massachusetts, 1992, 759.
14. US Food and Drug Administration, *Mercury in fish: cause for concern?*, available at: http://www.fda.gov/opacom/catalog/mercury/html, 1995.
15. P. Grandjean, *Public Health Rep.*, 1999, **114**, 512.
16. T. Clarkson, *C.M.A.J.*, 1998, **158**, 1465.
17. G.J. Meyers and P.W. Davidson, *Environ. Health Persp.*, 1998, **106**, 841.
18. L. Magos, *J. Appl. Toxicol.*, 2001, **21**, 1.
19. R.R. Lauwerys and P. Hoet, *Industrial Chemical Exposure: Guidelines for Biological Monitoring*, Lewis Publishers, Boca Raton, Florida, 2001.
20. M.E. Pichichero, E. Cernichiari, J. Lopreiato and J. Treanor, *Lancet*, 2002, **360**, 1737.
21. WHO Environmental Health Criteria 101 Methylmercury, World Health Organization, Geneva, 1990.
22. E. Cernichiari, T.Y. Toribara, L. Liang, *et al.*, *Neurotoxicology*, 1995, **16**, 613.
23. Agency for Toxic Substance and Disease Registry, *Am. Fam. Physician*, 1992, **46**, 1731.
24. J. Forman, J. Moline, E. Cernichiari, *et al.*, *Environ. Health Persp.*, 2000, **108**, 575.
25. M. Wilhelm, F. Muller and H. Idel, *Toxicology Lett.*, 1996, **88**, 221.
26. K.A. Ritchie, W.H. Gilmour, E.B. Macdonald, *et al.*, *Occup. Environ. Med.*, 2002, **59**, 287.
27. D. Echeverria, *Occup. Environ. Med.*, 2002, **59**, 285.
28. P.W. Davidson, G.J. Myers, C. Cox, *et al.*, *J.A.M.A.*, 1998, **280**, 701.
29. P. Grandjean, P. Weihe, R. White, *et al.*, *Neurotoxicol. Teratol.*, 1997, **19**, 417.
30. C. Dumont, M. Girard, F. Bellavance and F. Noel, *C.M.A.J.*, 1998, **158**, 1439.
31. J.C. Smith, P.V. Allen and R. Von Burg, *Arch. Environ. Health*, 1997, **52**, 476.
32. A.H. Stern and A.E. Smith, *Environ. Health Persp.*, 2003, **111**, 1465.
33. P. Grandjean, P. Weihe, P.J. Jørgensen, *et al.*, *Arch. Environ. Health*, 1992, **47**, 185.
34. E. Cernichiari, R. Brewer, G.J. Myers, *et al.*, *Neurotoxicology*, 1995, **16**, 705.
35. A.H. Stern, M. Gochfield, C. Weisel and J. Burger, *Arch. Environ. Health*, 2001, **56**, 4.
36. S.E. Schober, T.H. Sinks, R.L. Jones, *et al.*, *J.A.M.A.*, 2003, **289**, 1667.
37. D. Airey, *Sci. Total Environ.*, 1983, **31**, 157.
38. Swedish Expert Group, *Nord Hyg Tisdskr*, 1971, **4(suppl)**, 19.
39. Health Canada, *Information on Mercury Levels in Fish*, available at: http://www.hc-sc.gc.ca/english/archives/warnings/2001/2001_60e.htm, 2001.
40. A.D. Matthews, *Environ Res.*, 1983, **30**, 305.
41. G.J. Myers, P.W. Davidson, C. Cox, *et al.*, *Lancet*, 2003, **361**, 1686.

42. T. Kjellstrom, P. Kennedy, S. Wallis and C. Mantell, *National Swedish Environmental Protection Board Report 3080*, Solna, Sweden, 1986.
43. J.W. Mitchell, T.E. Kjellstrom and R.L. Reeves, *N. Z. Med. J.*, 1982, **95**, 112.
44. E.C.O. Santos, I.M. Jesus, E.S. Brabo, *et al.*, *Environ. Res.*, 2000, **84**, 100.
45. E.C.O. Santos, V.M. Câmara, I.M. Jesus, *et al.*, *Environ. Res.*, 2002, **90**, 6.
46. F. Bakir, S.F. Damluji, L. Amin-Zaki, *et al.*, *Science*, 1973, **181**, 230.
47. C. Cox, T.W. Clarkson, D.O. Marsh, et al., *Environ. Res.*, 1989, **49**, 318.
48. T.W. Clarkson and J.J. Strain, *J. Nutr.*, 2003, **133**, 1539S.
49. J.M. Hightower and D. Moore, *Environ. Health Persp.*, 2003, **111**, 604.
50. L. Knobeloch, M. Ziarnik, H.A. Anderson and V.N. Dodson, *Environ. Health Persp.*, 1995, **103**, 604.
51. T.M. Bellanger, E.M. Caesar and L. Trachtman, *J. La State Med. Soc.*, 2000, **152**, 64.
52. G.V. Stajich, G.P. Lopez, H.W. Sokei and W.R. Sexson, *J. Pediatr.*, 2000, **136**, 679.
53. S. Seidel, R. Kreutzer, D. Smith, *et al.*, *J.A.M.A.*, 2001, **285**, 67.
54. S.J. Steindel, and P.J. Howanitz, *J.A.M.A.*, 2001, **285**, 83.
55. T.G. Rossman in *Environmental and Occupational Medicine*, W. N. Rom (ed), Lippincott-Raven, Philadelphia, Pennsylvania, 1998, 1011.
56. M.M. Rahman, U.K. Chowdhury, S.C. Mukherjee, *et al.*, *J. Toxicol. Clin. Toxicol.*, 2001, **39**, 683.
57. K. Chandra Sekhar, N.S. Chary, C.T. Kamala, *et al.*, *Environment International*, 2003, **29**, 601.
58. T. Lin, Y, Huang and M. Wang, *J. Toxicol. Environ. Health*, 1998, **53**, 85.
59. Y.C. Chen, Y.L. Guo, H.J. Su, *et al.*, *J. Occup. Environ. Med.*, 2003, **45**, 241.
60. M. Ford in *Goldfrank's Toxicologic Emergencies*, L.R. Goldfrank, N.E. Flomenbaum, N.A. Lewin *et al.* (ed), McGraw-Hill, New York, NY, 2002, 1183.
61. J.T. Hindmarsh, *Clin. Biochem.*, 2002, **35**, 1.
62. M. Vahter, E. Marafante and L. Dencker, *Sci. Total Environ.*, 1983, **30**, 197.
63. E. Marafante, M. Vahter and L. Dencker, *Sci. Total Environ.*, 1984, **34**, 223.
64. B.K. Mandal, Y. Ogra and K.T. Suzuki, *Toxicol. Appl. Pharm.*, 2003, **189**, 73.
65. J.M. Harrington, J.P. Middaugh, D.L. Morse, *et al.*, *Am. J. Epidemiol.*, 1978, **108**, 377.
66. H. Smith, *J. For. Sci. Soc.*, 1964, **4**, 192.
67. A.J. Van den Berg, J.J.M. de Geoji and J.P.W. Hortman in *Modern Trends in Activation Analysis*, Vol. 1, J.R. DeVoe (ed), NBS, Washington DC, 1968, 272.
68. E.S. DiPietro, D.L. Phillips, D.C. Paschal and J.W. Neese, *Biol. Trace Elem. Res.*, 1989, **22**, 83.
69. A.L. Hindwood, M.R. Sim, D. Jolley *et al.*, *Environ. Health Persp.*, 2003, **111**, 187.
70. V. Bencko, *Toxicology*, 1995, **101**, 29.
71. A. Shraim, X. Cui, S. Li, *et al.*, *Toxicol. Lett.*, 2003, **137**, 35.

72. D. Chakraborti, S.C. Mukherjee, S. Pati, *et al.*, *Environ. Health Persp.*, 2003, **111**, 1194.

73. J. Mahata, A. Basu, S. Ghoshal, *et al.*, *Mutation Research*, 2003, **534**, 133.

74. D. Chakraborti, S.C. Mukherjee, K.C. Saha, *et al.*, *J. Toxicol. Clin. Toxicol.*, 2003, **41**, 963.

75. J.T. Hindmarsh, D. Dekerhove, G. Grime, *et al.*, in *Arsenic Exposure and Health Effects*, W. R. Chappell, C. O. Abernathy and R. L. Calderon (ed), Elsevier, Amsterdam, 1999, 41.

76. A. Fischbein in *Environmental and Occupational Medicine*, W. N. Rom (ed), Lippincott-Raven, Philadelphia, Pennsylvania, 1998, 973.

77. J. Pirkle, D.J. Brody, E. Gunter, *et al.*, *J.A.M.A.*, 1994, **272**, 284.

78. World Bank Group in *Pollution Prevention and Abatement Handbook*, Washington DC, 1998.

79. Y. Cheng, J. Schwartz, D. Sparrow, *et al.*, *Am. J. Epidemiol.*, 2001, **153**, 164.

80. F. Gerr, R. Letz, L. Stokes, *et al.*, *Am. J. Indus. Med.*, 2002, **42**, 98.

81. H. Hu, A. Aro, M. Payton, *et al.*, *J.A.M.A.*, 1996, **275**, 1171.

82. D.E. Burger, F.L. Milder, P.R. Morsillo, *et al.*, *Basic Life Sci.*, 1990, **55**, 287.

83. H. Hu, F.L. Milder, D.E. Burger, *et al.*, *Arch. Environ. Health.*, 1990, **45**, 335.

84. A.C. Todd and F.E. McNeill in *Human Body Composition Studies*, K. J. Ellis and J. D. Eastman (ed), Plenum Press, New York, NY, 1993, 299.

85. D.A. Paschal, E.S. DiPietro, D.L. Phillips and E.W. Gunter, *Environ. Res.*, 1989, **48**, 17.

86. H. Vishwanathan, A. Hema, D. Edwin and M.V. Rani, *Environ. Monit. Assess.*, 2002, **77**, 149.

87. E. Esteban, C.H. Rubin, R.L. Jones and G. Noonan, *Arch. Environ. Health*, 1999, **54**, 436.

88. T. Niculescu, R. Dumitru, V. Botha, *et al.*, *Brit. J. Indus. Med.*, 1983, **40**, 67.

89. S.C. Foo, N.Y. Khoo, A. Heng, *et al.*, *Int. Arch. Occup. Environ. Health*, 1993, **65**, S83.

90. B. Revich, *Arch. Environ. Health*, 1994, **49**, 59.

91. A.J. Newman-Taylor in *Environmental and Occupational Medicine*, W.N. Rom (ed), Little, Brown, Boston, Massachusetts,1992,767.

92. K. Hoffman, K. Becker, C. Friedrich, *et al.*, *J. Expos. Anal. Environ. Epidemiol.*, 2000, **10**, 126.

93. E. Hać, M. Krzyżanowski and J. Krechniak, *Sci. Tot. Environ.*, 1998, **224**, 81.

94. A.M. Bernard *et al.*, *Brit. J. Indus. Med.*, 1990, **47**, 559.

95. H.A. Roels *et al.*, *Brit. J. Indus. Med.*, 1990, **47**, 331.

96. H.A. Roels *et al.*, *Brit. J. Indus. Med.*, 1993, **50**, 37.

97. A.M. Bernard and C. Hermans, *Sci. Tot. Environ.*, 1997, **199**, 205.

98. K.A. Bustueva, B.A. Revich and L.E. Bezpalko, *Arch. Environ. Health*, 1994, **49**, 284.

99. J. Chlopicka, Z. Zachwieja, P. Zagrodzki, *et al.*, *Biol. Trace Elem. Res.*, 1998, **62**, 229.

100. I. Eržen and L.Z. Kragelj, *Croat. Med. J.*, 2003, **44**, 538.

101. D. Mergler and M. Baldwin, *Environ. Res.*, 1997, **73**, 92.

102. D.G. Barceloux, *J. Toxicol. Clin. Toxicol.*, 1999, **37**, 293.

103. M.M.A. Boojar and F. Goodarzi, *J. Occup. Environ. Med.*, 2002, **44**, 282.

104. A. Woolf, R. Wright, C. Amarasiriwardena and D. Bellinger, *Environ. Health Persp.*, 2002, **110**, 613.

105. M. Bader, M.C. Dietz, A Ihrig and G. Triebig, *Int. Arch. Occup. Environ. Health*, 1999, **72**, 521.

106. S. Loranger and J. Zayed, *Int. Arch. Occup. Environ. Health*, 1995, **67**, 101.

107. K. Ono, K. Komai and M. Yamada, *J. Neurol. Sci.*, 2002, **199**, 93.

108. M. Mercurio and R.S. Hoffman in *Toxicologic Emergencies*, L. Goldfrank (ed), McGraw-Hill, New York, NY, 2002, 1272.

109. R.S. Hoffman, *Toxicol. Rev.*, 2003, **22**, 29.

110. S. Galván-Arzate and A. Santamaría, *Toxicol. Lett.*, 1998, **99**, 1.

111. M. Hirata, K. Taoda, M. Ono-Ogasawara, *et al.*, *Indus. Health*, 1998, **36**, 300.

112. I. Tromme, D. Van Neste, F. Dobbelaere, *et al.*, *Brit. J. Dermatol.*, 1998, **138**, 321.

113. F. Herrero, E. Fernandez, J. Gomez, *et al.*, *J. Toxicol. Clin. Toxicol.*, 1995, **33**, 261.

114. D.E. Rusyniak, R.B. Furbe, and M.A. Kirk, *Ann. Emerg. Med.*, 2002, **39**, 307.

115. J.B. Sullivan in *Clinical Environmental Health and Toxic Exposures*, J.B. Sullivan and G.R. Krieger (ed), Lippincott Williams & Wilkins, Philadelphia, PA, 2001, 954.

116. J. McCormack and W. McKinney, *Postgrad. Med.*, 1983, **74**, 239.

117. J.P. Mulkey and F.W. Oehme, *Vet. Hum. Toxicol.*, 1993, **35**, 445.

118. J. Schoer in *Handbook of Environmental Chemistry*, O. Hutchinger (ed), Springer-Verlag, Berlin, 1984, 143.

119. P. Lazar, *J.A.M.A.*, 1974, **229**, 1908.

120. S. Barrett, *J.A.M.A.*, 1985, **254**, 1041.

121. D.J. Fletcher, *Postgrad. Med.*, 1982, **72**, 79.

122. M. Frisch and B.S. Schwartz, *Environ. Health Perspect.*, 2002, **110**, 433.

123. L.M. Klevay, B.R. Bistrian, C.R. Fleming and C.G. Neumann, *Am. J. Clin. Nutr.*, 1987, **46**, 233.

124. American Medical Association *Hair Analysis: A Potential for Abuse*, Policy No. H-175.995, Chicago, 1994.

125. T.L. Guidotti, *J. Occup. Med.*, 1983, **25**, 693.

126. ATSDR, *Hair Analysis Panel Discussion: Exploring the State of the Science: Summary Report*, available at http://www.atsdr.cdc.gov/HAC/hair_analysis/, 2001.

127. R. Yamamoto and T. Suzuki, *Int. Arch. Occup. Environ. Health*, 1978, **42**, 1.

128. M.I. Greenburg, Pre-meeting Comments, ATSDR, *Hair Analysis Panel Discussion: Exploring the State of the Science*, available at http://www.atsdr.cdc.gov/HAC/hair_analysis/, 2001.

129. T. Clarkson, Pre-meeting Comments, ATSDR, *Hair Analysis Panel Discussion: Exploring the State of the Science*, available at http://www.atsdr.cdc.gov/HAC/hair_analysis/, 2001.

130. S. Salmela, E. Vuori and J.O. Kilpiö, *Anal. Chim. Acta*, 1981, **125**, 131.

131. G. Chittleborough, *Sci. Tot. Environ.*, 1980, **14**, 53.
132. J. Sen and A.B. Das Chaudhuri, *Amer. J. Phys. Anthropol.*, 2001, **115**, 289.
133. IAEA, *Activation analysis of hair as an indicator of contamination of man by environmental trace element pollutants*, 1977, 10.
134. P. Bermejo-Barrera, O. Muniz-Naveiro, A. Moreda-Pineiro and A. Bermejo-Barrera, *Forensic Sci. Int.*, 2000, **107**, 105.

Hair and Exposure to Environmental Pollutants

VLADIMÍR BENCKO

7.1. Monitoring of Environmental Pollution

The environmental pollution in a number of industrial agglomerations continues to be high and so assessment of the degree of health risk involved is increasingly becoming a major public health concern in Europe, the US, Canada, Japan and several other developed countries and regions.

Underlying the growing interest among public health authorities in biomarkers of human exposure to environmental pollutants and the potential health risk related to this exposure, is the difficulty to assess qualitatively and quantitatively the full extent of environmental pollution. Furthermore, analyses of air and surface water samples if collected non-systematically yield virtually worthless data in this respect, as the degree of environmental contamination may be highly variable. For example, concentrations of air pollutants are influenced by actual weather conditions, local air movements or by thermal inversions causing critical accumulation of emissions in given areas during smog episodes. Moreover, the quality of surface waters, especially in rivers and streams, is generally dependent on flow rate, *i.e.* the degree to which the incoming discharges are diluted by the stream flow. There may be fluctuation in the quality of surface water, influenced by the nature and amount of discharge of industrial effluents depending on the actual production technology used.

Ideally, a continuous measurement of environmental pollution can be effected through the use of a network of automated monitoring systems, *e.g.* a hexagonal scheme of air pollution monitoring networks; or line systems for the monitoring of, for example, water stream pollution. These networks and systems should be capable of automatic sampling, analysis, registration and evaluation of data. Although ideal, these automated monitoring systems are not easily accessible at present, both technically and economically, and their use in the near future is expected to remain limited to localities suffering from the greatest degree of environmental pollution.

As an alternative to the technical approach to this problem, biological indicators could be used to monitor pollution of the environment. This strategy would appear to be particularly well-suited to demonstrating environmental pollution by potentially toxic trace elements, including trace metals.[1–3]

The use of animals in the monitoring of noxious substances present in a working environment has a long tradition, beginning with canaries or mice in coal mines, used as indicators of the presence of, for example, carbon mono- or dioxide, up to the current wide range of biological exposure tests. A further, well-characterised, example of the exploitation of animals to monitor pollution is to count losses or even extinction of honey bee populations in areas affected by emissions containing arsenic, first described in the late 1930s.[4,5]

Thus, the examination of animals can provide important data that may complete the information obtained by the examination of the region's human inhabitants. It may be even assumed that the pollution effects may be detected in animals earlier than those in humans, because animals tend to be exposed to the impact of contamination more directly, by all routes including local food chains. Thus, free-living animals might signal in advance the danger threatening humans. Interestingly, haematological changes found in hares living in areas polluted by industrial emissions were comparable with those encountered in local children. Similarly, we found virtually identical concentrations of arsenic in the hair of children and the lot of rabbits living in the same locality.[6]

The advantages of investigations of environmental pollution effects on animals are not limited to the above-mentioned toxic metals and non-metal inorganic pollutants. After the Seveso accident, Lombardy, North Italy in 1976, besides the associated cutaneous manifestations in children (chloracne), it was the massive death of small animals that drew attention to the dioxine leakage. In the same way, cattle and domestic animals were dying in Bhopal, India in 1984 following an inhalation of the organic poison methylisocyanate. In this context, a primary concern of environmental health (or more specifically of environmental toxicology) is the assessment of human exposure. For this the examination of suitable human material would appear to be more appropriate than the analysis of plant or animal materials as are currently used in ecological studies to demonstrate environmental pollution. Human biological materials that are accessible for sampling usually include blood and urine. Successful attempts have also been made to measure the accumulation of metals, such as lead,[7–9] and non-metal pollutants, such as fluorine, in deciduous teeth[10] demonstrating non-occupational exposure to these materials in children. In addition, two ectodermal derivatives, the nail and the hair, are becoming increasingly valuable as highly accessible tissues for the purpose of monitoring human exposure to environmental pollutants.[9,11]

7.2 The Hair Fibre as a Biomarker of Human Exposure to Metals and Inorganic Substances

The human hair fibre represents an easily accessible material for the non-invasive sampling of individuals or population groups, to assess criminal, occupational or environmental exposure to toxic elements. Although originally proposed in 1974, utilisation of hair for these purposes has recently received a great deal of attention in the literature.[12–19] The current trend is towards the increasing use of the human hair fibre as the bio-resource of choice for monitoring excessive exposure to trace elements. This development however, is closely linked with the availability of suitable analytical procedures, sensitive enough to quantify the content of the respective pollutant in the biological specimen tested. The first steps towards this end are represented by the classic studies aimed at determining the arsenic content of human hair using the neutron activation (NAA) technique of investigation.[20–22] The power of this method is illustrated by the fact that by 1976 there were as many as 50 reports published on the analysis of pollutants in hair using NAA.[23] This activation analysis approach remains one of the most appropriate analytical techniques, especially for cases requiring non-destructive determination of a wide spectrum of trace elements simultaneously using instrumental NAA (INAA).[24–27] Its disadvantage is that it is less sensitive in determining particular elements *e.g.* lead, cadmium and nickel. In such cases the use of atomic absorption spectrophototmetry (AAS) is preferable. Other analytical procedures used successfully today for hair analysis purposes include X-ray fluorescence and proton-induced X-ray emission (PIXE) techniques[19,28–30] and inductively coupled plasma mass spectrometry (ICP MS).[31] To ensure accuracy of determination, certified reference materials (CRM) are now used as a quality control (QC) standard (*e.g.* commercially available CRM Human Hair GBW 07601).

Our study conducted in the mid 1960s demonstrated that determination of arsenic in human hair has potential utility as a biomarker. This study examined the environmental contamination with this noxious agent due to the burning of locally mined coal with high arsenic content (*i.e.* 900 to 1500 g As per tonne of dry coal) in a local power plant situated in the Upper Nitra Valley, Central Slovakia.[32] A later study in the same region revealed a correlation between arsenic content of hair and the expected degree of arsenic contamination in environment[33] (Figure 7.1).

In groups of 10-year-old boys from selected localities, a relatively high correlation was observed for residential distances from the source of emissions and hair arsenic content. Moreover, levels of arsenic in urine correlated with that observed in the children's hair, suggesting the feasibility of such an approach for the demonstration of human exposure to arsenic. It is necessary, however, to emphasise the importance of group-wide procedures whenever such values are evaluated. The comparison of individual values found in the hair and urine of the subjects, both in exposed and control areas, revealed a low degree of correlation or none at all. Data obtained during the examination of sufficiently large panels of probands are nevertheless only of descriptive value (see below). However, this is not an exceptional situation in environmental toxicology. For example, a similar situation is seen when assessing chromosomal aberrations of the peripheral lymphocytes as biomarkers of exposure to genotoxic substances.[34,35] Only by examining a group can one provide an evaluative answer (Figure 7.2).

The relationship between the arsenic content in blood, urine and hair is illustrated in Table 7.1. Arsenic content in blood determined by a destructive separation NAA technique in the early 1970s, appeared to be the most appropriate and still relevant today. The urine and hair samples were analysed using a classical colorimetric method.[36] In order to achieve sufficient accuracy using this test, not

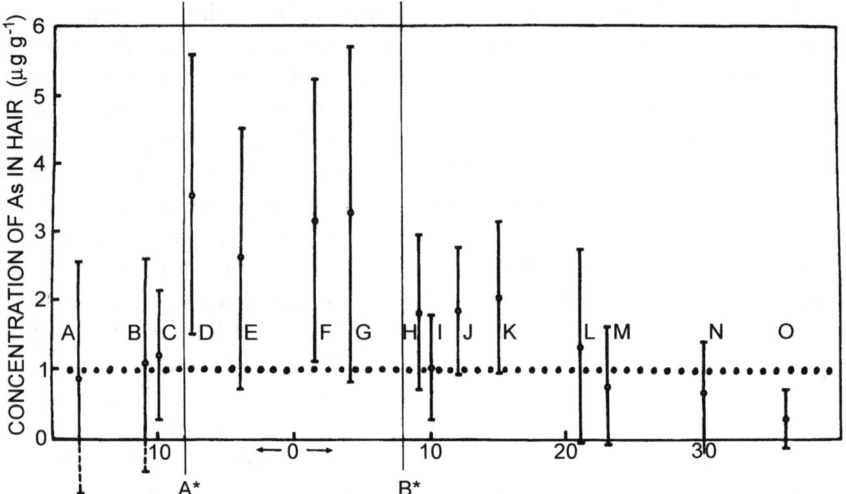

A*, B*: Vertical Lines Delimitating a Most Exposed Part of the Prievidza District:
7.5 Km Circle Round the Source of Emission Marked 'O'.
Dotted Line: Limit Value of a 'Normal' Arsenic Hair Content. Point C indicates position of Prievidza

Figure 7.1   *Values of arsenic in the hair of 10-year-old boys (mean value and standard deviation in µg g⁻¹)*

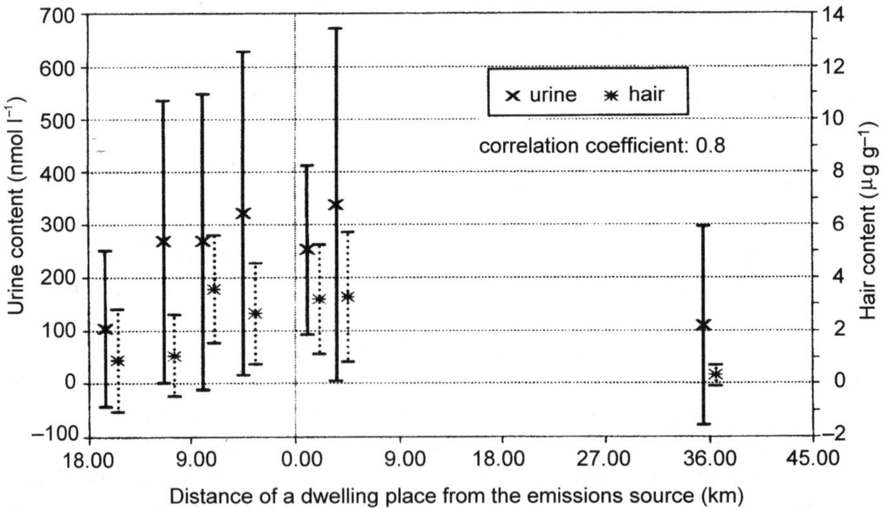

Figure 7.2   *Total arsenic content in the hair and in the urine of 10-year-old boys living in a vicinity of a power plant burning coal of high arsenic content. Adapted from reference 46*

Table 7.1 *Comparison of arsenic values (g g⁻¹)* — *in blood, urine and hair of 10-year-old boys residing in polluted (G) and control (O and P) communities. Adapted from reference 33*

	Community	No. of samples	Mean value	Standard deviation	t-test		Significance
Blood	G	10	0.00453	0.002028	G-O	4.261	S
	O	10	0.00145	0.000768	O-P	1.020	NS
	P	10	0.00188	0.001005	G-P	3.512	S
Urine	G	25	0.0253	0.025	G-O	2.893	S
	O	20	0.0082	0.013	O-P	0.617	NS
	P	24	0.0109	0.015	G-P	2.382	S
Hair	G	23	3.261	2.431	G-O	5.642	S
	O	23	0.295	0.413	O-P	1.653	NS
	P	44	0.152	0.279	G-P	5.978	S

All values are given in g g⁻¹, in urine after adjustment to standard density. G, the most heavily polluted part of the area; O, group residing 36 km from the source of emission; P, control group residing outside of the polluted area in a large industrial city.

less than a 1 g (usually about 2 g) hair sample and a 200 ml 24 hour urine sample were needed from each boy for analysis. These data show that the tests for arsenic content of hair, used for years to demonstrate arsenic exposure for the purpose of forensic or industrial toxicology,[21–23,37] are equally applicable to environmental settings if the method of group examination is applied. Today, extensive data is available in the literature, including the monographs quoted above, which document the advantages of this approach. Furthermore, the usefulness of hair analysis for the monitoring of exposure to toxic metals in humans has also been repeatedly confirmed in a series of our studies concerned with toxic trace elements such as manganese, lead, nickel, cobalt and antimony.[6,38–42]

It is evident that hair samples are far simpler to collect, transport and store than samples of blood and urine, and similarly their processing prior to analysis, provided that the content of potentially toxic elements is determined by adequate analytical techniques.[19,43–45] Table 7.2 illustrates the feasibility of hair analysis for the monitoring of excessive exposure to trace metals under different conditions.[40,46] These data were obtained from control and exposed groups of children, from control groups of adults as well as from occupationally exposed cohorts. Except for arsenic and antimony determinations, carried out spectrophotometrically by the use of silver diethyl-dithiocarbamine or phenylfluorone, all the remaining hair element concentration data were obtained by the method of classical flame atomic absorption spectrometry using a Varian Techtron AA 6-D spectrometer. In these studies, all the hair samples for analysis in environmental settings were obtained from boys as we felt that analogous hair sampling in girls might raise objections from their parents especially their mothers. The weight of hair samples collected was at least 1 g and usually about 2 g as mentioned above. In the case of antimony

Table 7.2 *Element concentrations in hair in children and adults (occupationally exposed) in regions polluted by selected toxic metals, metalloids and control groups in $\mu g\ g^{-1}$. Adapted from reference 40*

Element	Cohort		n	Median	min.	max.	Geometric mean	Standard error	P
Mn	Children								
	Control	A	100	0.60	0.25	3.20	0.747	0.0226	
		B	63	0.90	0.0	3.50	0.913	0.0436	
		C	106	1.20	0.0	4.16	1.243	0.0320	
	Exposed	A	22	2.74	0.33	19.2	3.953	0.1318	<0.01
		B	95	10.0	2.02	101.0	12.14	0.0651	<0.01
	Adults								
	Control	A	102	1.40	0.0	25.1	1.456	0.1468	
		B	32	1.00	0.03	9.25	1.076	0.0800,	
	Welders		12	2.90	0.79	7.25	2.900	0.1286	<0.05
	Cu		18	2.55	1.0	8.7	2.949	0.0991	<0.01
	Ni		33	2.60	1.04	15.2	4.120	0.0962	<0.01
	Co Smelters		30	3.85	0.64	11.2	4.551	0.0885	<0.01
	Fe		20	4.23	0.68	78.4	4.827	0.2273	<0.01
	Mn		36	19.0	1.47	117.0	17.28	0.1345	<0.01
Sb	Children								
	Control	A	130	2.88	0.0	55.0	2.698	0.0727	
		B	40	2.22	0.0	61.0	2.952	0.1668	
	Adults								
	Control	A	20	7.21	3.0	33.9	7.969	0.1308	
	Sb Smelters		39	266.0	122.0	710.0	292.9	0.0649	<0.01
As	Children								
	Control		125	0.24	0.0	3.9	0.331	0.0439	
	Exposed	A	100	1.31	0.18	4.42	1.277	0.0349	<0.01
	Adults								
	Control	A	20	0.0	0.0	0.93	0.153	0.0481	
	Exposed	A	58	0.51	0.06	4.62	0.613	0.0468	<0.01
		B	40	2.53	0.35	19.63	3.124	0.1059	<0.01
	Sb Smelters		39	44.0	0.34	151.3	36.2	0.1432	<0.01
Cr	Children								
	Control	A	100	0.1	0.0	2.6	0.209	0.0169	
		B	63	0.35	0.0	0.9	0.336	0.025,7	

Table 7.2 *(Continued)*

Element	Cohort				min.	max.	Geometric mean	Standard error	P
	Exposed	A	95	0.2	0.0	13.9	0.733	0.0777	<0.01
		B	106	0.8	0.0	5.9	0.911	0.0590	<0.01
	Adults								
	Control	A	62	0.0	0.0	1.0	0.124	0.0219	
		B	40	0.3	0.0	2.0	0.369	0.0337	
		C	31	0.1	0.0	14.0	0.370	0.1187	
Cr	Welders		12	2.47	0.84	5.56	1.468	0.1311	<0.01
	Mn		35	0.15	0.0	9.0	0.358	0.0727	<0.05
	Fe Smelters		20	1.56	0.0	7.77	1.610	0.1252	<0.01
	Ni Smelters		33	2.33	0.0	5.0	3.117	0.0851	<0.01
	Co		30	3.10	0.0	12.2	4.326	0.0955	<0.01
Ni	Children								
	Control	A	106	0.6	0.0	3.4	0.711	0.0469	
		B	100	0.6	0.2	5.5	0.721	0.0292	
	Exposed		95	1.3	0.0	207.9	2.088	0.0879	<0.01
	Adults								
	Control		102	0.0	0.0	13.0	0.286	0.0481	
	Welders		12	2.4	1.6	3.5	2.385	0.0378	<0.01
	Cu		18	1.45	0.1	3.3	1.407	0.0771	<0.01
	Co Smelters		30	29.2	10.0	136.0	34.78	0.1226	<0.01
	Ni		33	224.0	42.7	2140.0	222.5	0.1438	<0.01
Co	Children								
	Control	A	106	0.0	0.0	0.8	0.032	0.0087	
		B	100	0.0	0.0	0.5	0.046	0.0062	
	Exposed		95	0.1	0.0	0.9	0.179	0.0172	<0.01
	Adults								
	Control		102	0.6	0.0	1.6	0.525	0.0204	
	Cu		18	0.3	0.2	0.7	0.355	0.0231	>0.05
	Ni Smelter		33	3.4	0.0	37.2	3.862	0.1165	<0.01
	Co		30	109.5	26.6	631.8	97.3	0.1731	<0.01
Cu	Children								
	Control	A	95	9.5	2.7	212	9.321	0.0638	
		B	100	10.5	4.5	200	14.807	0.0842	

Table 7.2 *(Continued)*

Element	Cohort				*Limit values*		*Geometric mean*	*Standard error*	*P*
					min.	*max.*			
	Adults								
	Control	A	62	7.1	1.5	63.0	7.70	0.0723	
		B	40	9.7	4.7	60.0	10.4	0.0720	
	Fe Smelters		20	10.4	6.9	23.08	11.1	0.0613	>0.05
	Cu Smelters		18	13.7	6.8	130.0	24.4	0.2260	<0.01

Results of analysis of hair samples from groups of control and exposed children and from groups of control and occupationally-exposed adults. Except for As and Sb determinations, which were carried out spectro-photometrically using silver diethyldithiocarbamine or phenylfluorone, all other hair element concentration data were obtained by classical flame AAS (atomic absorption spectrometry) using a Varian Techtron AA 6-D spectrometer); 0.0 = value below the detection limit of the method used.

and copper content of hair we found no suitable non-occupationally exposed child populations. The significantly increased levels of the two metals namely antimony and copper, reported here are from groups of occupationally exposed adults.

7.3 Advantages and Limitations of Hair Fibre Analysis as a Biomarker of Human Exposure to Trace Elements

Major advantages and limitations of the use of human hair analysis as a biomarker of excessive exposure to trace metals can be summarised as follows: The extent of human exposure to pollutants in the general environment does not as a rule reach the level of exposure in occupational settings, and varies greatly from individual to individual, thus leading to relatively great intra-group differences in values. The only rational approach that might help to overcome this problem is to use a group approach when assessing the risk of environmental exposure. Our experiences to date suggest that there should be at least 20 individuals per population group to be sampled, to ensure that the differences in element content of hair samples in these groups can be qualified, quantified and analysed statistically.[40,42]

When comparing the trace element content of human hair in 'exposed' and 'control' groups it is preferable, according to our experience, to use the geometric mean and standard error of the mean of the values found. In the case of great inter-group differences, the use of non-parametric tests, *e.g.* the Wilcoxon test, is commonly advisable.[42,46] Where only a smaller difference in inter-group values exists, the use of the t-test to compare the means (for two groups), or the F-test to compare the variances (for more groups than two) is of advantage for assessing the significance of inter-group differences.

The frequency distribution pattern for the intra-group values of hair element content is, as a rule, asymmetrical. Thus, a logarithmic transformation should precede the employment of comparative tests. Just for illustrating the inter-group differences it is, as a rule, preferable to use the median and range, especially for ease of calculation. Although the median may differ in its absolute value from the respective geometric mean, it still constitutes the tool of choice for the orientation assessment of inter-group differences.

The accumulation of trace elements in hair might be, at least partially, dependent on age and sex. To overcome this problem we have used since the mid 1960s,[32] groups of 10-year-old boys as the most suited representatives of non-occupationally exposed populations under surveillance.

To date there has been no reliable data that would allow the establishment of generally applicable limits for normal content of individual trace elements in human hair. The elemental content of hair tends to vary from one geographical region to another, depending on natural background conditions, including composition of soil, element concentration in water and food, and eating habits.[12–15,47] From this, it follows that all findings obtained in the area under surveillance be compared against the values in suitable control groups of 'unexposed' human populations. One of the quite rare examples of a systematic effort to obtain reliable background trace element data on a national basis – deserving appreciation in this context – is the current project *System of Monitoring the Environmental Impact on Population Health* conducted with the full co-operation of the Czech Republic since 1994.[48,49] Within this project four selected regions of the Czech Republic are regularly assessed and blood, urine and hair samples of the child population collected. Scalp hair samples were collected from 3556 school children (1741 boys and 1815 girls) of average age 9.9 years (boys 10.1 and girls 9.7 years) over the period 1994–2001. For details of washing procedure and mineralisation see the original reference[48] and the Section 7.4 in this chapter dealing with the washing procedure applied. Individual elements were determined by means of electrothermal or flame atomic absorption spectrometry (ETAAS, FAA). For determinations the following instruments were used: Perkin Elmer AA spectrometer 4100 ZL, Perkin Elmer AA spectrometer 3300 (USA). Cadmium, chromium and lead were determined by means of ETAAS using $NH_4H_2PO_4$ or Pd/Mg modifiers. Copper was determined by ETAAS (modifier $Mg(NO_3)_2$) or by FAAS, zinc by the FAAS method only. Selenium was determined by ETAAS or by the technique of hydride generation (with HCl and $NaBH_4$ as reducing agents) using instrumentation by Perkin Elmer and FIAS 400 (USA). Mercury was determined using the single-purpose spectrophotometer AMA 254 (Altec, CZ). For all elements standards were prepared from commercially available standard solutions for AAS [Merck]. Accuracy was checked by means of reference material CRM Human Hair GBW 07601. Control samples and blanks were prepared and measured simultaneously with each series of analysed samples. Statistical evaluation was made with the t-test, χ^2 and the Kruskal Wallis test at 0.05 significance levels.[49] Results of determinations of cadmium, chromium, copper, mercury, lead, selenium and zinc concentrations in hair of children (of all children, of boys only, and of girls only) are given in Tables 7.3–7.5, respectively. The results are given in $\mu g \ g^{-1}$.

Table 7.3 *Summary results of metal concentrations in hair of children (~10-year-olds) collected in four selected areas of the Czech Republic from 1994–2001. Adapted from reference 48*

	Cd	Cr	Cu	Hg	Pb	Se	Zn
N	3554	3427	3553	3427	3555	3330	3556
Av	0.23	0.40	20	0.27	2.05	0.36	128
SD (mean)	0.31	0.89	27	0.57	2.02	0.46	54
Geomean	0.15	0.25	14	0.20	1.54	0.25	117
Median	0.14	0.22	12	0.19	1.60	0.22	124
Qv0.25	0.07	0.14	9	0.12	1.04	0.15	94
Qv0.75	0.28	0.32	19	0.31	2.50	0.34	155
Qv0.90	0.47	0.56	39	0.49	3.65	0.59	190
Qv0.95	0.66	1.27	65	0.68	4.76	1.60	214
Min	0.01	0.01	1	0.01	0.01	0.01	8
Max	5.45	25.0	337	28.00	44.8	4.92	642

The results found in this study do not differ significantly from the data published in different European regions,[50] which are analogous to the Czech Republic from the point of view of geological, geographical and demographic conditions.[51–53] Distribution analysis has proved that concentrations of cadmium, copper, mercury, lead and selenium approximate the log-normal distribution. Selenium and zinc concentrations have normal distribution. The correlation analyses have shown,

Table 7.4 *Summary results of metal concentrations in hair of boys of average age 10 years in $\mu g\ g^{-1}$. Adapted from reference 48*

	Cd	Cr	Cu	Hg	Pb	Se	Zn
N	1741	1674	1740	1667	1741	1623	1741
Avg	0.20	0.45	16	0.27	2.03	0.38	117
SD (mean)	0.26	1.12	23	0.39	1.95	0.47	47
Geomean	0.13	0.29	12	0.20	1.49	0.26	107
Median	0.12	0.22	10	0.19	1.56	0.23	115
Qv0.25	0.06	0.14	8	0.12	1.00	0.16	85
Qv0.75	0.24	0.32	15	0.31	2.50	0.35	142
Qv0.90	0.44	0.61	28	0.50	3.72	0.64	172
Qv0.95	0.62	1.72	46	0.72	4.98	1.70	191
Min	0.01	0.01	1	0.01	0.01	0.01	8
Max	3.17	25.0	337	10.60	25.7	4.92	570

Table 7.5 *Summary results of metal concentrations in hair of girls of average age 9.7 years in* μg g^{-1}. *Adapted from reference 48*

	Cd	Cr	Cu	Hg	Pb	Se	Zn
N	1813	1753	1813	1760	1814	1707	1815
Avg	0.26	0.35	24	0.28	2.07	0.34	140
SD (mean)	0.35	0.59	29	0.70	2.08	0.44	59
Geomean	0.18	0.31	15	0.24	1.57	0.27	105
Median	0.17	0.22	13	0.2	1.64	0.22	135
Qv0.25	0.09	0.14	10	0.13	1.10	0.15	104
Qv0.75	0.30	0.31	24	0.32	2.50	0.31	167
Qv0.90	0.49	0.52	50	0.47	3.58	0.53	203
Qv0.95	0.68	1.00	81	0.65	4.51	1.50	236
Min	0.01	0.01	1	0.01	0.01	0.01	10
Max	5.45	9.73	327	28.00	44.8	3.14	642

Tables show arithmetic means (Avg), geometric means (Geomean), medians (Me), quantiles (Kv) and limit values (Min, Max). N is the number of hair samples. From Table 7.3 it follows that normal values (medians) of trace elements in hair of the Czech child population are as follows: 0.14 μg g^{-1} for cadmium, 0.22 μg g^{-1} for chromium, 12 μg g^{-1} for copper, 0.19 μg g^{-1} for mercury, 1.6 μg g^{-1} for lead, 0.22 μg g^{-1} for selenium and 124 μg g^{-1} for zinc.

except for the pair cadmium-chromium, no significant correlation between selected pairs of the trace elements under our examination. Tables 7.4 and 7.5 further show results of determinations in boys' and girls' hair separately. It can be seen from the Tables that statistically significant differences exist between boys and girls in the concentrations (medians) of cadmium (0.12 μg g^{-1} vs. 0.17 μg g^{-1} p $<$ 0.01), copper (10 μg g^{-1} vs. 13 μg g^{-1} p $<$ 0.01) and zinc (115 μg g^{-1} vs. 135 μg g^{-1} p $<$ 0.01). Significantly lower concentrations of copper and zinc which appear in the boys' hair were not conclusively explained. The differences could hardly be caused by differences in nutritional habits, which do not play an important role in juveniles. Perhaps it could be caused by as yet unknown possible initial hormonal influences. A similar tendency was found also in the blood.[49] The increased level of cadmium in girls' hair is quite unclear. Detailed knowledge of reference values of the trace elements studied in humans shall constitute, therefore, a basis to further studies in the field.

Sampling, transport and storage of hair samples is easier than the sampling, transportation and processing of blood or urine samples, the most common specimens used today to demonstrate human exposure to a variety of noxious, environmental agents.[2,54,55] Parts of human hair samples are also very easy to preserve for later control re-analyses.

With the exception of mercury and selenium,[56–58] no biological limit values (or exposure indexes *i.e.* concentrations which, if surpassed, would indicate overexposure and a potential public health emergency), have been established as yet for exposures to toxic trace elements in the general environment. Consequently, their determinations in hair can hardly be interpreted as representing exposure tests in the classic sense, but may substantially help in monitoring excessive exposure or, in other words, may help to monitor pollution in general or occupational settings.

7.4 Washing of Hair Samples

In contrast to blood and urine specimens, hair should be cleared of external contaminants prior to elemental analysis. Different authors have preferred different removal techniques and the question of washing methods has received a great deal of attention. Because the metallic cations released into the washing solution might form bonds with sulfur in the keratin matrix of the hair,[59–61] it is obvious that a non-polar solvent should be used as a first-step wash to remove, together with hair surface grease, the attached dust particles containing metals in un-dissolved form. In order to ensure the comparability of test results, a standardised washing procedure has been recommended by the International Atomic Energy Agency in Vienna.[25,28,62]. According to this standard, each hair sample to be analysed should first be subjected to washing in a non-polar solvent (double distilled acetone), followed by two washings in a polar solvent (de-ionised or redistilled water) and finally washed in acetone. It is recommended that this scheme of washing be repeated twice, and that each washing step should last about 10 min. Despite the fact that there is no general consensus on how the washing procedure should be done,[44,63–66] this IAEA-recommended procedure should be preferred over other techniques that are not standardised. In this context it should be noted, however, that excessive external contamination of hair usually reflects massive environmental pollution and resulting human exposure, which is actually what is to be demonstrated by these tests in most instances,[46] at least in the pilot stage of the exposure assessment. For example, relatively high levels of dioxins (TCDDs) and dibenzofurans (TCDFs) have been discovered in hair samples taken from people scavenging for recyclable waste from Cairo's municipal waste dumping sites. The TCDD/TCDF congeners pattern observed in GC/MS-SIM fragmentograms of hair extracts was almost identical to that observed in emissions from municipal waste incinerators.[67]

The data provided by human hair analysis techniques, carried out in compliance with the above described principles, may well serve as a basis for identifying areas excessively contaminated with emissions containing toxic trace elements, including their geographical delimitation.

7.5 Discussion

The method of biological monitoring using hair analysis appears to be ideally suited for use in pilot prospective studies in the context of biological monitoring of human exposure to metals and metalloids. If an excessive exposure is detected it is recommended that the epidemiological examination be extended to the analyses of other biological materials, most often blood and urine, in order to obtain a closer assessment of the degree of exposure in the respective population (see Table 7.1 and Figure 7.2). Attempts to use hair analyses in clinical diagnostics[6,68,69] is perhaps one further application of this investigative technique. In this context, the method of multi-elemental analysis, enabling comparison of the content of various trace elements, is likely to be more useful that isolated monitoring of just one micro-element content in hair.

This approach, based on the determination of the selected spectrum of essential and toxic trace metals and their mutual proportion in hair, is also of course of advantage if biological markers of exposure in the general environment are the subject of interest.

Nevertheless, taking into account limitations in our current knowledge that result from insufficient data, the analysis of hair fibre samples does not permit prediction of human health implications of environmental/occupational exposures. Thus, the presence of essential or toxic metals and metalloids or some organic xenobiotics in hair samples indicates exposure only. This information does not however, inform us of the route of association *e.g.* internally by inhalation or digestion, or externally.[70,71] Thus, a panel of experts was convened by the Agency for Toxic Substances and Diseases Registry (ATSDR), Atlanta, USA, and expressed its concern about the misuse of hair analysis to justify and support unnecessary and so unethical medical therapy.[72] This view is consistent with the 1984 policy statement of the American Medical Association (AMA), which was reaffirmed in 1994. The AMA stated: 'The AMA opposes chemical analysis of the hair as a determinant of the need for medical therapy and supports informing the American public and appropriate governmental agencies of this unproven practice and its potential for health care fraud'.[73]

The science and utility of hair analysis can only advance through well-designed studies. Future research should reflect a more complex approach by respecting fundamental principles of epidemiological study. A model of such an approach may be the *System of Monitoring the Environmental Impact on Population Health* project underway in the Czech Republic, in which analysis of hair samples constitutes just one aspect of a more complex approach.

Perhaps today a similar pattern of study in selected regions of Europe for example the newly started European Commission 6th Framework Programme can balance and/or complete the current efforts of the ATSDR in Atlanta.

7.6 Conclusions

Hair samples are far simpler to collect, transport and store than samples of blood and urine, and similarly their processing prior to analysis, with the exception of problems with removal of external contamination which can be neglected when exposure comes from the food supply or *via* drinking water. While blood and urine concentrations reflect a recent exposure well, but usually vary within a relatively wide range, hair as well as nails, both derivates of ectoderm, reflect a long-term or past exposure. Unlike bone or tooth, hair growth is very time-resolved, *i.e.* 1 cm growth per month on scalp, thus exposures will make contemporaneous contributions rather than 'averaging' as in bone and tooth.[74]

Detection of toxic and trace metals in human tissue samples could help to overcome problems associated with the detection of multimedia integrative exposure to metals and metalloids as well as some selected organic xenobiotics (*e.g.* TCDDs/Fs and PCBs). The data provided by human hair analysis may well serve as a basis for identifying populations at specific risk, *e.g.* those exhibiting a deficit in essential trace elements like selenium and zinc, and/or populations in excessively contaminated areas, including the latter's geographical delimitation.

One of the essential conditions for ensuring the realistic evaluation of a population's excessive exposure in both occupational and environmental settings is the need to examine sufficiently large population groups and also the use of a group diagnostic approach on a cross-section or follow-up basis in optional cases fully respecting principles of random sampling of pertinent populations.

7.7 References

1. UNEP/WHO, Report of a Meeting of Government-Designated Experts on Health-Related Monitoring, Annex III, *Recommendations for the Future Work Related to the UNEP/WHO Pilot Project on Assessment of Human Exposure to Pollutants Through Biological Monitoring (Metals Component)*, Geneva, 8–12 March, 1982.
2. M. Vahter, in *Assessment of Human Exposure to Lead and Cadmium Through Biological Monitoring*, UNEP and WHO, Stockholm, 1982.
3. E. Subramarian in *Biological Monitoring of Exposure to Chemicals and Metals*, H.K. Dillon and M.H. Ho (ed), Wiley-Interscience Publications, 1991, 255.
4. J. Svoboda, *Czechoslovak Acad. Agricul. Sci.*, 1936, **12**, 589.
5. H. Prell, *Arch. Gewerbepath. Gewerbehyg.*, 1937, **7**, 656.
6. V. Bencko in *COST 6.2 Report Series on Air Pollution Epidemiology Report No 3.*, P. Rudnay (ed), 1992, 141.
7. G. Fosse and N.P. Berg-Justensen, *Arch. Environ. Health*, 1978, **33**, 166.
8. M. Cikrt, P.Lepší, J. Handzel, J. Kratochvíl and H. Hanáková, *Cs. Hyg.*, 1983, **28**, 525.
9. Z. Zaprianov in *COST 613/2 Report Series On Air Pollution Epidemiology*, No. 3, P. Rudnay (ed), 1992, 146–150.
10. G. Balážová, P. Macúch and A. Rippel, *Fluoride*, 1969, **2**, 33.
11. A. Agahain, J.S. Lee, J.H. Nelson and R.E. Johns, *Am. Ind. Hyg. Assoc. J.*, 1990, 51, **12**, 646.
12. A. Chattopadhyay and R.E. Jervis in *Proc. VIII Symposium, Trace Substances in Environmental Health*, D.D. Hemphill (ed), Univ. of Missouri, Columbia, 1974, pp.31.
13. J.J. Hefferre, *J. Am. Dental Assoc.*, 1976, **92**, 1213.
14. D.W. Jenkins in *EPA 68–03–0443 Biological Monitoring of Environmental Pollutants*, N.E.R.C., Las Vegas, Nevada, 1977.
15. D.W. Jenkins, *Biological monitoring of Toxic Metals in Human Populations Using Hair and Nails*, International Workshop on Biol. Specimen Coll., Luxembourg, 1977, 18.

16. V. Valkovic, in *Trace Elements in Human Hair*. Garland Publ. Inc., New York, 1977.

17. M. Stoeppler and K.Brandt, *On the Importance of Hair as an Indicator of Environmental and Occupational Exposure to Toxic Trace Elements*, Report to IAEA Vienna, 1978.

18. M. Anke and M. Risch in *Haaranalysse und Spurenelement- Status*, Gustav Fischer Verlag, Jena, 1979.

19. T. Suzuki in *Biological Monitoring of Toxic Metals*, T.W. Clarson, L. Friberg, G.F. Nordberg and, P.R. Sager (ed), Plenum Press, New York, 1988, 623.

20. A.A. Smales and B.D. Pate, *Analyst*, 1952, **77**, 196.

21. H. Smith, S. Forschfvud and A. Wassen, *Nature*, 1962, **194**, 725.

22. H. Smith, *J. Forensic Sci.*, 1964, **4**, 192.

23. IAEA, Advisory Group Meeting Report, *Application of nuclear methods in environmental research, Attachment 2*, 22–26, March, 1976.

24. IAEA, *Activation Analysis of Hair as an Indicator of Contamination of Man by Environmental Trace Element Pollutants*, IAEA/RL/41 H, Vienna, 1977.

25. O. Obrusník and V. Bencko, *Radiochem. Radioanal. Lett.*, 1979, **38**, 189.

26. S. Yu. Ryabukhin, *J. Radioanal. Chem.*, 1980, **60**, 7.

27. I. Obrusník, *J. Hyg. Epidemiol. (Praha)*, 1986, **30**, 11.

28. E.C. Montenegro, G.P. Baptista, L.V. De Castro Faria and A.S. Paschoa, *Nucl. Instr. Methods*, 1980, **168**, 479.

29. A.E. Pillay and M. Peisach, *J. Radioanal. Chem.*, 1981, **63**, 85.

30. T. Badica, C. Ciorka, V. Cojocaru, M. Ivascu, A. Petrovici, A. Popa, I. Popescu, M. Salagean and S. Spiridon, *Nucl. Instr. Math. Physics Res.*, 1984, **3**, 288.

31. R. Knight, S.J. Haswell, S.W. Lindow and J. Barry, *J.A.A.S.*, 1999, **14**, 127.

32. V. Bencko, *US Publ. Hlth. Service*, 1966, **3**, 948 (1969).

33. V. Bencko and K. Symon, *Environ. Res.*, 1977, **13**, 378.

34. P. Rössner, M. Černá, H.Bavorová, A. Pastorková and D. Očadlíková, *Centr. Europ. J. Publ. Hlth.*, 1995, **3**, 219.

35. P. Rössner, R.J. Šrám, H. Bavorová, D. Očadlíková, M. Černá and E. Švandová: *Toxicol. Lett.*, 1998, **96–97**, 137.

36. V. Vašák and V. Šedivec, *Chem. Listy*, 1952, **46**, 341.

37. I. Porázik, V. Legáth, K. Puchá and I. Kratochvíl, *Prac. Lek.*, 1966, **10**, 352.

38. V. Bencko, D. Arbetová, V. Skupeňová and A. Pápayová in *Proc. Industrial and Environ. Xenobiotics*, Springer-Verlag, Berlin, Heidelberg, 1981, 69.

39. V. Bencko, K. Erben, K. Zmatlíková, L. Filková and M. Tichý, *Cs. Hyg.*, 1982, **27**, 206.

40. V. Bencko, T. Geist, D. Arbetová, D.M. Dharmadikari and E. Švandová, *J. Hyg. Epidemiol. (Praha)*, 1986, **30**, 1.

41. V. Bencko, V. Wagner, M. Wagnerová and V. Zavázal, *Environ. Res.*, 1986, **40**, 399.

42. V. Bencko In *Biological Monitoring of Exposure to Chemicals and Metals*, H.K. Dillon and M.H. Ho (ed), Wiley-Interscience Publications, 1991, 243.

43. H.G. Petering, D.W. Yeager and S.O. Whitherup, *Arch. Environ. Health.*, 1973, **27**, 327.

44. G. Chittelborough, *Sci. Total Environ.*, 1980, **14**, 53.

45. R.R. Lauwerys, *Industrial Chemical Exposure: Guidelines for Biological Monitoring*. Biomedical Publications, Davis, California. 1993.
46. V. Bencko, *Toxicology*, 1995, **101**, 29.
47. H. Teraoka, *Jpn. J. Hyg.*, 1981, **36**, 712.
48. B. Beneš, J. Sladká, V. Spěváčková and J. Šmíd, *Cent. Eur. J. Publ. Health* 2003, **11(4)**, 184.
49. V. Spěváčková, B. Beneš, J. Šmíd and V. Spěváček, *Centr. Eur. J. Publ. Hlth* 1996, **4(2)**, 102.
50. M. Wilhelm and H. Idel, *Zbl. Hyg.*, 1996, **198**, 485.
51. A. Stararo, G. Parvoli, L. Doretti, *Biol. Trace Elem. Res.*, 1994, **40**, 1.
52. O.M. Paulsen, J.M. Christiansen and E. Sabbioni, *Sci. Total Environ.*, 1994, **158**, 191.
53. R. Cornelis, E. Sabbioni and M.T. Van den Vanne, *Sci. Total Environ.*, 1994, **141**, 191.
54. J.P. Buchet, H. Roels, R. Lauwers, P. Bruaux, F. Clayes-Thoreau, A. Lafontaine and G. Verduyn, *Environ. Res.*, 1980, **22**, 95.
55. V. Bencko, M. Tichý, J. Pekárek, V. Wagner and M. Wagnerová, in, *Proc. III Spurenelement-Symposium, Nickel*, K.M.U. Leipzig u. F. SCH. U., Jena, 1980, 332.
56. H. Al-Sharistani, K.M. Shibal and I. K. Al-Haddad, in *Conf. on Intoxication Due to Alkyl Mercury-Treated Seeds*, Baghdad, Iraq, 9–13 Sept. 1974, *Bull. WHO, Suppl.*, **53**, 105.
57. S.A. Katz and R.B. Katz, *J. Appl. Toxicol.*, 1992, **12(2)**, 79.
58. K. Kratzer, P. Beneš and V. Spěváčková, *J.A.A.S.* 1994, **9**, 303.
59. M.A. Rota and C. Zorzi *Zacchia*, 1961, **2**, 505.
60. P. Blanquet, M. Croizet, P. Bourbon and G. Broussy, *Ann. Biol. Clin. (Paris)*, 1966, **23**, 59.
61. K. Nishiyama and G.F. Nordberg, *Arch. Environ. Health*, 1972, **25**, 92.
62. S. Yu, Ryabukhin in *Activation Analysis of Hair as an Indicator of Contamination of Man by Environmental Trace Element Pollutants*, Report IAEA (RL) 50, Vienna, 1978.
63. L.C. Bate, *J. Forensic Med.*, 1965, **10(1)**, 60.
64. D.C. Hildebrand and D.H. White, *Clin. Chem.*, 1974, **20(2)**, 148.
65. A.N. Clarke, D.J. Wilson and N. Tenn, *Environ. Res.*, 1974, **6**, 247.
66. S. Salmela, E. Vuori, D. Jukka and J.O. Kilpio, *Anal. Chem. Acta*, 1981, **125**, 131.
67. A. Kočan, V. Bencko and W. Sixl, *Toxicol. Environ. Chem.*, 1992, **36**, 33.
68. W. Weisener, W. Goerer and S. Niese in *Proc. Int. Symp. on Nuclear Activation Techniques in the Life Sciences*, IAEA, Vienna, 22–26 May, 1978.
69. E.L. Rees, *J. Orthomol. Psychiatry*, 1979, **8**, 37.
70. M. Wilhelm, F.K. Ohnesorge, *Sci. Total Environ.*, 1990, **92**, 199.
71. V. Senofonte, N. Violante and S. Caroli, *J. Trace Elem. Med. Biol.*, 2000, **14(1)**, 6.
72. D.K. Harkins and A.S. Susten, *Environ Health Perspect*, 2003, **111**, 576.
73. AMA, *Hair Analysis: A Potential for Abuse*, Policy No. H-175.995, American Medical Association, Chicago, 1994.
74. V. Neste and D.J. Tobin, *MICRON* 2004, **35(3)**, 193.

Hair and Nutrient/Diet Assessment

TAMSIN O'CONNELL

8.1 Hair and Diet

The appearance of an individual's hair has long been known to reflect their diet and nutritional status, and with the advent of sophisticated scientific analyses, we have realised that diet and nutritional information can also be gleaned from the chemical and elemental composition of hair. Since 'you are what you eat', atoms and molecules consumed as food are incorporated into the consumer's body tissues and therefore if a chemical signal can be tracked either unchanged or altered in a quantifiable fashion from food into the body, this can reveal what an individual has consumed.

Hair can yield dietary and nutritional information at several levels: at the physical and structural level; from incorporated atoms and molecules; and from chemical signals within constituent proteins. Dietary and environmental signals that can be measured within constituent proteins are primarily the stable isotopic ratios of the light elements carbon, nitrogen, sulphur, hydrogen and oxygen. Variation in the distribution of stable isotopes of carbon, nitrogen and sulfur throughout different global ecosystems, and the corresponding isotopic variation in different foodstuffs, makes it possible to use these elements as natural dietary tracers. Furthermore, oxygen and hydrogen isotopic distributions reflect environmental temperature and water sources and so can provide information as to an individual's place of origin. Atoms and molecules incorporated within the hair include organic metabolites and trace elements (particularly metals). Most organic metabolites that are detectable in hair are related to ingestion of substances other than food, mainly drugs and alcohol (see Chapter 4). As yet, no metabolites from dietary sources have been measured in hair, such as vitamin breakdown products, but there has been little research in this area. Metals detectable in hair can be considered to be either nutritionally necessary or toxic, and can be related to both dietary and non-dietary exposure (see Chapters 6 and 7, elsewhere in this volume). Nutritionally desirable metals in hair are primarily derived from diet.[1] Ingestion of toxic metals can be *via* non-dietary (*e.g.* occupational exposure or accidental ingestion) and dietary sources (*via* consumption of foods containing the metal *e.g.* mercury in fish). For nutritional assessment, the elemental concentration of incorporated metals is usually measured. However, isotopic ratios of metals can also be measured, but this approach provides information as to an individual's place of origin, rather than about nutrition, since the relative amounts of radiogenic isotopes (*e.g.* strontium, lead and rubidium) relate to the local geological environment. The physical structure of hair is altered primarily *via* nutritional deficiencies and genetic disorders, leading to changes in size, shape, texture, pigmentation and growth rates.

8.2 Diet and Nutritional Investigations Using Hair – a Brief Outline

This chapter focuses on dietary and nutritional assessment of hair from modern humans, but the research covered is closely related to a number of research fields, including archaeology, forensics and toxicology. The techniques used are similar and the conclusions drawn are based on the same underlying principles.

The assessment of dietary intake and the assessment of nutritional status are two very different issues and it is essential to distinguish between them. Diet can be defined as that which is intentionally consumed by an individual, including both food and water. Nutritional status is the degree to which an individual has a certain level of nutrients for optimal metabolic function. It is possible for individuals to consume wildly differing diets, and yet be equal/similar in their nutritional status.

Chemical analysis of hair has been applied to studies of both diet and nutritional status. Light stable isotope analysis of hair has been used to identify aspects of an individual's long-term diet type (such as the importance of animal products, vegetables and fish). This work has had most application in anthropology and links closely to the field of archaeological science known as palaeodiet. A combination of dietary and environmental indicators have been used in forensic science for identification of an individual's place of origin, again tying in with archaeological studies of human ecology and migration.

The nutritional status of an individual or a population has been assessed by measurement of the nutritionally necessary elements including zinc, copper, manganese, chromium, selenium, calcium, sodium and magnesium amongst others. This work is related to investigations of occupational/non-occupational exposure to toxic metals including cadmium, lead, mercury, arsenic, aluminium and chromium (see Chapter 6 elsewhere in this volume).

Visual inspection and physical examination of hair is used primarily as a clinical diagnostic tool for identifying certain metabolic or genetic conditions, and as such shall only be touched on briefly towards the end of this chapter.

8.3 Advantages of Using Hair as a Study Tissue

The advantages of dietary or nutritional analysis of an individual using a hair sample are manifold. Hair is an ideal biopsy material, since it is available from almost all individuals, and is easy and painless to sample. Its physical and chemical stability means that no special storage between sampling and analysis is required, unlike, for example, blood or urine, and its external growth in a linear fashion provides a longitudinal measure of body tissue synthesis and therefore potentially of variation in diet or nutrition over a long period of time.

Rapid developments in the sophistication of analytical techniques, together with the relatively high concentrations of elements (metals and non-metals) within hair mean that most analyses can be performed on samples of 1–2 mg. Hair proteins contain carbon, nitrogen, oxygen, hydrogen and sulfur at similar or higher percentages than other tissue proteins, and the hair shaft contains most of the trace metals at relatively high concentrations compared to the rest of the body.[2,3]

In addition to non-invasive tissue sampling from living individuals, the tissue preserves well post-mortem, making it a good forensic biopsy material. It has been more widely used in this context as a source of DNA evidence (see Chapter 5 elsewhere in this volume), but can also be used as a source of information on diet and geographical origin.

8.4 Problems Associated with the Use of Hair as a Study Tissue

Unfortunately, despite the inherent advantages of hair as a tissue for analysis, the problems associated with it are also many and widely documented. Trace metal analysis for nutritional assessment in particular is now widely discredited within the medical profession.[4-9]

The associated difficulties vary, depending on the type of analysis being performed. While the analysis of hair proteins is relatively trouble-free, great difficulties are associated with regards trace metal analysis of hair. These problems are due to:

- the association between the chemical signal measured and an individual's diet or nutritional status (*i.e.* the correlation issue relating to diet and body pools)
- endogenous variability in the chemical signal being measured
- hair growth rates and its effect on the measured chemical signal
- hair contamination.

8.4.1 Correlation with Diet and Body Pools

The assumption underlying hair analysis for dietary or nutritional assessment is either that signals measured in hair can be correlated with diet, or that the signals in hair are representative of those in other body tissues and/or the 'metabolic pool'. These are two significantly different statements, and a number of the problems of hair trace element analysis can be traced to confusion between the two.

8.4.1.1 Isotope Ratios of Light Elements

The relationship between the light stable isotope ratios of diet and of hair are fairly well documented in animals using controlled feeding experiments.[10-19] A strong correlation is observed between the carbon, nitrogen and sulfur isotopic values of the dietary protein and those of hair keratin. This is so despite our incomplete understanding of the mechanisms underlying alterations of isotopic ratios between those seen in diet and those seen in a consumer's different body tissues. For example, there is a ^{15}N enrichment between diet and body (the 'trophic level effect'[15,20]), and also an effect of dietary protein routing on body tissue carbon isotopic values.[17,21] The number of studies correlating hair isotope ratios with those in other tissues is small,[10–17,22,23] but these are however fairly consistent and so may enable comparison with different body pools, particularly bone collagen.

8.4.1.2 Trace Metal Analysis in Hair

The relationship between trace metal concentrations in hair and that of the diet is not a simple one, and studies of the link between the two have produced conflicting results. One study on rats found a positive correlation between dietary zinc and that in hair, bone and testes,[24] yet a second study, also on rats, found no such relationship for either zinc, copper or manganese.[25] Moreover, dietary supplementation with selenium dramatically increased hair selenium, with no effect on blood selenium concentrations.[26] However, supplementation with zinc and copper had little or no effect on hair metal concentrations.[27–29] It is important to recognise in this context that trace metal concentrations in the body are not solely dependent on the metal concentration in the diet, but are dependent on other dietary constituents.[25] For instance, dietary phytate and fibre levels are known to affect zinc absorption.[30] Ascorbic acid increases the intake of inorganic iron but reduces copper absorption, and competitive absorption in the gut between iron and manganese and between copper and zinc will affect the body levels of these pairs of elements.[1,31,32] Results of analyses of trace element levels in hair relative to the individual's nutritional status have also produced less than clear-cut results. For example, marginally zinc-deficient children from affluent backgrounds in Denver had lower hair zinc than those judged zinc-sufficient,[33] yet the hair of seriously malnourished children contained greater amounts of zinc than did a comparable healthy population.[34] A review of studies comparing hair zinc with anthropometric indices found that the vast majority of studies found no correlation between the two.[35] Moreover, hair copper levels were similar for well-nourished and malnourished individuals.[28] Hair concentrations of zinc, copper, iron and manganese in healthy Thai children did not correlate with either growth indicators or with morbidity,[25] leading the researchers to conclude that for individuals with an adequate dietary metal intake, hair trace metal concentrations were 'of little value in predicting health status'.

For nutritional assessment in a clinical setting, blood concentrations are generally considered as representative of the 'metabolic pool', and therefore, for hair to act as a proxy diagnostic tissue, hair concentrations must ideally correlate with blood concentrations. This has not been conclusively demonstrated with several studies reporting a lack of a significant correlation between hair and blood (serum or plasma) for a number of elements including zinc, copper, iron, calcium and aluminium.[24,26,36–43] The exception is selenium, however, which does appear to be highly correlated in blood (serum) and hair.[44] However, some researchers consider that blood concentrations of trace metals are of 'uncertain significance'[36] in trace element nutrition, since blood concentrations of trace metals such as zinc, copper and chromium may be chemically buffered, thus not representative of the body pool, and so potentially irrelevant for nutritional assessment.[36,39,45–47] Therefore, the absence of a relationship between hair and blood trace element concentrations does not, of itself, render trace element analysis of hair for nutritional status as flawed.

The association between trace metal levels in hair and other body tissues, is again not simple. There are few specific comparative data concerning the extent to which hair trace element concentrations relate to those in other body tissues, and there is little consistency between studies. In one survey, no correlation between hair and other human body organs (liver, kidney, spleen, heart, cerebellum) was found for magnesium, calcium, iron, copper, zinc or selenium.[48] In another however, a weak positive correlation was found for selenium in human hair and kidney, though not for either iron or zinc.[49] In a study using rats on controlled zinc and copper intakes, hair, bone and testes zinc levels were strongly correlated, as were levels of copper in hair, liver, kidney and heart. However, hair, kidney and liver zinc concentrations were either weakly correlated or not correlated at all.[24,50,51]

The many factors that control hair trace element concentrations make definitive relationships between diet and body, or between hair and nutritional adequacy, very difficult to describe. In addition, examining all the studies discussed above, it becomes apparent that there is little consistency in sample collection methods and analytical techniques, and as such it is perhaps not surprising that no coherent relationships have been identified. A number of the studies suggest that hair can potentially record exposure to, or the body burden of, some trace metals rather than the metabolic pool or nutritional status. However, a more rigorous consideration of analytical variables is required to assess this. In the light of the current understanding, it appears that hair trace metal analysis cannot be used to assess the nutritional status of trace metals at the level of the individual under conditions of adequate nutritional status. However, this does not preclude the use of hair as an indicator of nutritional status or dietary exposure at the population level.[47]

8.4.2 Endogenous Variability in Chemical Signals from Hair

8.4.2.1 Isotope Ratios of the Light Elements Carbon, Nitrogen, Sulfur, Oxygen and Hydrogen

Although diet is the primary influence on carbon and nitrogen in tissue, secondary influences including environmental and physiological factors have been shown to affect isotopic values. For nitrogen, the 'trophic level enrichment' (*i.e.* the increase in nitrogen isotopic values between diet and body tissues[12]) between diet and hair is quite variable under a range of environmental and climatic conditions (temperature, altitude, aridity), as well as being affected by physiological factors such as water stress, starvation and growth, and diet macronutrient composition.[14,15,20,52–58] The magnitude of the these influences is unknown, primarily because of our lack of understanding of the origin of the $\delta^{15}N$ enrichment between diet and body, and how it is transformed by metabolic processes. However, the general trends are that stress (water stress, heat, starvation) elevates $\delta^{15}N$, whereas growth depresses $\delta^{15}N$ values.

Hair colour and location on the body appear to have no discernable effect on isotopic values: no difference was found between the carbon and nitrogen isotopic values. Similarly, C/N ratios of pigmented and non-pigmented (grey) hair were the same as were isotopic values of hair on different locations on the body.[59,60]

Other than that due to diet, physiology or environment, there appears to be little intrinsic variability in the carbon, nitrogen, sulphur, oxygen or hydrogen isotopic variations within an individual's keratin protein.[18,59] This is likely to be a result of the body protein pool acting as an isotopic 'buffer', reducing the response to short-term fluctuations. Despite the lack of variability within individuals, there is always some degree of intra-population isotopic variation. Some degree of isotopic variation between individuals has been observed within animal and human populations fed on diets of limited variability (and thus assumed to be isotopically uniform). This result suggests that personal dietary preferences can result in long-term isotopic differences and so a degree of caution must therefore be exercised when interpreting isotopic data based on analyses of a few individuals.

8.4.2.2 Trace Metals

Hair trace element concentrations have been shown to vary widely due to factors such as sex, age, race, hair colour and anatomic location of hair sampled. These parameters can cause as much variation in concentrations as can diet. The data are sometimes conflicting, and a cause and effect relationship has seldom been established. As for the studies investigating hair trace metal correlation with diet and other body pools, there is generally no consistency in sample collection methods and analytical techniques. Most worryingly, the sampling strategy is often *ad hoc*, and the majority of studies have, surprisingly, not even controlled for variation in dietary intake (both food and water), which is self-defeating when attempting to quantify endogenous variability.

Difference in trace element concentrations between hair from males and females has been reported in several studies, yet again there is little consistency here.[34,61–65] Various studies report changes in hair metal concentrations with age, with significant variations observed during early infancy, puberty, pregnancy and lactation, but yet again no clear patterns emerge.[33,35,43,63–66] However, pregnancy and the oral contraceptive pill are well-documented to alter hair copper levels.[37,38]

Both variation and lack of variation in a range of trace metals, including iron, magnesium, manganese, chromium, copper and zinc, has been reported in hair of differing intrinsic hair colour (black, brown, blond, red) and in comparisons of pigmented and non-pigmented (grey) hair.[61,62,67,68] Within the same individual, manganese levels have been shown to be higher in black hair than in white, whilst zinc and copper levels were reported to be similar, leading researchers to suggest that manganese plays a role in pigmentation.[69]

Differences in calcium, manganese, strontium, iron, copper, selenium and nickel were found between Caucasian, Negroid and Asiatic samples, however, some of this difference may simply be due to differences in hair colour, or indeed to differences in diet.[62,68] Anatomic variation across the body may also be a factor in variable metal concentrations, although again the little available data are contradictory. DeAntonio and colleagues found no overall concentration differences in five metals between scalp and pubic hair from one population, yet there were differences at an individual level.[70,71]

8.4.3 Hair Growth Rates

Hair typically grows at a rate of 1 cm per month,[72] but there are variations in this, for example for hair type and anatomic location on the body, and hormone levels. Where analytical techniques are independent of sample size (for instance with protein isotope ratios), then the rate of hair growth is largely irrelevant. In such situations, the rate of hair growth can be used as a means of calibrating the longitudinal record of diet, and small variations due to season and other physiological factors will merely cause a slight error on the time period calculated. However, where the analytical technique is not independent of sample size, then if hair growth rates are substantially altered, this has major implications for interpretation of data. Where the diet is adequately supplied with macro- and micro-nutrients, hair growth will be unaffected by diet, but in severe cases of malnutrition or specific micronutrient deficiencies (particularly zinc), growth will be substantially slowed.[73,74] This will have the effect of 'elevating' the elemental concentration per unit mass of hair, and can lead to spurious results. Thus, severely nutrient deficient individuals can exhibit similar hair elemental concentrations to those whose diet is nutritionally adequate.[34,46]

8.4.4 Hair Contamination

As an externally growing tissue, hair is subject to contamination, either endogenous, including sweat and sebaceous secretions, or exogenous, such atmospheric dust and cosmetic treatments (shampoos, colouring *etc.*), and this is exacerbated by hair's hygroscopic nature.[75]

The further the hair is from the scalp, the older it is, and the greater exposure time it will have had to contaminants. Such exposure primarily affects metal concentrations, rather than changes in hair protein, but as yet there is no consensus as to universal relationships. Correlations between trace element concentration and distance from the scalp has been observed in some studies and not in others for copper, zinc, chromium, manganese, selenium and calcium.[62,66,76–78]

The proximal hair shaft (*i.e.* closest to the scalp surface) is constantly bathed in sweat and sebum, both of which contain trace metals which adsorb onto and into the hair.[79,80] Specific cosmetic treatments are also known to alter hair shaft composition. Shampooing has been shown to alter zinc, copper, calcium and magnesium concentrations,[81,82] and use of shampoos containing selenium, copper and zinc greatly elevate hair levels of these elements.[83] Colouring, bleaching and permanent wave treatments have all been shown to increase or decrease specific trace elements, including copper and zinc.[76,83]

The hair protein itself, and thus its isotopic values, is less susceptible to contamination. No effects of contamination by sweat, sebum, atmosphere or shampoo have been demonstrated, but strong cosmetic treatments can alter the protein, for example bleaching by hydrogen peroxide (a strong denaturing agent) has been shown to remove certain amino acids, such as cysteine,[84] and to alter the C/N ratio and carbon and nitrogen isotopic values of hair.[59]

What is obvious is that care must be taken to ensure that hair is sampled from consistent locations, with full details of sampling location and the distance from the scalp noted, though this has, unfortunately, rarely been done in most studies (for example reference 68). Cleaning procedures prior to analysis ideally should remove all exogenous contamination, whilst leaving the endogenous signal unaltered. This has been shown to be an extremely complex matter, with evidence to suggest that cleaning may cause more problems than it solves (see later discussion).

8.5 Analysis and Data Interpretation

Analytically, hair is a challenging material, because of the unavoidable issue of contamination after its emergence from the follicle. Ideally, it is important to distinguish between the biogenic signal (*i.e.* that which occurs during hair formation) and that occurring after contamination. This is relatively simple for isotopic analysis of hair protein, since there are diagnostic criteria, independent of the isotopic analysis, to judge whether or not the protein is degraded or altered.[59,60] Having said that, carbon and nitrogen isotopic analysis have been shown to be relatively insensitive to degradation, remaining unaltered when the hair was judged to have been degraded.[85] Whilst this could be seen as problematic, it can also be seen as a measure of the robusticity of the analysis.

Trace metal analysis is however, more prone to difficulties. This is because metals can be lodged or adsorbed within the hair structure, but are themselves not an intrinsic component of it. Thus, without spatial resolution, it is impossible to distinguish between endogenous and exogenous material.

8.5.1 Analytical Methods

Light element stable isotope analysis is performed on small quantities of bulk hair fibre (1–2 mg) using continuous-flow isotope-ratio mass spectrometry.[86] Trace element concentrations are usually high enough for most elements of interest to permit multi-element analyses on such small samples using atomic absorption spectrophotometry (AAS) or inductively-coupled plasma mass spectrometry (ICP-MS).[87] For such bulk analytical methods, there are no independent criteria to judge whether or not the metals measured are endogenous or are from superficial contamination. However, positionally-sensitive analytical methods such as particle induced X-ray emission (PIXE) and secondary ion mass spectrometry (SIMS) can identify the cross-sectional and longitudinal distribution of the metals within the hair shaft, and thus distinguish between surface and internal metals.[88–91]

8.5.2 Pre-treatment of Hair for Analysis

Cleaning of hair fibres for protein isotopic analysis is usually done *via* a simple solvent extraction followed by an aqueous rinse.[59,60,92] It is important to avoid the use of detergents, since these can damage the hair shaft surface.[93]

A wide variety of treatments have been used to clean hair prior to trace element analysis, most using a variant of an aqueous detergent wash, in contrast to the accepted cleaning protocol for protein isotopic analysis. Research has demonstrated that most cleaning processes fail to remove all externally-adsorbed trace elements,[76] and that some techniques have the doubly disastrous effect of removing endogenous but not exogenous metals. In one study the pre-treatment method did not remove the external contaminant, but did leach internally deposited zinc.[75] In the widest ranging study conducted to date, Chittleborough compared 24 different pre-treatment methods and none were completely effective. While he concluded that a 'no-wash' policy was the best option,[79] this recommendation has not been taken up by most researchers. Research into metal distributions within the hair shaft have shown that endogenous adsorption pattern varies between elements, and the effect of various cleaning procedures differs for different elements. These findings therefore suggest that no single pre-treatment is appropriate for all elements.[88]

8.5.3 Data Interpretation

Aside from the technical difficulties of analysis, other hurdles must also be overcome when interpreting the data produced. A major constraint to accurate interpretation of trace metal data is that we have no real idea what constitutes normal ranges for trace element concentrations within different body pools in humans fed on diets of varying macro- and micro-nutrient composition.[2,3] The lack of adequate reference data must be addressed before the technique will achieve wide applicability. Similarly, reference data are also lacking for light elemental isotopes, from potential food sources. As light element stable isotope ratios of the consumer are relative to diet, to gain accurate dietary information from an individual's isotopic values, we must have isotopic data on available foods or at least on consumers of known diets. What little reference data are available are piecemeal, with most researchers measuring control subjects and likely food items as part of their on-going research projects. Lack of reference data is the primary limitation on the application of this technique to forensic cases.

8.6 Applications

The successful study of diet and nutritional status using hair is possible only where there are sufficient controls, either within population studies or where there are changes over time within an individual. Thus, studies of the hair of single individuals for diet or nutritional status have little or no academic or clinical value, since there is no framework for the interpretation of the results.

8.6.1 Dietary Intake and Hair Signals

The majority of studies measuring light isotopic values in hair have attempted to reconstruct dietary intake from analysis of body tissues, with the aim of utilising the technique for forensic and archaeological applications.

Webb and colleagues compared the carbon isotopic values of hair from omnivores and ovo-lacto-vegetarians (individuals eating secondary but not primary animal protein) in Australia and New Zealand,[94] and found little difference between the two. This is to be expected, since all animal-derived proteins from the same individual are isotopically equivalent in both carbon and nitrogen and can all be classed as animal protein for the sake of isotopic dietary reconstruction. However, O'Connell and Hedges compared omnivores, ovo-lacto-vegetarians and vegans,[59] and showed that vegans had lower nitrogen isotopic values. This reflects their lower position on the food chain: since they consume plant and not animal products, their diet has an overall lower nitrogen isotopic value. The higher nitrogen isotopic values of omnivores and ovo-lacto-vegetarians correlated with their increased frequency of animal protein ingestion (Figure 8.1), and so demonstrated the technique's ability to quantify animal protein intake.

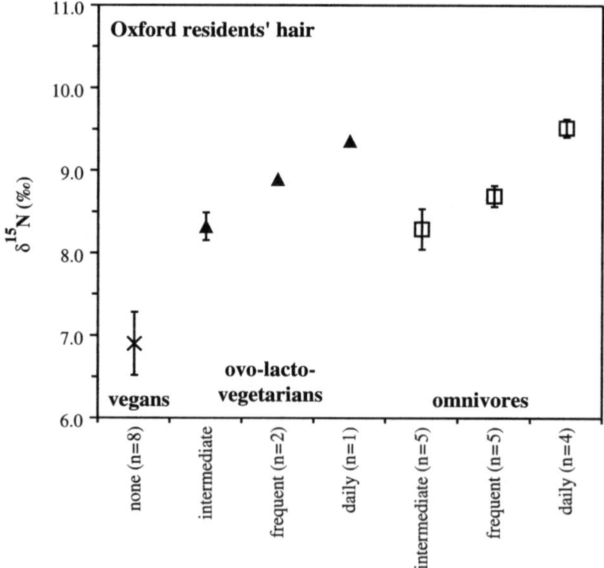

Figure 8.1 *Variation of nitrogen isotopic values of individuals depending on their reported frequency of animal protein consumption[59]*

Minagawa adopted a large-scale approach to reconstruct contemporary Japanese diet from carbon and nitrogen isotopic values found in the hair of Tokyo residents.[60] He modelled the expected dietary input, using a stochastic approach that compared the carbon and nitrogen isotopic values of typical food sources with the carbon and nitrogen isotopic values of the hair. The best-fit dietary pattern, detailing energy/protein proportions, and the relative amounts of plants, animal products and fish consumed, was in good agreement with the diet predicted from national statistical surveys of household consumption. A smaller study by Schoeller and co-workers produced similar results for a group of Chicago residents. In this way, carbon, nitrogen and hydrogen isotopic values of human head hair produced a good estimate of total dietary composition, and carbon isotopic values were particularly useful to assess C_4 dietary inputs, both directly as maize and corn syrup and indirectly *via* consumption of maize-fed animals.[92] Comparing all published studies of this kind, distinct isotopic differences in carbon, nitrogen, sulfur and hydrogen are apparent between populations living in different continents. This is most likely a result of consuming differing diets together with isotopic variation in water sources.[95,96] Based on these isotopic differences, there is increasing interest in the use of this technique in a forensic context, for the identification of an individual's place of origin or of residence. As yet, applications of this technique have been few and there is little published work.[97,98]

In the most comprehensive study so far, the link between hair composition, and food and water intake for four villages in different locations in Papua New Guinea has been studied.[99–103] Here, hair carbon and nitrogen isotopic values reflected the variation in dietary intake of the four villages, with the coastal and riverine villagers consuming more marine and aquatic organisms than the two inland villages. However, there was some considerable discrepancy between the diets as predicted from the isotopic values and those calculated from dietary reports by the villagers. A possible reason is that hair isotopic values average medium- to long-term diet, whereas dietary reports reflect current diet, so a comparison of the two may not allow for seasonal variation. In the comparison of hair trace metal content of hair with that of food and water sources, a strong relationship was found between trace metal profiles in hair and that of the local water source for each village. This correlation was lost, however, when hair was cleaned before analysis. Hongo and colleagues suggested this was caused by contamination of the hair by the water source during everyday activities such as hair washing, and so recalled the merits of Chittleborough's proposed 'no-wash' policy, considering that in this case the information about an individual's geographical origin is as important as their diet.[79,99] High manganese, iron and aluminium was also found in the hair of these subjects, corresponding to high dietary intakes. Conversely, low levels of zinc were found, which could not be solely related to low zinc intake, since high dietary fibre, phytate and iron were reported, all of which could interfere with zinc absorption.

8.6.2 Nutritional Status and Hair Signals

Under-nutrition has long been recorded as having a dramatic effect on both appearance and composition of hair. In early guidelines from the World Health

Organisation on identifying protein-energy malnutrition in children, four of the eleven indicators were to do with hair appearance.[104] Insufficient protein intake results in hairs of reduced diameter but with little other change in growth pattern, whereas caloric deficiency reduces apparent hair growth by increasing the number of follicles in the telogen or resting phase, and can result in hypochromotricia or depigmentation.[104,105] Specifically, protein-energy malnutrition (PEM) results in reduction in hair diameter, as well as changes in pigmentation and growth patterns.[74] Although hair condition and appearance is no longer used as a diagnostic tool of malnutrition, Bradfield and Jelliffe suggest that hypochromotricia can be a useful monitor for PEM during community survey work, since it can provide an indication of nutritional status over the previous year(s).[105] Whilst hair growth and appearance can be within normal ranges on very low protein and energy intakes, crash dieting and its resulting protein and energy inadequacy, can result in substantial hair loss.[106]

Hypochromotricia has also been reported as a side-effect of a number of specific nutritional conditions. Copper and iron deficiencies can cause hypomelanosis.[107] Pernicious anaemia resulting from a deficiency in Vitamin B_{12} can result in the sufferer's hair turning white, which is reversible after vitamin supplementation.[108] Hills records two rare cases of prematurely white-haired individuals whose hair, after diagnosis of coeliac disease in middle age and subsequent avoidance of dietary gluten, reverted to brown over the following few years.[109]

In an study of iron nutritional status and cooking practices in Southern Africa, Baumslag and Petering compared !Kung Bushman women with Bantu and American inner-city mothers.[110] Bushman women, who cook all their food in iron pots, had much higher levels of iron in their hair than the two urban groups, had no incidence of anaemia and their haematological indices showed high intakes of zinc and iron. In contrast the American women were all anaemic and the Bantu women were deficient in iron and folic acid. The authors suggested that the replacement of iron cooking pots with aluminium or enamel pots within urban societies has led to a decrease in iron and zinc intake, and the concomitant health problems.

McKenzie *et al.* studied Polynesian islanders, known to have low zinc and copper intake, but failed to find any correlation between diet and hair metal levels,[39] suggesting that the islanders were nutritionally adequate in these two metals.

One of the most successful applications of hair analysis for assessing nutritional status has been in monitoring selenium levels. Keshan's disease, a myocardial disease resulting in cardiovascular failure with high degree of fatality, was observed to be widely prevalent in certain areas of China in the early and middle twentieth century. A large-scale study by the Keshan Disease Research Group of the Chinese Academy of Sciences showed that residents of Keshan Disease Areas had very low selenium levels in both blood and hair, as well as very low selenium intake.[44] Subsequent work demonstrated that the nutritional deficiency of selenium, whilst not the cause of Keshan's disease, played an important role in the disease aetiology. Supplementation with selenium, as a standard practice, in areas at risk has now greatly reduced the incidence of the disease.

8.6.3 Diseases and Illnesses Associated with Changes in Hair Composition

The measurement of variations in metal composition of hair in disease or illness has, generally, not been used for diagnostic purposes. However, it has been used as a means of investigating the underlying pathomechanisms and causes. Its usefulness and applicability have been assessed for a wide variety of conditions, including metabolic dysfunction of macro- and micro-nutrients, neurological function and genetic metabolic disorders.

Chromium has long been known to be involved in carbohydrate metabolism and chromium deficiency is correlated with impaired glucose tolerance.[45] Hair from juvenile diabetic patients contained lower levels of both chromium and zinc than for age-matched normal controls.[111,112] A large-scale study of over 40,000 patients found significant age-related decreases in the level of chromium in hair, sweat and serum,[65] leading the researchers to suggest that age-related decreases in chromium levels may increase the risk of developing age-related impairment of glucose metabolism, such as in type II diabetes.

Endocrine disorders, such as hyperthyroidism, and hyper- and hypoparathyroidism, as well as disorders of bone metabolism, such as osteomalacia, affect calcium and phosphorus balance within the body. Investigations of women with such conditions indicated that concentrations of various trace elements including calcium and phosphorus in hair were influenced characteristically by these diseases, suggesting that hair analysis could be used as a complementary diagnostic tool.[113]

Ahlskog and colleagues used hair analysis to test whether the high incidence of neurodegenerative disease on the Pacific island of Guam was a result of abnormalities in calcium metabolism, resulting in secondary hyperparathyroidism and subsequent heavy metal absorption into the brain.[114] However, there was no difference in the concentration of heavy metals in hair, nail or serum of patients and controls, thus providing no support for the suggestion of calcium deficiency as a cause or contributing factor in this pathology. High manganese levels in hair have been reported for Aborigines living in Groote-Eylandt in Australia, supporting the view that high manganese intake may be partly responsible for the high prevalence of neurological dysfunction amongst this population.[115,116]

A suggestion that autism may be a result of trace element imbalance has been tested by a comparison of hair trace metal levels. No difference in calcium, magnesium, zinc, copper, lead or cadmium concentrations was found between autistic and non-autistic children in one study,[117] but recent work found that autistic children had lower hair, plasma and erythrocyte zinc levels than normal controls, suggesting that autistic children may have either a metabolic abnormality of zinc, or chronic zinc deficiency.[118]

Two studies examined epileptics in Venezuela and in Pakistan, and showed that these individuals had double the concentration of magnesium in their hair compared to normal individuals. These authors proposed as a diagnostic criterion the ratio of magnesium/zinc in hair, since the ratio was always less than one for non-sufferers, and always greater than one in epileptics, with higher ratios correlating with increased fit frequency.[119,120] For a single individual whose hair

Mg/Zn was 3.35, reduced intake of salt (containing 7% magnesium) reduced the hair Mg/Zn ratio to 1.3 in a short space of time, together with a decrease in fit frequency. The authors postulate that the high excretion of magnesium in hair could be related to anomalies of magnesium uptake or utilisation.

There is now increasing interest in the effect of diet and dietary constituents on disease development. However, the debate centres on how best to assess diet, since dietary recall methods are frequently inaccurate.[121] In a study of patients with Alzheimer's disease, nitrogen isotopic values in hair, used as a proxy for medium-term dietary fish and meat intake, were compared to dietary records and cognitive function.[122] Cognitive function and hair nitrogen isotopic values did not correlate in controls. In patients however, cognitive scores related directly to the reported frequency of eating fish and to hair nitrogen isotopic values, and inversely to the reported frequency of eating beans. The potentially controversial conclusion of this study is that a diet rich in fish may ameliorate Alzheimer's disease, whereas more vegetarian diets may not.

Inborn errors of protein and amino acid metabolism have been reported to cause alteration in the amino acid composition of hair proteins. In phenylketonuria, the excessively high circulating levels of phenylalanine are reflected in abnormally high concentrations of phenylalanine in hair keratin, whereas haemoglobin proteins are unaffected.[123] In patients with homocystinuria, keratin cystine levels were normal but again phenylalanine was elevated relative to controls.[123]

A number of genetic disorders can cause substantial changes to hair growth colour, texture and shape, often resulting from errors in metabolism of trace metals. One such example is Menkes' syndrome, associated with severe copper deficiency.[124] Hair analysis has also been suggested as a diagnostic technique for Wilson's disease – a genetic condition of systemic copper overload – where sufferers have low total serum copper levels, but high levels of circulating non-caeruloplasmin-bound copper causing the metal to accumulate in many tissues. Diagnosis is generally *via* a liver biopsy, as that is the initial concentration site. Using hair as an alternative non-invasive biopsy had been suggested, but results show that copper is not sequestered in hair, as sufferers do not have elevated hair copper concentrations compared to normal controls.[90,125]

Complications in trace metal status as a side effect from ongoing medical procedures has also been reported. Marumo and co-workers compared body burdens of aluminium, calcium and zinc in controls and patients with chronic renal failure.[126] Controls had lower aluminium and calcium concentrations in their hair and blood than did the patients. Instead controls had higher plasma zinc levels, indicating that the use of aluminium-rich and zinc-poor dialysis substitution fluid can have lasting effects on body stores.

8.7 Summary

Hair analysis has not yet proved itself as useful a tool for diet and nutrient assessment as it has for toxicity studies or forensic DNA analysis. However, if one acknowledges the limitations of our current interpretations, these analyses can still yield valuable information. Its advantages, in terms of availability and ease of sampling, ensures that many will continue to investigate how far we can reconstruct an individual's dietary and nutritional history from studying their hair.

8.8 References

1. E.J. Underwood in *Trace Elements in Human and Animal Nutrition*, Academic Press, New York, 1977.
2. G.V. Iyengar, W.E. Kollmer and H.J.M. Bowen in *The elemental composition of human tissues and body fluids: a compilation of values for adults*, Weinheim, New York, Verlag Chemie, 1978.
3. G.V. Iyengar in *Chemical Toxicology and Clinical Chemistry of Metals*, S. S. Brown and J. Savory (ed), Academic Press, London, 1983.
4. S. Seidel, R. Kreutzer, D. Smith, S. McNeel and D. Gilliss, *Journal of American Medical Association*, 2001, **285**, 67.
5. K.M. Hambidge, *American Journal of Clinical Nutrition*, 1982, **36(5)**, 943.
6. S. Barrett, *Jama-Journal of the American Medical Association*, 1985, **254(8)**, 1041.
7. P. Manson and S. Zlotkin, *Canadian Medical Association Journal*, 1985, **133(3)**, 186.
8. L.M. Klevay, B.R. Bistrian, C.R. Fleming and C.G. Neumann, *American Journal of Clinical Nutrition*, 1987, **46(2)**, 233.
9. J. Dormandy, *British Medical Journal*, 1986, **293**, 975.
10. S.H. Ambrose in *Biogeochemical approaches to Paleodietary Analysis*, S. H. Ambrose and M.A. Katzenberg (ed), Kluwer Academic/Plenum, New York, 2000, 243.
11. M.J. DeNiro and S. Epstein, *Geochimica et Cosmochimica Acta*, 1978, **42**, 495.
12. M.J. DeNiro and S. Epstein, *Geochimica et Cosmochimica Acta*, 1981, **45**, 341.
13. A. Nakagawa, A. Kitagawa, M. Asami, K. Nakamura, D.A. Schoeller, R. Slater, M. Minagawa and I.R. Kaplan, *Biomedical Mass Spectrometry*, 1985, **12**,502.
14. M. Sponheimer, T. Robinson, L. Ayliffe, B. Roeder, J. Hammer, B. Passey, A. West, T. Cerling, D. Dearing and J. Ehleringer, *International Journal of Osteoarchaeology*, 2003, **13**, 80.
15. M. Sponheimer, T. Robinson, B. Roeder, L. Ayliffe, B. Passey, T. Cerling, D. Dearing and J. Ehleringer, *Journal of Archaeological Science*, 2003, **30**, 1649.
16. L.L. Tieszen, T.W. Boutton, K.G. Tesdahl and N.A. Slade, *Oecologia*, 1983, **57**, 32.
17. L.L. Tieszen and T. Fagre in *Prehistoric Human Bone – Archaeology at the Molecular Level*, J. B. Lambert and G. Grupe (ed), Springer-Verlag, Berlin. 1993, 121.
18. M.P. Richards, B.T. Fuller, M. Sponheimer, T. Robinson and L. Ayliffe, *International Journal of Osteoarchaeology*, 2003, **13**, 37.
19. L.K. Ayliffe, T.E. Cerling, T. Robinson, A.G. West, M. Sponheimer, B.H. Passey, J. Hammer, B. Roeder, M.D. Dearing and J.R. Ehleringer, *Oecologia*, 2004, **139(1)**, 11.
20. S.H. Ambrose, *Journal of Archaeological Science*, 1991, **18**, 293.

21. S.H. Ambrose and L. Norr in *Prehistoric Human Bone – Archaeology at the Molecular Level*, J.B. Lambert and G. Grupe (ed), Springer Verlag, Berlin, 1993, 1.

22. T.C. O'Connell and R.E.M. Hedges, *Journal of Archaeological Science*, 1999, **26**, 661.

23. T.C. O'Connell, M.A. Healey, R.E.M. Hedges and A.H.W. Simpson, *Journal of Archaeological Science*, 2001, **28**, 1247.

24. S. Deeming and C. Weber, *American Journal of Clinical Nutrition*, 1977, **30**, 2047.

25. S.N. Gershoff, R.B. McGandy, A. Nondasuta, U. Pisolyabutra and P. Tantiwongse, *American Journal of Clinical Nutrition*, 1977, **30**, 868.

26. M. Gallagher, P. Webb, R. Crounse, J. Bray, A. Webb and E. Settle, *Nutrition Research*, 1984, **4**, 577.

27. R.S. Gibson, F. Yeudall, N. Drost, B.M. Mtitimuni and T.R. Cullinan, *Journal of Nutrition*, 2003, **133(11)**, 3992S.

28. R. Bradfield, A. Cordano, J. Baertl and G. Graham, *Lancet*, 1980, **2**, 343.

29. S. Davies, *Science of the Total Environment*, 1985, **42**, 42.

30. J. Reinhold, B. Faradji, P. Abadi and F.I. Beigi, *Journal of Nutrition*, 1976, **106**, 493.

31. C. Hill in *Trace Elements in Human Health and Disease, Volume II*, A. Prasad and D. Oberleas (ed), Academic Press, New York, 1976.

32. S. Skoryna and D. Waldron-Edwards in *Intestinal absorption of metal ions, trace elements and nuclides*, S. Skoryna, and D. Waldron-Edwards (ed), Pergamon Press, New York, 1972.

33. K.M. Hambidge, C. Hambidge, M. Jacobs and J. Balm, *Pediatric Research*, 1972, **6**, 868.

34. J. Erten, A. Arcasoy, A.O. Çavdar and S. Cin, *American Journal of Clinical Nutrition*, 1978, **31**, 1172.

35. J. Dorea and P. Paine, *Human Nutrition: Clinical Nutrition*, 1985, **39C**, 389.

36. L. McBean, M. Mahloudji, J. Reinhold and J. Halstead, *American Journal of Clinical Nutrition*, 1971, **24**, 506.

37. S. Deeming and C. Weber, *American Journal of Clinical Nutrition*, 1978, **31**, 1175.

38. S.C. Vir and A.H.G. Love, *American Journal of Clinical Nutrition*, 1981, **34(8)**, 1479.

39. J. McKenzie, B. Guthrie and I. Prior. *American Journal of Clinical Nutrition*, 1978, **31**, 422.

40. T.B. Haddy, D.M. Czajkanarins, H.H. Skypeck and S.L. White, *Public Health Reports*, 1991, **106(5)**, 557.

41. M. Baer and J.C. King, *American Journal of Clinical Nutrition*, 1984, **39**, 556.

42. J.M. Huddle, R.S. Gibson and T.R. Cullinan, *British Journal of Nutrition*, 1998, **79(3)**, 257.

43. L.M. Klevay, *American Journal of Clinical Nutrition*, 1970, **23**, 1194.

44. G. Yang, J. Chen, Z. Wen, K. Ge, L. Zhu, X. Chen and X. Chen, *Advances in Nutrition Research*, 1984, **6**, 203.

45. K.M. Hambidge, *American Journal of Clinical Nutrition*, 1974, **27**, 505.

46. K.M. Hambidge, *Pediatric Clinics of North America*, 1980, **27(4)**, 855.
47. M. Hambidge, *Journal of Nutrition*, 2003, **133**, 948S.
48. J. Yoshinaga, H. Imai, M. Nakazawa, T. Suzuki and M. Morita, *Science of the Total Environment*, 1990, **99(1–2)**, 125.
49. Y. Muramatsu and R. M. Parr, *Science of the Total Environment*, 1988, **76**, 29.
50. L.M. Klevay, *Nutrition Reports International*, 1981, **23(2)**, 371.
51. R. Jacob, L.M. Klevay and G. Logan, *American Journal of Clinical Nutrition*, 1978, **31**, 477.
52. P.L. Koch, J. Heisinger, C. Moss, R.W. Carlson, M.L. Fogel and A.K. Behrensmeyer, *Science*, 1995, **267**, 1340.
53. T.H.E. Heaton, J.C. Vogel, G. von la Chevallerie and G. Collett, *Nature*, 1986, **322**, 822.
54. D.R. Gröcke, H. Bocherens and A. Mariotti, *Earth and Planetary Science Letters*, 1997, **153**, 279.
55. K.A. Hobson, R.T. Alisauskas and R.G. Clark, *The Condor*, 1993, **95**, 388.
56. H.P. Schwarcz, T.L. Dupras and S.I. Fairgrieve, *Journal of Archaeological Science*, 1999, **26**, 629.
57. C.M. Scrimgeour, S.C. Gordon, L.L. Handley and J.A.T. Woodford, *Isotopes in Environmental and Health Studies*, 1995, **31**, 107.
58. J.C. Vogel, B. Eglington and J.M. Auret, *Nature*, 1990, **346**, 747.
59. T.C. O'Connell and R.E.M. Hedges, *American Journal of Physical Anthropology*, 1999, **108**, 409.
60. M. Minagawa, *Applied Geochemistry*, 1992, **7**, 145.
61. H. Schroeder and A. Nason, *Journal of Investigative Dermatology*, 1969, **53**, 71.
62. H.H. Sky-Peck and B.J. Joseph in *Chemical Toxicology and Clinical Chemistry of Metals*, S. S. Brown and J. Savory (ed), Academic Press, London, 1983, 159.
63. G. Gordon, *Science of the Total Environment*, 1985, **42**, 133.
64. J. Creason, T. Hinners, J. Bumgarner and C. Pinkerton, *Clinical Chemistry*, 1975, **21**, 603.
65. S. Davies, J.M. Howard, A. Hunnisett and M. Howard, *Metabolism-Clinical and Experimental*, 1997, **46**, 469.
66. M. Yukawa, M. Suzukiyasumoto and S. Tanaka, *Science of the Total Environment*, 1984, **38**, 41.
67. E. Eads and A. Lambdin, *Environmental Research*, 1973, **6**, 247.
68. H.H. Sky-Peck, *Clinical Physiology and Biochemistry*, 1990, **8**, 70.
69. O. Guillard, J. Gombert, M. Brierre, D. Reiss and A. Piriou, *Clinical Chemistry*, 1985, **31**, 1251.
70. S.M. DeAntonio, S.A. Katz, D.M. Scheiner and J.D. Wood, *Clinical Chemistry*, 1982, **28**, 2411.
71. R. Thatcher and M. Lester, *Clinical Chemistry*, 1983, **29**, 16912.
72. M. Saitoh, M. Uzuka, M. Sakamoto and T. Kobori in *Hair Growth*, W. Montagna and R.L. Dobson (ed), Pergamon Press, Oxford, 1969, 183.
73. M. Ryder in *The Biology of Hair Growth*, W. Montagna and R. Ellis (ed), Academic Press, New York, 1958.
74. S. Desai, R. Sheth and P. Udani in *Hair research: status and future aspects*, C. Orfanos, W. Montagna and G. Stuttgen (ed), Springer-Verlag, Berlin, 1981, 257.

75. R. Buckley and I. Dreosti, *American Journal of Clinical Nutrition*, 1984, **40**, 840.

76. J. McKenzie, *American Journal of Clinical Nutrition*, 1978, **31**, 470.

77. K. Hambidge, *American Journal of Clinical Nutrition*, 1973, **26**, 1212.

78. I. Obrusnik, J. Gislason, D. Meas, D. McMillan, J. D'Auria and B. Pate, *Journal of Radioanalytical Chemistry*, 1973, **15**, 115.

79. G. Chittleborough, *Science of the Total Environment*, 1980, **14**, 53.

80. H. Hopps, *Science of the Total Environment*, 1977, **7**, 71.

81. D. Hilderbrand and D. White, *Clinical Chemistry*, 1974, **20**, 148.

82. J. Abraham, *Lancet*, 1982, **2**, 554.

83. P. Clanet, S.M. DeAntonio, S. Katz and D.M. Scheiner, *Clinical Chemistry*, 1982, **28**, 2450.

84. N. Baba, Y. Nakayama, F. Nozaki and T. Tamura, *Journal of Hygiene and Chemistry*, 1973, **19**, 47.

85. A.S Wilson and T.C. O'Connell, unpublished data.

86. I.T. Platzner in *Modern Isotope Ratio Mass Spectrometry*, 1997, John Wiley and Sons.

87. I. Rodushkin and M.D. Axelsson, *Science of the Total Environment*, 2000, **250(1–3)**, 83.

88. A. Bos, C. van der Stap, V. Valkovic, R. Vis and H. Verheu, *Science of the Total Environment*, 1985, **42**, 157.

89. J.L. Campbell, S. Faiq, R.S. Gibson, S. Russell and C. Shulte, *Analytical Chemistry*, 1981, **53**, 1249.

90. F. Watt, J.P. Landsberg, J.J. Powell, R.J. Ede, R.P.H. Thompson and J.A. Cargnello, *Analyst*, 1995, **120(3)**, 789.

91. J.A. Cargnello, J.J. Powell, R.P.H. Thompson, P.R. Crocker and F. Watt, *Analyst*, 1995, **120(3)**, 783.

92. D.A. Schoeller, M. Minagawa, R. Slater and I.R. Kaplan, *Ecology of Food and Nutrition*, 1986, **18**, 159.

93. R.E. Taylor, P.E. Hare, C. Prior, D. Kirner, L. Wan and R. Burky, *Radiocarbon*, 1995, **37**, 319.

94. Y. Webb, D.J. Minson and E.A. Dye, *Search*, 1980, **11**, 200.

95. T.C. O'Connell, unpublished data.

96. M.A. Katzenberg and H.R. Krouse, *Canadian Society of Forensic Science Journal*, 1989, **22(1)**, 7.

97. T. Cerling, J. Ehleringer, A. West, E. Stange and J. Dorigan, *Forensic Science International*, 2003, **136**, 172.

98. Z.D. Sharp, V. Atudorei, H.O. Panarello, J. Fernández and C. Douthitt, *Journal of Archaeological Science*, 2003, **30(12)**, 1709.

99. T. Hongo, T. Suzuki, R. Ohtsuka, T. Kawabe, T. Inaoka and T. Akimichi, *Ecology of Food and Nutrition*, 1990, **24(3)**, 167.

100. T. Hongo, T. Suzuki, R. Ohtsuka, T. Kawabe, T. Inaoka and T. Akimichi, *Ecology of Food and Nutrition*, 1989, **23(1)**, 39.

101. T. Hongo, T. Suzuki, R. Ohtsuka, T. Kawabe, T. Inaoka and T. Akimichi, *Ecology of Food and Nutrition*, 1989, **23(4)**, 293.

102. J. Yoshinaga, M. Minagwa, T. Suzuki, R. Ohtsuka, T. Kawabe, T. Hongo, T. Inaoka and T. Akimichi, *Ecology of Food and Nutrition*, 1991, **26**, 17.

103. J. Yoshinaga, M. Minagawa, T. Suzuki, R. Ohtsuka, T. Kawabe, T. Inaoka and T. Akimichi, *American Journal of Physical Anthropology*, 1996, **100**, 23.
104. R. Bradfield in *Hair Research: Status and Future Aspects*, C. Orfanos, W. Montagna and G. Stuttgen (ed), Springer-Verlag, Berlin, 1981, 251.
105. R. Bradfield and D. Jelliffe, *Lancet*, 1974, **1**, 461.
106. D. Goette and R. Odom, *Journal of American Medical Association*, 1976, **235**, 2622.
107. J. Ortonne and J. Thivolet in *Hair Research: Status and Future Aspects*, C. Orfanos, W. Montagna and G. Stuttgen (ed), Springer-Verlag, Berlin, 1981.
108. N. Noppakun and D. Swasdikul, *Archives of Dermatology*, 1986, **122(8)**, 896.
109. L. Hill, *British Medical Journal*, 1980, **281**, 115.
110. N. Baumslag and H. Petering, *Archives of Environmental Health*, 1976, **31**, 254.
111. K.M. Hambidge, D. Rogerson and D. O'Brian, *Diabetes*, 1968, **17**, 517.
112. A. Amador and A. Gonzales, *Lancet*, 1975, **2**, 1146.
113. N. Miekeley, L. Fortes, C. da Silveira and M. Lima, *Journal of Trace Elements in Medicine and Biology*, 2001, **15**, 46.
114. J.E. Ahlskog, S.C. Waring, L.T. Kurland, R.C. Petersen, T.P. Moyer, W.S. Harmsen, D.M. Maraganore, P.C. Obrien, C. Estebansantillan and V. Bush, *Neurology*, 1995, **45(7)**, 1340.
115. J.L. Stauber and T.M. Florence, *Science of the Total Environment*, 1989, **83(1–2)**, 85.
116. J.L. Stauber, T.M. Florence and W.S. Webster, *Neurotoxicology*, 1987, **8(3)**, 431.
117. T.R. Shearer, K. Larson, J. Neuschwander and B. Gedney, *Journal of Autism and Developmental Disorders*, 1982, **12(1)**, 25.
118. O. Yorbik, C. Akay, A. Sayal, A. Cansever, T. Sohmen and A. Cavdar, *Journal of Trace Elements in Experimental Medicine*, 2004, **17**, 101.
119. W. Ashraf, M. Jaffar, D. Mohammed and J. Iqbal, *Science of the Total Environment*, 1995, **164(1)**, 69.
120. K.P. Shrestha and A. Oswaldo, *Science of the Total Environment*, 1987, **67**, 215.
121. J.A. Novotny, W.V. Rumpler, J.T. Judd, H. Riddick, D. Rhodes, M. McDowell and R. Briefel, *Journal of the American Dietetic Association*, 2001, **101(10)**, 1189.
122. J.H. Williams and T.C. O'Connell, *Journals of Gerontology Series A-Biological Sciences and Medical Sciences*, 2002, **57(12)**, M797.
123. M. van Sande, *Archives of Disease in Childhood*, 1970, **45**, 678.
124. Olsen, E.A. in *Disorders of Hair Growth (Diagnosis and Treatment)* E. A. Olsen (ed), McGraw-Hill, New York, 1999.
125. G. Sturniolo, A. Martin, G. Mastropaolo, G. Gurrieri and R. Naccarato, *Lancet*, 1982, **2**, 608.
126. F. Marumo, Y. Tsukamoto, S. Iwanami, T. Kishimoto and S. Yamagami, *Nephron*, 1984, **38**, 267.

Part 3
Chemistry and Toxicology of Personal Hair Care Products

Hair Colorant Chemistry

THOMAS CLAUSEN and WOLFGANG BALZER

9.1 Introduction

Our aging and increasingly ethnically mixed societies continue to spur the usage of hair care products containing hair colorants. While the principal application continues to be coverage of grey hair, another significant factor in this growth is the increasing 'cosmetification' of young adults, males as well as females, in western societies. Both target groups seek to either add colour to their hair (original or different), or to lighten/bleach their hair to generate a preferred lighter colour or permit subsequent addition of colour(s) shades lighter than their natural colour. The hair colour market worldwide is already massive, worth 9 billion euro per year (Source: *Euromonitor*), with some estimates of up to 50% of the adult Western population (men and women) using hair colorants, many to cover grey.

This chapter focuses on the chemistry of cosmetic colorants and agents that modify hair colour, though changes in hair colour can also be due to iatrogenic biochemical processes *i.e.* side effects of medication.[1] For example, the anti-malarial drug chloroquine and the spasmolytic drug mephenesin can lighten hair colour, though original hair colour returns upon discontinuation of drug treatment. Furthermore, melanin formation (*via* melanogenesis) can be inhibited by chemicals such as hydroquinone and its ether derivatives.[2] It should be emphasised that alterations in hair colour by these biochemical interventions should not be regarded as cosmetic treatments because of their mechanism of action and potential side effects (allergy, irritation, hair loss).

9.2 Hair Dyes: Bleaching

Hair can be bleached by both chemical and photochemical (*i.e.* sunlight) oxidative mechanisms. Hair pigments, hair proteins and hair lipids are susceptible to degradation by visible and ultraviolet radiation (UVR). Indeed, lipids of the hair cortex cohesive membrane complexes may be more sensitive to light radiation than to UVR[3] and degradation may cause fractures in the fibre. This may be particularly severe if the hair has already been pre-exposed to chemical bleaching (see below). Sunlight affects the amino acids of the hair fibre's cuticular covering more so than its underlying, and somewhat protected, bulk cortex. Hair proteins appear to absorb light principally between 254 and 350 nm and so most light-associated degradation may occur between these wavelengths[4] affecting cystine, methionine, phenylalanine, tryptophan, histidine, proline and leucine.[5–9] Interestingly, amino acids in darker hair may be more resistant to photodamage than paler shades, suggesting the eumelanin in these hairs provides some protection. Chronic exposure to sunlight is however, likely to cause fibre brittleness, that can be seriously exacerbated if further subjected to even relatively mild chemical bleach conditions (*e.g.* aqueous alkalinity and peroxides). Curiously, the oxidation of hair that lacks melanin (senile white hair) occurs much more slowly than does melanised hair fibres, suggesting that peroxide reacts preferentially with melanin pigment compared to hair proteins. These effects have been subjects for several studies over the past years.[5–9]

Photochemical bleaching of hair dye by sunlight or UV light has also been investigated with regard to bleaching of the natural hair colour,[10–12] although these techniques have never been applied to products.

Chemical bleaching of natural hair colour can be performed either as a single stand-alone cosmetic procedure or as part of oxidative hair colouring. Melanin pigments in hair are irreversibly destroyed, partially or completely, by oxidation leading to bleaching, blonding, and lightening of human hair. The degree of lightening is controlled by the type and the concentration of oxidant, temperature and duration of treatment. The maximum possible extent of hair colour lightening is dependent on hair quality, as melanin degradation also modifies hair keratins.

9.2.1 Chemistry of Bleaching

The structure and formation of melanin are discussed elsewhere[13] and in this volume (see Chapter 3). Human hair melanin is relatively resistant to reduction and, in acid solution, also to oxidation. In alkaline hydrogen peroxide, however, the melanin granules are quickly degraded. It appears that melanin's polymer structure is partly destroyed by oxidation resulting in the formation of carboxyl groups that facilitate dissolution of the granule under alkaline conditions.[13,14] If bleaching is complete, melanin is totally removed from site of the melanin granule.[15] Persulfates are generally added to hydrogen peroxide to intensify the bleaching action. Hydrogen peroxide can also be replaced by adding compounds with urea and carbamine peroxohydrate. Other oxidants have been proposed, *e.g.* sodium carbonate peroxohydrate ($2\ Na_2CO_3 \times 3\ H_2O_2$) or sodium peroxoborate tetrahydrate ($NaBO_3 \times 4\ H_2O$), though these have not been widely accepted and are instead only used in speciality products.

The detrimental reaction between bleach and hair keratin has been well documented.[16] Here disulfide bonds containing cystine are principally targeted which are oxidatively cleaved *via* a series of reaction intermediates[17] to form principally cysteic acid; this results in the higher alkali solubility of bleached hair. Indeed, normal bleaching may break up to 25% of the fibres' disulfide bonds, while bleaching from black to blond may break as many as 50% of these bonds.[18] Further hydrolysis destroys both hydrogen bonds and ionic bonds and the hydroxyl and amino side chains of various keratin amino acids offer further targets for oxidative degradation.[19] Strong bleaches not only alter the hair fibres' chemical properties but also modify its physical properties, *i.e.* creating a fibre with higher extensibility and so reduced mechanical strength with a rough, straw-like feel when dry and a spongy feeling when wet. Increased porosity causes the hair to swell more rapidly when hydrated. Moreover, a permanent wave can be produced using a more dilute solution than normal, and the dyeing behaviour is also modified.

9.2.2 Bleaches

The two components of a hair bleach are the oxidant and a vehicle, the latter used to prevent the product from running out of the hair. For formulation examples see reference[20]. Paste bleaches provide the most intensive bleaching action and these are prepared before use by mixing a 6–12 vol % hydrogen peroxide solution with a bleach powder. Components of bleach powder include a peroxodisulfate, an alkalinising agent, stabilisers, thickeners and other additives. The common peroxodisulfates used include sodium, potassium or ammonium peroxodisulfate. The ammonium salt is most effective when combined with an alkalinising component like sodium carbonate or sodium silicate, which results in the production of ammonia. This base readily penetrates the hair fibre and promotes bleaching. The decomposition of hydrogen peroxide in the alkaline preparation is suppressed by the addition of stabilisers such as sodium pyrophosphate or sodium oxalate[21,22] resulting in an enhancement of the bleaching action.

Similarly, the decomposition of bleach by the presence of trace amounts of heavy metals can be retarded by the addition of complexing agents, also called sequestrants (*e.g.* ethylenediaminetetraacetic acid). Carboxymethyl celluloses, xanthine derivatives and synthetic polymers can be included as thickening additives. Given the toxicity of inhaled peroxodisulfates, the dusting of bleaching powders has to be prevented by the addition of oils to bind fine particles,[23] the granulation or enlargement of powder particle size[24] or by the formulation of water-free oil-containing pastes.[25]

Despite the importance of bleaches in oxidation dyeing, unfortunately very few technical improvements, apart from formulation aspects, have occurred recently in this field. The selective adsorption of metal ions,[26] especially of iron(II) salts,[27] on to melanin has been proposed for gentler bleaching of human hair. However, this process has not been widely accepted to date.

9.3 Hair Dyeing

Despite its very ancient origin, hair-colouring technology has struggled to keep up with demand and recent changes in consumer lifestyles and expectations. Advances in the field of hair dyeing by oxidation have related mainly to the areas of toxicology and application techniques. Hair dyeing includes the use of permanent, semi-permanent and temporary dyes. People may choose to change their hair colour in a non-permanent way and hair colorants are available that can be removed after about 2–10 shampoos.[28] By contrast, temporary hair colorants can be washed out of the hair with a single shampoo. However, a permanent dye lasts through continuous washings as well as permanent waving.

9.3.1 Permanent Hair Dyes

9.3.1.1 Dye Precursors

The requirements of a effective hair colorant include colour range, masking of white hair and permanence. Only oxidation hair dyes can achieve these adequately and satisfactorily and these represent approximately 80% of the hair colour market.[29] The principle of oxidation dyes is over a century old[30] and is used in a similar form for dyeing furs.[12] In hair dyeing, however, toxicological considerations are more important (see Chapters 10 and 11 in this volume). The dyes are produced inside the hairs from colourless precursors by oxidation with hydrogen peroxide in an alkaline solution.[31] The dyes themselves and the kinetics of their formation have been studied intensively.[32]

9.3.1.2 Dye Oxidation

A primary intermediate, an aromatic *para* compound, such as a derivative of 1,4-diaminobenzene or 4-aminophenol, is oxidised to a quinonediimine or quinonemonoimine, respectively. The imine then reacts with a secondary intermediate (coupler), which is a *meta* compound such as a derivative of 1,3-diaminobenzene, 3-aminophenol or resorcinol. A further oxidation step yields an indo, phenazine or oxazine dye that is two to three times the size of the precursors.[12] The mechanism of this coupling reaction is shown in Figure 9.1.

This increase in the size of the dye is largely responsible for fixing the dye in the hair. Altering the choice of substituents can yield a diverse range of colours from yellow to blue[19,33] (Figure 9.2).

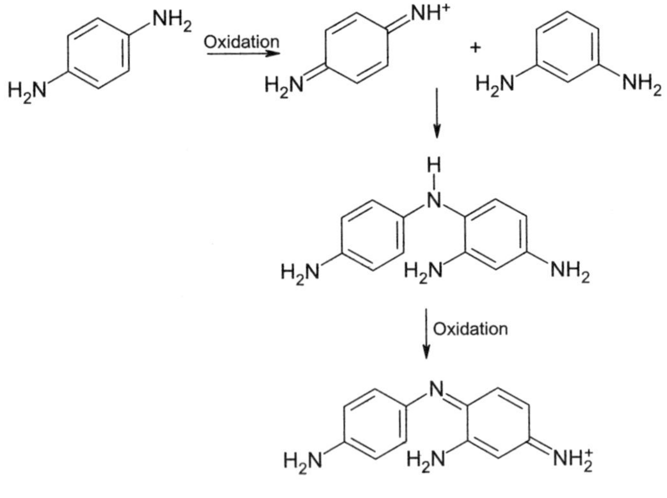

Figure 9.1 *Oxidative formation of hair dyes*

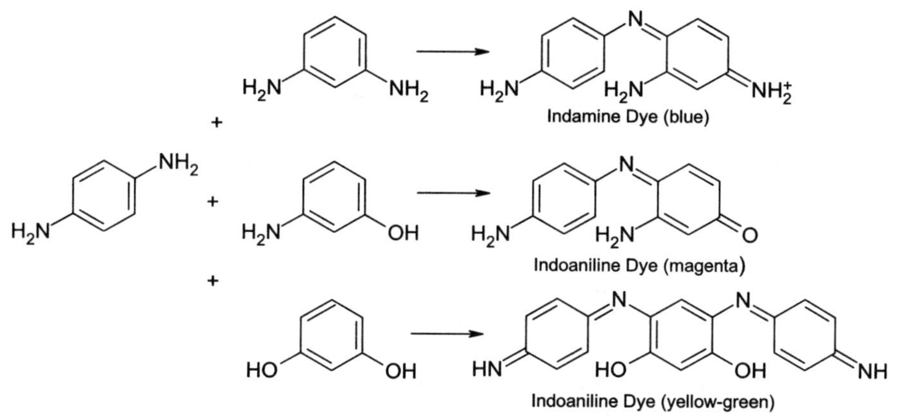

Figure 9.2 *Formation of colour using different couplers*

Thus, the mixture of one or two primary intermediates with various (usually 2–5) couplers, can generate a range of colours in the hair to give the desired shade as an overall impression. Oxidants include hydrogen peroxide (H_2O_2) or its addition compounds and in some special cases atmospheric oxygen can also be used. Enzyme systems have also been proposed.[34]

Ammonia is the alkali of choice to accelerate dyeing and swell the hair fibre, though mono-ethanolamine some sometimes also used. Under these conditions, lightening of the natural melanin pigment depends on the concentration of the oxidant and the alkaliser. It is desirable to lighten the hair before dyeing as this produces a more even background colour whereby the synthetic dye is prevented for giving rise to a range of different shades when applied to 'grey' hair resulting for an admixture of naturally pigmented and white hair. Thus, a simultaneous bleaching process allows hair to be dyed to a lighter color with only small amounts of dye intermediates needed to offset the shade produced by bleaching. The degree of lightening is limited; the colour corresponds to that obtained by bleaching without peroxodisulfate (see above).

9.3.1.3 Primary Intermediates

The main primary intermediate used worldwide in permanent hair dyes for the last 100 years or so has been 1,4-diaminobenzene (*p*-phenylenediamine) or 2,5-diaminotoluene (*p*-toluenediamine). Other primary intermediates include tetra-aminopyrimidine and *N,N*-dialkylated, ring-alkylated or ring-alkoxylated 1,4-diaminobenzene derivatives (*e.g. N,N*-bis(2′-hydroxyethyl)-*p*-phenylene-diamine and 2,5-diamino(hydroxyethylbenzene). 4-Aminophenol and its derivatives, such as 4-amino-3-methylphenol, yield red/violet and orange indoanilines and indophenols with the usual couplers; they are therefore important for red shades, whereas the diaminopyrazoles and their derivatives yield bright red, red-violet and orange colors with a broad range of different couplers.[35]

9.3.1.4 Couplers

Couplers can be classified into three groups: blue, red, and yellow-green couplers, according to the colour obtained with the aromatic 1,4-diamines. The usual blue couplers are 1,3-diaminobenzene derivatives with the ability to couple at the 2,4-positions relative to the amino groups. Examples in use include *N*-hydroxyethyl derivatives[36] and *O*-hydroxyethyl derivatives,[37] dimeric couplers[38] and heterocyclic compounds.[39] The principal red couplers are 3-aminophenol, 5-amino-2-methylphenol and 1-naphthol. Among yellow/green couplers, resorcinol, 4-chlororesorcinol and benzodioxoles should be mentioned, as well as 2-methylresorcinol and its derivatives. The importance of the yellow/green couplers lies in the broadband absorption of the dyes produced, which makes natural–looking hair shades possible. Table 9.1 shows a range of colours obtained by reaction of primary intermediates with different couplers.

Table 9.1 *Combinations of primary intermediates and couplers used in hair colorants*

Primary intermediate	Coupler	Colour on hair
2,5-diaminotoluene	m-phenylenediamine	blue
2,5-diaminotoluene	2,4-diaminophenoxyethanol	violet/blue
2,5-diaminotoluene	m-aminophenol	magenta/brown
2,5-diaminotoluene	5-amino-2-methylphenol	magenta
2,5-diaminotoluene	1-naphthol	purple
2,5-diaminotoluene	resorcinol	greenish brown
N,N-bis-(2-hydroxyethyl)-p-phenylene-diamine	m-phenylenediamine	green/blue
N,N-bis-(2-hydroxyethyl)-p-phenylene-diamine	2,4-diaminophenoxyethanol	green/blue
N,N-bis-(2-hydroxyethyl)-p-phenylene-diamine	m-aminophenol	green/blue
N,N-bis-(2-hydroxyethyl)-p-phenylene-diamine	5-amino-2-methylphenol	violet/blue
N,N-bis-(2-hydroxyethyl)-p-phenylene-diamine	1-naphthol	blue
N,N-bis-(2-hydroxyethyl)-p-phenylene-diamine	resorcinol	greenish brown
4-amino-3-methyl-phenol	m-phenylenediamine	magenta
4-amino-3-methyl-phenol	2,4-diaminophenoxyethanol	magenta
4-amino-3-methyl-phenol	m-aminophenol	pale orange/red
4-amino-3-methyl-phenol	5-amino-2-methylphenol	orange/red
4-amino-3-methyl-phenol	1-naphthol	red
4-amino-3-methyl-phenol	resorcinol	yellow/grey
4,5-diamino-1-methyl-pyrazole	m-phenylenediamine	magenta/brown
4,5-diamino-1-methyl-pyrazole	2,4-diaminophenoxyethanol	purple/brown
4,5-diamino-1-methyl-pyrazole	m-aminophenol	bright orange/red
4,5-diamino-1-methyl-pyrazole	5-amino-2-methylphenol	intense red
4,5-diamino-1-methyl-pyrazole	1-naphthol	purple
4,5-diamino-1-methyl-pyrazole	resorcinol	red

9.3.1.5 Other Dye Intermediates

Besides the aforementioned hair dye precursors, other compounds may be used as dye intermediates *e.g.* 2-aminophenol. The compounds trihydroxybenzenes, dihydroxyanilines, diphenylamines are used chiefly in 'auto-oxidative' systems, *i.e.* in those systems using oxidation with atmospheric oxygen without an added oxidant. The diphenylamines are oxidised to indo dyes – an generic term to include indamine, indoaniline and indophenol dyes. There have been intensive research efforts made, since the first patent appeared,[40] to generate indole precursors that mimic biological melanogenesis in melanocytes by using 5,6-dihydroxyindole or 5,6-dihydroxyindoline. Until now, the impact of these products has been low. This may in part be due to the fact that the latter dyes result in a 'progressive' colouration whereby the colour intensity increases gradually with each treatment. Also the range of available shades is limited.

There are excellent reviews available of the hair colorant patent literature, as well as general progress in the field of hair dyeing generally.[41–43]

Some nitro dyes (see below) are used in oxidation hair dyes as well, especially for the shading of brilliant (mainly red) fashion colours. However, they need to be stable to the oxidative conditions during use, as well to antioxidants needed to stabilise the respective formulas.

Recently some classes of cationic azo-dyes have been found to be stable too under these conditions and add brilliance and shine, mainly for use in fashion shades achieving simultaneously high lift.[44]

9.3.2 Semipermanent and Temporary Hair Dyes

Direct dyes, unlike oxidation dyes, contain a dye (and not its precursor) that when applied to the hair fibre imparts a semi-permanent or temporary colour that lasts for a variable time. Thus, lightening is not possible with these dyes. A review of historical texts reveals that the first known hair dyes were semi-permanent ones. The ancient Egyptians and Romans dyed hair and fingernails with: a) henna containing lawsone (2-hydroxy1,4-naphthoquinone), a red/orange dye; b) walnut shells containing juglone (5-hydroxy1,4-naphthoquinone) which gives a yellow/brown colour and c) indigo. They also employed combinations or mordants with metal salts. To a very limited extent, hair is still dyed with henna, juglone, indigo and extracts such as chamomile (containing apigenin, 4′,5,7-trihydroxyflavone, which gives a yellow colour). The use of these products is low however, because of poor selection of shades, uneven colouring and laborious method of application.[19] In addition natural dyes may be combined with oxidation or direct dyes to achieve a higher colour intensity or stability.[45] Nearly all the chromophore systems common in dye chemistry (nitro, azo, anthraquinone, triphenylmethane, and azomethine) are now currently used.[43,46]

9.3.2.1 Nitro Dyes

Nitro dyes are the most important class of direct hair dyes. They are substituted derivatives of nitrobenzene or nitrodiphenylamine.[12] A spectrum of dyes from

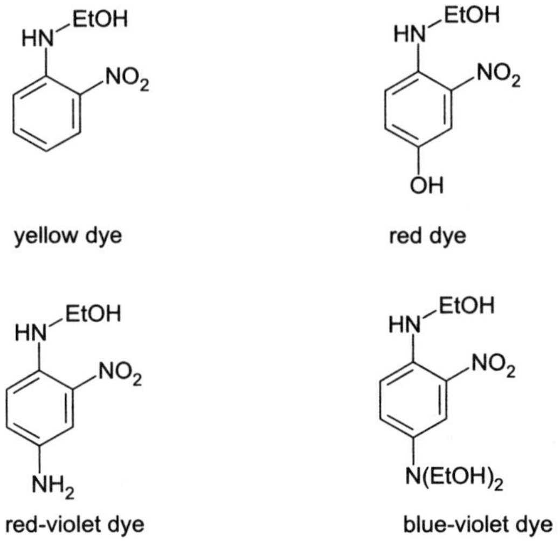

yellow dye

red dye

red-violet dye

blue-violet dye

Figure 9.3 *Influence of substituents on the colour of nitrodyes*

yellow to blue/violet can be prepared by proper selection of donor groups and substitution sites on the benzene ring[19,47] (Figure 9.3).

Nitro dye molecules have some important characteristics, most particularly their small size, that allow them to penetrate into the hair fibre and impart colour throughout the hair or in an annular pattern.[48] In this way, intense colours can be obtained even on hair with a large proportion of white shafts. However, these dyes are not fixed into the hair by size enlargement or ionic bonds (as in oxidation dyes) and so they slowly wash/leach out. Rinse-out rates of these dyes correlate with size of the dye molecules used, but this also depends on the region of the hair fibre. For example, dye is removed most readily from weathered hair fibre tips than the scalp end and so accommodation needs to be built into the formulations to ensure evenness of tones. Furthermore, some dyes, especially small mononuclear dye molecules, may be washed out more readily from previously bleached hair.

Thus, the physical and chemical properties of dyes are very important for even colouring. First, a hair dye should have comparable affinities for the roots, damaged areas and tips. Second, the combined yellow-to-blue dyes used for shading must have similar properties so that colour shifts will not take place when the hair is washed, for example. The importance of these problems is illustrated by the increasing number of patent applications disclosing new, 'custom-tailored' dyes[41] and suitable dye mixtures.[49] The relatively good colourfastness and stability of some nitro dyes allow them to be used in oxidation hair dyes as well, especially for the shading of brilliant (mainly red) fashion colours.

9.3.2.2 Cationic (Basic) Dyes

Methyl violet or methylene blue are used in dye rinses. Cationic azine or azo dyes are also used in tints. Their molecular size and charge prevent them from penetrating the hair, so that the colour pattern in cross-section is annular.[31] The charge on the dyes, however, leads to a relatively stable ionic bonding of the dye to acid groups of the hair and so the colour lasts for several washings.

9.3.2.3 Anionic (Acidic) Dyes

Anionic dyes are usually azo dyes and employed only in special cases. Because the skin is more basic than the hair, these dyes are adsorbed preferentially on the skin, so that contact with the dyeing product leads to severe scalp staining. For this reason, the mainly anionic food colourings are only occasionally used as hair dyes for special applications. Attempts to mitigate these disadvantages by using appropriate vehicles have not been successful.

9.3.2.4 Other Dyes

'Disperse' dyes of the azo and anthraquinone types, on the other hand, are used in hair tints. They also give annular colour patterns, which are relatively durable because of the poor solubility of the dyes in aqueous systems. They can be combined with other classes of dyes.

'Metal complex' dyes are of minor importance.[12] Other methods for dye formation *in situ*[50] or with reactive dyes have not been accepted because of technical drawbacks or of toxicological concerns.

9.3.3 Dyeing with Inorganic Compounds

Historically, salts of metals like silver, bismuth, cobalt, copper, iron, mercury and lead have been used to dye hair. However, only lead salts (*e.g.* lead acetate with sulfur) are still in use. These are often chosen by men with greying hair as the darkening of the hair shaft occurs very gradually, mostly likely *via* the formation of lead-sulfur complexes in the periphery of the hair shaft. Users should be aware that these complexes maybe unstable, especially when exposed to other chemicals and dyes. Permanent hair dyes consisting of metal salts produce a finely dispersed metal deposit on the hair. The addition of sulfur compounds can lead to the formation of metal oxides or sulfides. A metallic grey, brown or black colouration is obtained depending on the concentration and type of metal ion used. Metal salts for permanent hair dyeing can be used in two ways. First, pre-treatment with a tri-hydroxybenzene compound (*e.g.* pyrogallol) is followed by treatment with an ammoniacal silver salt solution. The formation of metallic silver and oxidation products of the trihydroxybenzene derivative permit rapid dyeing. If pretreatment is performed with thiosulfate instead of a benzene derivative, the process yields unstable silver thiosulfate and finally black silver sulfide. In the second method, a metal salt solution (*e.g.* silver, lead or bismuth; less often nickel, cobalt or manganese) is applied, sometimes together with colloidal sulfur. Dyeing is based on the reaction of the metal salts with the added sulfur and the sulfur in the hair keratin, to yield metal sulfides. Moreover, there may also be a deposition of finely divided metals or metal oxides. A 'progressive' colouration is obtained with these products. It should be emphasised that metal salt dyes are no longer important in professional hairdressing, largely because their use may involve toxicological problems. However, the absorption of lead from hair-colorant use has been reported to represent about 0.5% of the lead absorption from average daily environmental lead intake.

Metal salt-based hair dyes cannot be used with permanent-wave neutralisation, bleaching or oxidation dyeing as the metal salts catalyse the decomposition of hydrogen peroxide used as oxidant. When the temperature rises, the liberated oxygen can make the hair brittle. Finally, the selection of colours is very limited with metal dyes and the shades inevitably look metallic and unnatural. Dyes based on silver nitrate are still used.[51] Dyes based on lead and bismuth salts are sold in small quantities as 'colour restorers'.

9.4 Product Forms

The dyes discussed in the above sections can be used in a variety of colouring products.[52] Many proposed formulations for the preparations discussed below can be found in Reutsch *et al.*,[9] while a more application-related review is provided by Williams and Schmidt.[53]

The easiest way to alter hair colour is by using colour rinses, whereby the hair is washed with a dilute aqueous or aqueous/alcoholic dye solution. As the dyes are generally cationic they are adsorbed by the hair surface and can mask the yellow colour that results from bleaching or if the proper shade is selected, can impart an attractive accent to the hair. However, only slight colour change is possible due in large part to the short contact time with the hair. The same holds for coloured or tint setting lotions[12] that contain added dyes, usually cationic, disperse and/or nitro dyes. The coloured film-forming polymers remaining on the hair also contribute to colour impression. Changes are limited to refreshing the colour, masking individual white hairs and producing colour effects as well as silver or grey shades in white or greying hair. The dye molecules used here tend to be larger than those used for semi-permanent colorants and are chosen for their maximum water solubility and minimum penetration, thereby facilitating easy rinse out.[54]

More pronounced colour changes are possible with tints. Tints are formulated with direct dyes especially with nitro dyes allowing intense colours to be obtained. Depending on the desired colouring effect, tints are left on the head for 3–20 min. Therefore, thickened products are generally employed. In foam tints, a surfactant solution is dispensed as foam from an aerosol. The foam is distributed easily and does not drip off the head. Tints can also be thickened with cellulose derivatives, natural mucilage or synthetic polymers. Concentrated solutions with intense colouring action can be obtained by using co-solvents (*e.g.* alcohols and ethylene glycol ethers) and vehicles (*e.g.* urea derivatives or benzyl alcohol).

Permanent hair colours can also be achieved with tint shampoos, but here the shampoo base is adjusted to an alkaline pH and contains oxidation dye inter-mediates. Before application, it is mixed with hydrogen peroxide or a hydrogen peroxide addition compound. In comparison with oxidation hair dyes, tint shampoos employ lower concentrations of base and oxidant. This suppresses the simultaneous bleaching process that occurs during dyeing (see above). As a result, damage to the keratin in hair is diminished, but the uniform colouring action is lost.

For ease of use, oxidation hair dyes are now sold in cream- or gel-based formulas. The oil-in-water emulsions commonly used can be supplemented with auxiliary ingredients, such as polymers to improve combing ability, as well as other conditioning additives.[41] Gel formulations may be based on alcoholic solutions of non-ionic surfactants or fatty acid alkanolamide solutions, which form a gel when mixed with the oxidant. The type (emulsion or gel) and the basic composition of the preparation strongly influence dyeing[55] and so different base formulations with the same dye content yield varying colour depths and shading due to the distribution of the dye between the different phases of the product, interaction with surfactants, and diffusion from the product into the hair. For use, the mixed product is applied to the hair with a brush or applicator. The processing time is about 30 min at room temperature. However, there may be the need to prolong the processing time to 40 min to achieve better lightening of the natural pigment. To prevent 'delayed' oxidation, which may result from residual hydrogen peroxide in the hair, antioxidants such as ascorbic or glyoxylic acid may be added to the conditioning agent. If the conditioner is adjusted to an acidic pH, alkali residues are neutralised and the hair swelling is reversed.

9.5 Dye-removal Preparations

Recently, there has been an increasing desire to be able to completely remove oxidation hair colours. Older, more intense dyes, in particular, were difficult to remove entirely. Now special reductive dye-removal products, mostly based on sulfites, reduce indo dyes to the leuco bases (diphenylamines), which are easier to remove by washing. However, there are some caveats with this approach. For example, the process is not complete and remaining diphenylamine in the hair fibre is reoxidised. An oxidative treatment such as bleaching has, therefore, often been used for the removal of dyes, although certain colours are quite stable to oxidative cleavage. Recently mixtures of reductive agents (*e.g.* ascorbic acid, sulfites and/or cysteic acid) together with organic acids have been described,[56] which may be used for the complete removal of a broad range of oxidation dyes. By contrast, metal salt dyes cannot be removed with these preparations because of the catalytic decomposition of hydrogen peroxide and their resistance to reducing agents.

On the other hand, direct dye tints can be removed by washing with shampoo (the rate of removal depending on the type of dye) or, in some cases, by rubbing the hair with ethanol.

9.6 Testing Hair Dyes

Hair dyes must meet a number of conditions in order to be fit for use. The colour itself can be assessed by colorimetry,[57] though this will depend greatly on the substrate (*i.e.* the hair type) used for the measurements. Perhaps surprisingly, studies are not usually conducted using test subjects as there are difficulties relating to uneven natural hair colour and the background colour, (*i.e.* of the scalp). Scalp hair tresses are difficult to prepare to constant high quality levels. While measurements on wool cloth can give reproducible results, shades generated using oxidation dyes are not identical to those produced on human hair. Colorimetric methods are therefore useful only for comparative measurements on the same object, for example, in light-fastness tests. The fastness of the colour produced in hair does not need to be as great as that required for textiles, in large part because hair must be re-dyed after four to six weeks due to growth. However, colour stability is still a concern with many dyes especially the indo dyes. Some of these fade even in the absence of light while others exhibit poor light,[58] sweat, or acid fastness.[59–61]

9.7 Toxicology

Cosmetic products are used extensively by most humans, at least in the economically developed countries and so new dyes and formulations must be tested for toxicological acceptability.[12] Moreover, many hair cosmetics are used daily or frequently over long periods of time and so consumer exposure to their ingredients may be considerable. Current legislation (see Chapter 13 elsewhere in this volume) defines far reaching safety requirements for cosmetic products. Safety requirements from a regulatory point of view and consumer expectations are adequately summarised by the corresponding text within the EU Cosmetics Directive: 'A cosmetic product put on the market within the community must not cause damage to human health when applied under normal or reasonably foreseeable conditions of use...'[62]

The history of cosmetology, past experience with the safety of cosmetic products and in particular statistics published by intoxication advisory centres show that the cosmetic industry in general has met these criteria.[63] This is achieved by continuously evolving safety assessment strategies, which take into account the actual state of the art in safety or toxicity testing.

9.8 References

1. F. Herrmann, H. Ippen, H. Schaefer and G. Stüttgen in *Biochemie der Haut*, Thieme Verlag, Stuttgart, 1973, 100.
2. S.S. Bleehen, M.A. Pathak, Y. Hori and T.B. Fitzpatrick, *J. Invest. Dermatol.*, 1968, **50**, 103.
3. E. Hoting and M. Zimmermann, *J. Soc. Cosmet. Chem.*, 1997, **48**, 79.
4. J. Arnaud, *Int. J. Cosmet. Sci.*, 1984, **6**, 71.
5. C. Dubief, *Cosmet. Toiletries*, 1992, **107**(10), 95.
6. S. Kanetaka, K. Tomizawa, H. Iyo and Y. Nakamura, in *Preprints Platform Pres.*, **Vol. 3**, IFSCC Intern. Congress, Yokohama, 1992, 1059.
7. S.B. Ruetsch, Y. Kamath and H.D. Weigmann, *J. Cosmet. Sci.*, 2000, **51**, 103.

8. L.J. Wolfram and L. Albrecht, *J. Soc. Cosmet. Chem.*, 1987, **38**, 179.

9. T. Takahashi and K. Nakamura, *J. Cosmet. Sci.*, 2004, **55**, 291.

10. Clairol, US 4 792 341, 1986 (S.D. Kozikowski, J. Menkart and L.J. Wolfram).

11. Wella, EP 451 261, 1990 (R.E. Godfrey, T. Clausen, W.R. Balzer, G. Mahal and R. Rau); L'Oréal, EP 682 937, 1995 (C. Caisey, C. Monnais and H. Samain).

12. *Ullmann's Encyclopedia of Industrial Chemistry*, 6th edn, 2000, electronic release.

13. C.R. Robbins, *Chemical and Physical Behaviour of Human Hair*, 4th edn, Springer-Verlag, New York, 2002.

14. C.R. Robbins, *J. Soc. Cosmet. Chem.*, 1971, **22**, 339.

15. I.J. Kaplin, A. Schwan and H. Zahn, *Cosmet. Toiletries.*, 1982, **97**, 22.

16. J. Cegarra and J. Gacen, *Wool Sci. Rev.*, 1982, **59**, 1.

17. H. Zahn, *J. Soc. Cosmet. Chem.*, 1966, **17**, 687.

18. C. Robbins and C. Kelly, *J. Soc. Cosmet. Chem.*, 1969, **20**, 555.

19. C. Zviak, in *The Science of Hair Care*, Marcel Dekker, New York, 1986.

20. K. Schrader, in *Grundlagen und Rezepturen der Kosmetika*, Hüthig Verlag, Heidelberg, 1979, 575.

21. J. Cegarra, J. Ribé and J. Gacén, *J. Soc. Dyers Colour.*, 1964, **80**, 123.

22. V. Böllert and L. Eckert, *J. Soc. Cosmet. Chem.*, 1968, **19**, 275.

23. Goldwell, EP-A 560 088, 1993 (H. Lorenz and F. Kufner); Goldwell, EP684 036, 1994 (H. Lorenz and W. Eberling); Clairol US 5 698 186, 1997 (G. Weeks).

24. L'Oréal, DE 2 023 922, 1970 (F.A. Vorsatz and A.F. Risch); Wella, US 5 279 313, 1991 (T. Clausen and W.R. Balzer); Wella, EP-A 650 719, 1994 (T. Clausen, W.R. Balzer, V. Port and J. Kujawa); L'Oréal, EP 619 114, 1994 (J.-M. Millequand, C. Tricaud and A. Gaboriaud).

25. Schwarzkopf, DE 3 844 956, 1988 (T. Oelschläger and W. Wolff); Wella, DE 195 45 853, 1997 (M. Schmitt, H. Goettmann, W.R. Balzer and H. Schiemann); Wella, DE 197 23 538, 1998 (M. Schmitt, U. Lenz and W.R. Balzer).

26. A. Bereck, H. Zahn and S. Schwarz, *TPI Text. Prax. Int.*, 1982, **37**, 621.

27. Shiseido, KK and Kokai, JP 5 4129-134, 1978 (S. Kubo and I. Yamada); Deutsches Wollforschungsinstitut an der RWTH Aachen, DE 3 149 978, 1981 (A. Bereck).

28. J. Corbett, *Cosmet. Toiletries*, 1973, **24**, 103.

29. J.S. Anderson, *J. Soc. Dyers. Col.*, 2000, **116**, 193.

30. L'Oréal, FR 158 558, 1883 (H. Monnet).

31. H. Wilmsmann, *J. Soc. Cosmet. Chem.*, 1961, **12**, 490.

32. J.F. Corbett, *J. Soc. Cosmet Chem.*, 1979, **30**, 191 and references cited there in.

33. J.F. Corbett, in *The Chemistry of Synthetic Dyes*, K. Venkataraman (ed), Vol 5, Academic Press, New York, London, 1971, 475.

34. Kyowa Hakko Kogyo Kabushika Kaisha, EP 310 675, 1988 (Y. Tsujino, Y. Yokoo, K. Sakato and H. Hagino); Wella, DE 196 10 392, 1997 (M. Kunz and D. LeCruer); Novo Nordisk, WO 95/33 836, 1995 (R.M. Berka, S.H. Brown, F. Xu, P. Schneider, D.A. Aaslyng and K.M. Oxemboell).

35. Wella, DE 38 43 892, 1990 (T. Clausen, U. Kern and H. Neunhoeffer); Wella, DE 42 34 887, 1994 (H. Neunhoeffer, S. Gerstung T. Clausen and W.R. Balzer); L'Oréal, FR 9 505 422, 1995 (L. Vidal, A. Burande, G. Malle and M. Hocquaux).

36. I.G. Farben, DE 559 725, 1930 (E. Lehmann).

37. L'Oréal, FR 2 362 116, 1976 (A. Bugaut and J.J. Vandebossche).

38. Henkel, DE 2 852 156, 1978 (D. Rose, P. Busch and E. Lieske); DE-OS 3 235 615, 1982 (D. Rose and E. Lieske).

39. Henkel, EP 106 987,1983 (N. Maak, P. Flemming and D. Schrader).

40. L'Oréal, US 2 934 396, 1958 (R. Charle and C. Pigerol); Henkel, DE 4 016 177, 1991 (G. Konrad, I. Matzik and E. Lieske).

41. J.F. Corbett, *Rev. Prog. Color. Relat. Top.*, 1973, **4**, 3; 1985, **15**, 52.

42. J.C. Johnson, *Hair Dyes*, Noyes Data Corp., Park Ridge, N.J. 1973.

43. J.F. Corbett, *Cosmet. Toiletries*, 1991, **106**, 53.

44. Ciba-Geigy, EP 0 681 464, 1994 (P. Möckli); L'Oréal, EP 0 810 851, 1995 (H. Samin).

45. Schwarzkopf, DE 196 07 220, 1997 (M. Akram, W. Wolff, S. Schlagenhoff, S. Schwartz and A. Kleen); Wella EP 898 954, 1998 (A. Sallwey, M. Schmitt and U. Lenz).

46. J.B. Wilkinson and R.J. Moore, in *Harry's Cosmeticology*, George Golswin, London, 7th edn, 1982.

47. J.F. Corbett, *J. Soc. Dyers Colour.*, 1967, **83**, 273.

48. S.K. Han, Y.K. Kamath and H.-D. Weigmann, *J. Soc. Cosmet. Chem.*, 1985, **36**, 1.

49. L'Oreal, DE 3 131366, 1981 (R. de la Mettrie and P. Canivet).

50. Ciba-Geigy, DE 2 807 780, 1978 (A. Bühler, A. Fasciati and W. Hungerbühler).

51. F.E. Wall, in *Cosmetics, Science and Technology*, 2nd edn, **Vol. 2**, M.S. Balsam and E. Sagarin (ed), Wiley-Interscience, New York 1972, 328.

52. P. Greß, D. Hoch, M. Schmock and D. Wanke *Das, Färben des Haares*, Wella AG, Darmstadt, 1984.

53. D.F. Williams and W.H. Schmitt (ed), in *Chemistry and Technology of the Cosmetics and Toiletries Industry,* 2nd edn, Blackie Academic & Professional (Chapman and Hall), London, 1996.

54. F.E. Wall, in *Cosmetics Science and Technology*, E. Sagarin (ed), Interscience, New York, 1957, 486.

55. R.L. Goldemberg and H.H. Tucker, *J. Soc. Cosmet. Chem.*, 1968, **19**, 423; K. Schrader, *Parfuem. Kosmet.*, 1982, **63**, 649.

56. Wella, DE 196 47 493 1998 (M. Kunz and D. LeCruer); Wella, DE 196 47 494, 1998 (M. Kunz and D. LeCruer); Clairol, US 5 982 933, 1998 (G. Wis-Surel, A. Mayer and I.Tsivim).

57. R. Feinland and W. Vaniotis, *Cosmet. Toiletries.*, 1986, **101**, 63.

58. H.H. Tucker, *J. Soc. Cosmet. Chem.*, 1967, **18**, 609.

59. I. Schwartz, J. Kravitz and A. D'Angelo, *Cosmet. Toiletries*, 1979, **94**, 47.

60. J.F. Corbett, *J. Soc. Cosmet. Chem.*, 1984, **35**, 297.

61. M.Y.M. Wong, *J. Soc. Cosmet. Chem.*, 1972, **23**, 165.

62. Council Directive of 27 July 1976 on the approximation of the laws of the Member States relating to cosmetic products (76/768/EEC).

63. A. Hahn, H. Michalak, K. Noack and G. Heinemeyer, Ärztliche Mitteilungen bei Vergiftungen nach § 16e Chemikaliengesetz (Zeitraum 1990–1995), Zweiter Bericht der *Dokumentations- und Bewertungsstelle für Vergiftungen im Bundesinstitut für gesundheitlichen Verbraucherschutz und Veterinärmedizin*; A.J. Cohen and F.J. Roe, *Food Chem. Toxicol.*, 1991, **29**, 485.

Hair Dyes And Skin Allergy

G. FRANK GERBERICK and CINDY A. RYAN

10.1 Introduction

Hair colouring has been practiced for many years and today millions of consumers use hair dyes. It is well known that ingredients of oxidative and direct hair dyes possess an allergenic potential. For example, *p*-phenylenediamine (PPD) and derivatives of PPD are contact allergens as evidenced by laboratory animal studies and human studies as well as clinical experience. In the following chapter we highlight some of the key studies relating hair dyes and their role in skin sensitisation and allergic contact dermatitis. Considerations such as skin penetration, metabolism, protein reactivity and the immune response are discussed. Finally, many of the animal and human studies available on hair dyes are reviewed, including diagnostic patch test studies, to help position the relationship between hair dye ingredients and skin allergy.

10.1.1 Biology of Skin Sensitisation and Allergic Contact Dermatitis

Allergic contact dermatitis (ACD) is the clinical manifestation or 'disease state' resulting from skin sensitisation, a type IV, cell-mediated (delayed type) hypersensitivity response. The biological process of skin sensitisation occurs in two distinct phases. The first phase, called the induction or sensitisation phase, begins with the chemical hapten penetrating into the skin. A number of factors influence a chemical's ability to gain access into the viable epidermis, including molecular weight and lipophilicity (*i.e.* lipid solubility).[1] The majority of known chemical contact allergens have a molecular weight below 500 dalton.[2] Once in the skin, the chemical hapten reacts with a carrier protein or peptide to form a complete antigen.[3] In some instances, the chemical may require biotransformation, *e.g.* undergo metabolic activation, in order to become protein reactive.[1] The hapten-modified proteins/peptides are then taken up and processed by Langerhans cells (LC), the primary antigen presenting cells (APC) in the epidermis. The role of the LC is to recognise, internalise and process the antigen encountered in the skin and then transport it, *via* the afferent lymphatics, to draining regional lymph nodes where it will present the antigen to T-lymphocytes. In order to become fully activated and migrate out of the epidermis, LC require not only the interaction with the hapten-modified protein but also signals provided by cytokines produced by epidermal keratinocytes.[4] It is believed that some degree of dermal inflammation is necessary for sensitisation to develop as it enhances keratinocyte cytokine production.[5] This inflammation may be the result of an irritant response induced by the allergenic chemical itself or some other concurrent trauma such as physical insult.[5,6] When LC arrive in the lymph node they present the antigen to T-lymphocytes. Upon recognition of the allergen, the T-lymphocytes become activated and proliferate, resulting in an expanded population of antigen-specific memory T-cells, which are then disseminated throughout the body.[4]

The second phase of skin sensitisation, called elicitation, occurs upon a subsequent encounter with the inducing hapten. As in the induction phase, the chemical enters the epidermis and is processed by LC. But in the elicitation phase, the protein-bound hapten is presented by the LC to memory T-cells that have percolated into the skin. Upon recognition of the antigen, the memory T-cells become activated and produce a number of pro-inflammatory cytokines which trigger an inflammatory response, resulting in the clinical signs associated with an allergic skin response: erythema, edema, pruritis and vesiculation.[7]

10.1.2 Skin Penetration

For sensitisation to occur, the chemical must have physico-chemical properties that allow it to pass through the stratum corneum and gain access to the viable epidermis. Two of the properties which can influence skin penetration are size (molecular weight) and lipophilicity (lipid solubility).[1] Most oxidative dyes and couplers and direct dyes have molecular weights under 300 dalton. Some of the temporary dyes which deposit directly on the hair surface have molecular weights over 500 and are reported to cause few skin problems.[8] The octanol-water partition coefficient, log $P_{o/w}$ or log K_{ow}, is often used to describe the lipid solubility of a chemical. The higher the log K_{ow}, the more lipophilic the chemical. Molecules with a log K_{ow} close to 1.0 tend to have the highest skin absorption because they must be able to pass into both the lipophilic stratum corneum and the aqueous environment of the epidermis and dermis. Work conducted on several oxidative and direct dyes found no relationship between skin penetration and the octanol/water coefficient.[9] It has been suggested that the use of partition coefficients to predict skin absorption of hair dyes may be confounded by their ability to bind to skin proteins.[10] Other factors such as dose (concentration), vehicle or formulation, and contact time, are known to influence skin penetration.

The anatomical region and state of the skin barrier, (*e.g.* hydration, damage, occlusion, irritation) also influence percutaneous absorption. It is estimated that the scalp, jaw and forehead are about four times more permeable than the palm of the hand.[1] However, during actual use, the hair acts as a competitive absorptive surface for the dye, thus reducing the amount of dye available for skin penetration. In a series of studies conducted in rats, Howes and Black[11] demonstrated that the absorption of 2-nitro-*p*-phenylenediamine (2NPPD) by skin with hair was approximately half that of skin which had been clipped.

The percutaneous absorption of hair dyes has been examined using both *in vitro*[12–18] and *in vivo* methods.[9,18–21] The *in vitro* assessment of hair dye skin penetration has been conducted primarily with human (cadaver or cosmetic surgery) or pig skin. For *in vitro* studies, the amount of dye that is considered to be bioavailable is defined as the amount which has penetrated and/or remains in the exposed skin. However, because hair dyes have a high affinity for keratin, they also bind to the uppermost layers of the stratum corneum. Since this layer consists of only dead cells with no access to the circulatory system, some *in vitro* methods exclude the amount of chemical adsorbed by the stratum corneum when determining bioavailable quantities. Data from *in vitro* work using pig skin has demonstrated that most of the applied dose (>90%) is not bioavailable; it is predominately rinsed off (85–89%) and about 3% is adsorbed by the stratum corneum.[17] With the stratum corneum excluded from analysis, the bioavailability of two representative oxidative hair dyes in representative standard cream formulations, PPD and bis-(5-amino-1-hydroxyphenyl)-methane, in the presence of hydrogen peroxide, was found to be less than 1% of the topically applied dose.[15,17] Similar skin penetration rates of 0.1% and 0.2% of the applied dose have been observed for PPD in studies using human skin.[22] *In vitro* studies with the non-oxidative dye 2NPPD found that approximately 9% of the applied dose was absorbed, with approximately 3% remaining in the skin (including the stratum corneum) and 6% passing through the tissue and into the receptor fluid.[16] Overall, data from *in vitro* studies suggest that only a small percentage of the applied amount of hair dye is capable of skin absorption.

Some of the more revealing data have been derived from studies conducted with human volunteers under actual hair dye use conditions.[9,18,20] In those studies ^{14}C-labelled hair dyes, incorporated into commercially available hair dye products, were applied to the hair then rinsed out. The level of radioactivity was measured in the rinse water, in digested and decolourised hair samples and in the subjects' urine and/or faeces collected for up to 120 to 144 hours following dye application. Data from such studies show that little PPD is absorbed during the hair dying process. Maibach and Wolfram[20] have reported absorption ranges of 0.07–0.21% of the applied dose and Goetz *et al.*[21] reported similar values of 0.04–0.25%.[10] The ranges have also been confirmed recently by Heuber-Becker *et al.*,[18] who reported absorption ranges of 0.32–1.02% of the applied dose. Skin penetrations of the oxidative couplers, resorcinol and 2,4-diaminoanisole, were reported to be 0.076% and 0.022% of the applied dose, respectively.[9] In the same study, slightly higher percentages of the applied dose, 0.143%, 0.235% and 0.151% were observed with the non-oxidative dyes, 2NPPD, 4-amino-2-nitrophenol and HC blue 1, respectively. Thus, it appears that in test methods that simulate use conditions, the skin penetration of hair dyes, both permanent (oxidative) and semi-permanent (non-oxidative) does not exceed 1% of the applied dose.[9] It is not surprising that only a small percentage of the applied dose would be available to penetrate the skin as most of the hair dye is rinsed off and, depending on the dye itself, another 5.9 to 20.3% is taken up by the hair.[10] While it would appear that the level of hair-dye actives present in the epidermis are low, they are obviously at sufficient levels to initiate an immune response *via* hapten interaction with LC.

10.1.3 Protein Reactivity and Skin Metabolism

While skin penetration is the first required event in the induction of sensitisation, not all chemicals that gain access to the epidermis are contact allergens. The ability to react with protein is another key characteristic of chemicals that function as contact sensitisers. The correlation of protein reactivity with skin sensitisation potential is well established.[23,24] Thus, if a chemical is capable of reacting with protein either directly or after appropriate biotransformation, then it has the potential to act as a contact allergen. The majority of chemical allergens has electrophilic properties and is able to react with various nucleophiles to form covalent bonds. In proteins, the side chains of many amino acids contain electron-rich groups, nucleophiles, capable of reacting with electrophilic allergens. Lysine and cysteine are those most often cited, but other amino acids containing nucleophilic heteroatoms, such as histidine, methionine, and tyrosine can also react with electrophiles.[23–25] Thus, electrophilic allergens have the capability to react with nucleophilic amino acids in proteins, forming extremely stable covalent bonds, and therefore are involved in the triggering of skin sensitisation responses. However, it is well known that some chemical allergens are prohaptens, and as such require biotransformation prior to initiating a skin sensitisation response *in vivo*.[1,26]

Aromatic amine dyes, such as PPD, are prohaptens that require conversion to haptens, although the exact mechanism of how this takes place is still under investigation. It was originally suggested these molecules (*e.g.* PPD) are converted to benzoquinones *via* protonated intermediates.[23] However, others believe that benzoquinone is neither the only, nor even the major, intermediate in cross-reactivity of *p*-amino groups.[27,28] Patch test studies have shown that less than half of PPD-sensitised patients cross-react with benzoquinone.[28,29] None of the PPD-sensitised patients were reported to cross-react with hydroquinone which also may be oxidized to benzoquinone. Similar cross-reactivity results with PPD and hydroquinone have been reported in the guinea pig.[27]

It has also been postulated that PPD sensitisation could result from the formation of reactive semiquinone imine radicals and be influenced by oxidative stress mechanisms.[30,31] PPD is capable of being oxidised to benzoquinone diimine, which, in turn may form the trinuclear dye, Bandrowski's base, one molecule proposed as the active metabolite responsible for sensitisation.[29,32,33] However, it has been demonstrated that PPD itself can be recognised by sensitised T-lymphocytes through a processing-independent pathway (*i.e.* T-cells were able to respond to PPD presented by 'fixed' APC), whereas its autooxidation product, Bandrowski's Base, required processing (*i.e.* viable APC) and possible metabolism to stimulate T-lymphocytes.[33,34] It has also been hypothesised that substitution at the *para* position of an aromatic amine compound can influence chemical reactivity, with NH_2, OH, and CH_3 being activating while NO_3, SO_3 and COOH are deactivating.[1] Of course, differences in sensitisation potential could also be influenced by other factors, including percutaneous penetration.

Studies have been performed to determine whether or not a correlation exists between PPD sensitisation in humans and the individuals' ability to potentially detoxify PPD in the skin by acetylation *via* the enzyme *N*-aceytyltransferase (NAT).[35–38] Kawakubo *et al.* [35] reported that there were similar numbers of slow and rapid acetylator phenotypes (using caffeine metabolism) in PPD-positive patients, but the non-PPD-sensitive patients were predominately rapid acetylators. In a comparative study, Schnuch *et al.*[36] observed a similar number of slow and rapid acetylators in PPD-sensitised subjects, but found a predominance of slow acetylator phenotypes in the non-allergic control group. More recently, Kawakubo *et al.*[37] demonstrated that PPD can be acetylated in human skin cytosols and keratinocytes with the presumably responsible enzyme being NAT1, one of two identified NAT isoenzymes. Investigators to date feel that the relationship of NAT polymorphism and the likelihood of becoming sensitised to PPD demonstrates at least a trend that slow-acetylators possess an increased risk to develop contact dermatitis; however, these data are not statistically significant.[39]

10.2 Predictive Testing of Hair Dyes for Skin Sensitisation

10.2.1 Animal Models

Historically, the sensitisation potential of hair dye ingredients has been examined largely in guinea pig assays. While a variety of such methodologies exist, this discussion will focus on the guinea pig maximisation test (GPMT) developed by Magnusson and Kligman[40,41] and the occluded patch test of Buehler[42,43] as they are perhaps the most well known of the guinea pig methods. More recently, the murine local lymph node assay (LLNA) has become the preferred method for assessing skin sensitisation potential. Unlike the traditional guinea pig methods, which are based on the visual assessment of erythema produced upon the elicitation of a contact allergic response in previously sensitised animals, the LLNA identifies potential skin-sensitising chemicals as a function of the lymphocyte proliferation response that is associated with the induction phase of contact sensitisation.[44] A substance is classified as a skin sensitiser if at one or more test concentrations it induces a three-fold or greater increase in lymph node cell proliferative activity compared with concurrent vehicle-treated controls. That is, sensitising chemicals by definition induce a stimulation index (SI) of three or more compared with vehicle controls.

P-Phenylenediamine (PPD), the major primary intermediate, has been studied widely for both its potential to induce sensitisation[41,45–49] and to cross-react with other hair dye materials.[27,50,51] In addition, due to the frequency and vigour of the allergic response to PPD, it has often been used as a benchmark contact allergen for evaluating the sensitivity and robustness of animal test methods.[45,48] PPD has been shown to be a strong sensitiser in a variety of guinea pig test methods, including the GPMT and the Buehler test, and in the LLNA. In the GPMT, challenge concentrations as low as 0.01% have elicited positive responses in over 80% of previously sensitised animals (Table 10.1). Sensitisation rates of 80% and greater are also seen with the Buehler Test, which does not employ adjuvant and relies

Table 10.1 *Guinea pig test results with hair dye chemicals*

Chemical name [CAS No.]	Test type	II[a] Concentration/ vehicle[b]	IP[a] Concentration/ vehicle	CP[a] Concentration/ vehicle	Results[c]	Reference
4-Amino-2-hydroxytoluene [2835–95–2]	GPMT	1% pg	25% pg	25% pg	1/10	99
4-Aminophenol (PAP) [123–30–8]	GMPT	0.1% oo	1% pet	0.1% A:dH$_2$O 0.01% 0.001%	5/6 2/6 0/6	51
3-Aminophenol (MAP) [591–27–5]	GPMT	1% sal.	10% A/S	5% A/S	100%	53
2-Methylresorcinol [608–25–3]	GPMT	1% aq	5% pg	5% pg	0/20	56
	GPMT	1% aq	25% pg	25% pg	0/12	56
1-Naphthol [90–15–3]	GPMT	0.1% aq	0.1% aq	0.1% aq 0.05% aq	0/19 0/19	55
p-Phenylenediamine (PPD) [106–50–3]	GPMT	0.25% sal	0.5%	0.5%	9/10	45
	GPMT	0.5% sal	10% pet	1% pet	20/20	46
	GPMT	0.25% sal	0.5% sal	0.5% sal	9/9	27
	GPMT	0.1% sal	1% pet	0.1% acetone 0.01% 0.001%	6/6 5/6 0/6	51
	GPMT	10% aq	10% aq	0.5% aq	16/20	100
	Buehler	not applicable	10%	5% 1% 0.2%	9/10 1/10 0/10	49
	Buehler	not applicable	2%	10% 8%	9/10 8/10	49
Toluene-2,5-diamine sulfate (PTD) [6369–59–1]	GPMT	0.1% sal	1% pet	0.1% A:dH$_2$O 0.01% 0.001%	6/6 5/6 0/6	51

[a] II injection induction; IP induction patch; CP challenge patch

[b] Vehicle abbreviations: oo olive oil; A/S acetone/saline; pet petrolatum; sal 0.9% saline; aq aqueous; pg propylene glycol; A:dH$_2$O a 1:1 mixture of acetone and distilled water; if no vehicle is listed then it was not given in the referenced paper

[c] Results shown are the number of positive responders (guinea pigs with an erythema grade >1 after challenge) over the total number of animals in the test group. When the number of animals is not given, the results are presented as the percentage of animals judged to be positive

solely on topical exposures. In the LLNA, sensitisation to PPD has been consistently induced at concentrations of 0.1%[48] (Table 10.2).

Toluene-2,5-diamine sulfate (PTD), another primary intermediate, displays a reactivity similar to that of PPD in preclinical models. GPMT data show an elicitation rate of approximately 80% in animals challenged with 0.01% PTD (Table 10.1). The lowest concentration of PTD reported to be tested in the LLNA, 0.5%, resulted in an SI of 4.4 (Table 10.2).

Other primary intermediates such as 2-aminophenol (OAP) and 4-aminophenol (PAP) have also been identified as skin allergens by preclinical test methods. LLNA results for OAP were similar to those of PTD, with an SI of 3.5 at a concentration of 0.5%.[52] PAP was positive in the GMPT though slightly less reactive than PPD and

Table 10.2 *Local Lymph Node Assay (LLNA) Data for Hair Dye Chemicals*

Chemical Name [CAS No.]	Vehicle[a]	LLNA Concentration, %	LLNA SI[b]	Reference
2-Aminophenol (OAP) [95–55–6]	AOO	0.5 1 2.5	3.5 5.0 7.4	52
3-Aminophenol (MAP) [591–27–5]	AOO	2.5 5 10	2.8 3.5 5.7	52
Hydroxyethyl-*p*-phenylenediamine sulfate [93841–25–9]	DMSO	0.5 1 2	2.7 4.5 7.0	101
p-Phenylenediamine (PPD) [106–50–3]	AOO	0.05 0.1 0.25 0.5 1	2.0 3.3 10.2 20.5 26.4	48
3-Phenylenediamine [108–45–2]	AOO	2.5 5 10	11.7 15.4 19.2	52
Resorcinol [108–46–3]	DMF	5 10 25	2.2 2.2 2.7	57
Toluene-2,5-diamine sulfate (PTD) [6369–59–1]	AAOO	0.5 1.5 2.8	4.4 10.4 19.4	101

[a] Vehicle abbreviations: AOO acetone/olive oil, 4:1; DMSO dimethylsulfoxide; DMF dimethylforma-dimethylformamide; AAOO aqua(distilled water)/acetone(1:1)/olive oil, 4:1
[b] SI stimulation index. An SI > 3 is considered a positive response.

PTD, as only 33% of the animals responded to a challenge concentration of 0.01% with similar induction concentrations (both injection and patch).[51]

Preclinical tests demonstrate a range of skin sensitising potential amongst the chemicals which function as couplers in hair dye formulations. 3-Aminophenol (MAP) has been shown to induce sensitisation in the GPMT, with 100% of the animals responding to a challenge concentration of 5%.[53] LLNA data for MAP were also moderately positive.[52] Based on results from several non-standard guinea pig test methods, the Cosmetic Ingredient Review on 1,3-phenylenediamine (*m*-phenylenediamine) concluded that it was a mild sensitiser.[54] However, based on LLNA data, 1,3-phenylenediamine appears to be a strong sensitiser.[52] The coupler 4-amino-2-hydroxytoluene has been shown to have weak sensitising ability, with only one out of ten animals responding in the GPMT to topical induction and challenge concentrations of 25%. However, for classification purposes, these data for 4-amino-2-hydroxytoluene would be considered negative as fewer than 30% of the animals responded. Negative GPMT results have been reported for 1-naphthol,[55] although the concentration used for induction and challenge, 0.1%, was somewhat low. Neither resorcinol[56,57] nor 2-methylresorcinol[56] have been shown to have skin sensitisation potential in preclinical test methods.

10.2.2 Predictive Testing of Hair Dyes in Humans

There are a limited number of reports concerning the predictive testing of hair dye ingredients in humans. The contact allergic response to PPD has been investigated using both the human maximisation test[58,59] and the modified Draize procedure.[60] In the human maximization test (HMT) conducted by Kligman,[58] 24 subjects were treated at a test site (typically the forearm or calf) with 10% PPD in petrolatum under occlusive patch for 48 h. This induction application was conducted for a total of five exposures. After a ten day rest period, the subjects were challenged on an alternative site (either the lower back or forearm) for 48 h with 0.5% PPD in petrolatum. Reactions were elicited in all of the subjects (24/24) and, based on this sensitisation rate of 100%, Kligman classified PPD as a grade 5, or extreme sensitiser. A lower sensitisation rate, 44% (15/34 subjects), was reported for the HMT conducted by Epstein and Taylor,[59] but a much lower concentration of PPD (2%) was used for induction. Interestingly, the same concentration of PTD, which displays reactivity similar to PPD in animal models, did not induce sensitisation (0/31 subjects). The HMT was also used to assess the sensitisation potential of resorcinol.[58] No responses were observed in 22 subjects induced with 15% and challenged with 5% resorcinol in petrolatum.

Marzulli and Maibach[60] utilised the modified Draize procedure to investigate dose response relationships of sensitisation to PPD. Subjects were treated with ten successive 48 or 72 h occlusive patch applications of 0.01%, 0.1% or 1% PPD in petrolatum. After an approximately two week rest period, subjects were then challenged with a single concentration of PPD, either 0.01% or 1%, in petrolatum for 72 h under occlusion. Only 7.2% (7/97) of the subjects responded to induction and challenge with 0.01% PPD. Responses were observed in 11.2% (11/98) of the subjects induced with 0.1% PPD and challenged with 1% PPD. The sensitisation rate increased to 53.4% (47/88) when 1% PPD was used for both induction and challenge. These data demonstrate that dose (*e.g.* amount of chemical per unit area of skin) plays a key role in the development of contact allergy.

10.3 Human Diagnostic Patch Testing
10.3.1 Dermatology Patient Population

As mentioned, the association of hair dyes and skin allergy has been known for many years. Skin allergy has been reported from occupational exposures, as well as from the use of hair dyes by consumers. PPD has been incorporated into standard patch test trays used around the world. For example, PPD is part of the International Contact Dermatitis Research Group (ICDRG), North American Contact Dermatitis Research Group (NACDG) and European standard series. PPD, as a free base, is most often used at 1% in petrolatum for patch testing patients with suspected contact dermatitis. However, in early years some investigators used PPD at 0.5%.[61] Generally, PPD was applied to patches (*e.g.* Finn Chambers from Epitest Ltd., Helsinki, Finland), which were fixed to the skin (usually the back) for 48 h. After two days, the patches are removed and examined for evidence of reaction at day 2, 3 and 6 or 7. Positive reactions were graded from + to +++, as recommended by the ICDRG. It should be noted that, amongst the large number of published patch test studies, dermatology clinics use slightly different procedures for diagnostic patch testing of their eczema patients (*e.g.* PPD concentration, patch type).

In Table 10.3, representative patch test results are listed to illustrate the frequency of positive patch test results to PPD from dermatology clinics in Europe, North America and Singapore. The % positive PPD individuals reported by many of the clinics are fairly similar with most values in the range of 3 to 8%. However, lower percentages (0.4 to 1.6%) are reported for Denmark and Sweden. Of great importance to clinicians and manufacturers is the need to understand whether the frequency of positive patch test reactions to PPD is increasing over time. Armstrong *et al.*[61] reported that the annual rate of positive reactions remained fairly constant over the study period 1982–1998. The NACDG and Netherlands PPD patch test results are strikingly similar with % values only ranging from 6.1 to 6.9% from 1972 to 1996[62–66] and 2.9 to 3.1% from 1995 to 2000,[67–69] respectively. The Information Network of Departments of Dermatology (IVDK) in Germany results indicate a very slight decrease in frequencies of positive PPD patch test results (4.8% to 4.3%) in two very large patient population groups tested from 1992–1996 versus 1999–2001.[70,71] However, in the period from 1992–1996 the yearly patch test results went from 4.8% up to 5.4% and then back down to 4.3%. The results reported from Denmark demonstrate a slight decrease in prevalence of PPD positive patients.[72] In Sweden, Edman and Möller[73] reported a decrease in PPD sensitisation in males, not females, in the period of 1969–1980. In a more recent Swedish study (1993–2000), the overall frequency of positive patch test results to PPD was reported to be lower than in the earlier Swedish study.[73,74] Only in Singapore do the percentages of PPD positive patients appear to be increasing over time with the percentages going from 4.5% in 1986–1990 to 8.1% in 1998–1999.[75,76] It is

Table 10.3 *Diagnostic patch test results for PPD from different clinics*

Country	Dates	Number of Patients Tested	% PPD Positive	Reference
Canada	1972–1981	4190	7.3	63
Denmark	1973–1977	3225	1.3	102
Denmark	1975–1980	3664	1.3	72
Denmark	1986–1990	6759	0.4	72
European Research Group	1967–1968	4825	4.9	103
Germany (IVDK)	1992–1996	45,250	4.8	70
Germany (IVDK)	1999–2001	25,451	4.3	71
Italy (GIRDCA)	1984–1993	42,839	3.5	104
Netherlands	1995–1999	2058	3.1	67
Netherlands	1996–1999	2375	3.0	68
Netherlands	1995–2000	1701	2.9	69
North America (NACDG)	1972–1974	3041	6.1	62
North America (NACDG)	1979–1980	2145	6.5	63
North America (NACDG)	1984–1985	1138	6.9	64
North America (NACDG)	1985–1989	3986	6.4	65
North America (NACDG)	1992–1994	3515	6.3	65
North America (NACDG)	1994–1996	3111	6.8	66
Singapore	1986–1990	5557	4.5	75
Singapore	1998–1999	406	8.1	76
Spain	1973–1977	4600	6.1	105
Sweden	1969–1980	8933	~4.0	73
Sweden	1993–2000	21,840	1.6	74
UK	1982–1998	26,706	2.5	61
USA	1998–1997	927	5.0	106
USA	1994–1999	1329	2.3	106

worth noting that the test population sample size was much smaller in the more recent study.

It is important to note that exposure to PPD can come from a number of different sources. In addition to hair dyes, PPD is also used in dyes for fur, leather, printer's ink, fax machines, photographic products, X-ray film fluids, lithography[67] and temporary tattoos.[77] For example, the high incidence of positive PPD reactions seen in rubber workers is proposed to reflect cross-reactivity between PPD and *N*-isopropyl-*N'*-phenyl-*p*-phenylenediamine or related antioxidants.[78] Thus, the potential for cross-reactions with other *para*-compounds can hinder establishing relevance in some PPD-positive individuals. Cross-sensitisations between PPD and other dyes have been frequently reported. For example, 128 PPD-positive patch test patients also reacted to disperse orange 3 and disperse yellow 3.[79] Little reactivity was noted for bismark brown, naphthol AS, disperse yellow 9, disperse blue 3 or disperse red 11. The authors speculate that simultaneous patch test reactions between PPD and the textile dyes are due either to cross-reactivity or to metabolic conversion of textile dyes to PPD in the skin. In a 1997 study of 236 azo-dye-sensitive subjects, Seidenari *et al.*[80] reported that co-sensitisation to PPD was present in 66% of subjects sensitised to disperse orange 3. It was also reported in this study that 20 subjects positive to PPD and disperse orange 3 claimed to have intolerance to hair dyes. It is worth noting that the molecular structures of PPD and disperse orange 3 are similar.[79] Hairdressers sensitised to PPD, and/or 2, 5-diaminotolulene sulfate and/or 2NPPD did not cross-react with the Food, Drug and Cosmetic (FD & C) dyes blue 1 and yellow 5, the D & C dyes red 33, yellow 10, orange 4, green 5 and EXT. violet 2 or the acid dyes black 1, red 52, red 14 and blue 3.[81]

PPD is regarded by some as a screening agent for contact allergy to *para* and azo dye compounds. Thus, investigators have performed numerous studies examining what other dyes can elicit responses in PPD-sensitised individuals. Recently, Koopmans and Bruynzeel[68] showed that positive responses to PPD correlated well with reactions to *para* compounds like *p*-aminoazobenzene and *p*-toluenediamine sulphate but not with disperse dyes. In a consumer complaint-based study, 16 patients with hair dye allergy were tested with a number of allergens associated with hairdressing. All 16 were patch test positive to PPD; 7 reacted to *p*-toluenediamine; 6 to aminoazobenzene; 5 to nitro-PPD; and 3 or fewer to 3-aminophenol, 4-aminophenol, ammonium persulfate, ammonium thioglycolate, disperse orange and diaminophenol.[82] Uter *et al.*[83] also report positive patch test reactions to developing and coupling agents such as 4-aminophenol, 3-aminophenol and hydroquinone in eczema patients suspected of having hair dye dermatitis from occupational or consumer exposure.

10.3.1.1 Occupational Skin Allergy to Hair Dyes

Occupational contact dermatitis among hairdressers is well documented.[84,85] A 'hairdressing series' of allergens exit for diagnostic patch testing and, besides PPD and other hair dye ingredients, it includes additional chemicals found in hair care products such as preservatives (Table 10.4). In addition to the risk of hand eczema from wet work, contact allergy to hair dye ingredients such as PPD, 2,5-diaminotoluene sulphate and 2NPPD has been reported.[81,83,86] For example, occupationally relevant sensitisations were found in 61% of 302 Italian hairdressers.[87] PPD showed the highest proportion of allergic reactions followed by the acid permanent wave ingredient glyceryl monothioglycolate. In a European centre study of 809 hairdressers and 104 clients,[86] the mean frequencies of sensitisation ranked as follows for these ingredients: glyceryl monothioglycolate (19%), PPD (15%), ammonium persulphate (8%) and 2,5-diaminotoluene sulphate (8%). Armstrong *et al.*[61] reported that hairdressers evaluated in their clinic had a 19% rate of PPD allergy. In many cases, hand eczema among hairdressers is due to a combination of chronic irritant and allergic dermatitis. In an occupational surveillance report from the UK, the rate of contact dermatitis for hairdressers and barbers was 116.3 in 100,000 (0.12%) workers with the majority of causative agents coming from hairdressing chemicals.[88] In Germany, the rate reported was 97.4 per 10,000 (0.97%) workers per year for the period 1990–1999.[84,85] Obviously, improved preventive strategies (*e.g.* increasing compliance of glove wearing) are crucial for reducing the rates of sensitisation for hairdressers and barbers.

Table 10.4 *Patch test allergens in the Chemotechnique Diagnostics' hairdressing series.[108] All allergens are in petrolatum unless specified otherwise*

Compound	Concentration % (w/w)
p-Phenylenediamine base	1.0
2,5-Diaminotoluene sulfate	1.0
2-Nitro-4-phenylenediamine	1.0
Ammonium thioglycolate	2.5 aqueous
Ammonium persulfate	2.5
Formaldehyde	1.0 aqueous
Nickel sulfate	5.0
Cobalt chloride	1.0
Resorcinol	1.0
3-Aminophenol	1.0
4-Aminophenol	1.0
Hydrogen peroxide	3.0 aqueous
Hydroquinone	1.0
Balsam Peru	25.0
Chloroacetamide	0.2
Glyceryl monothioglycolate	1.0
Cocamidopropylbetaine	1.0 aqueous
Cl+Me-isothiazolinone (Kathon CG, 200ppm)	0.02 aqueous
2-Bromo-2-nitropropane-1,3-diol (Bronopol)	0.25
Captan	0.5
4-Chloro-3-cresol (PCMC)	1.0
4-Chloro-3,5-xylenol (PCMX)	0.5
Imidazolidinyl urea (Germall 115)	2.0
Quaternium 15 (Dowicil 200)	1.0
Zinc pyrithione (Zinc omadine)	1.0
Diazolidinylurea (Germall II)	2.0

10.3.2 General Population

Most information on the prevalence of contact allergies to PPD arises from patch testing of patients with eczematous skin reactions who have been referred to specialised patch test clinics. Thus, it is difficult to extrapolate those data to the pattern of contact sensitisation in the general population. Epidemiological studies conducted on the general population are few in number with the range of individuals sensitised to PPD reported as being between 0.1 and 1%. In 1990 and 1998 15–41-year-old people were patch tested in two cross-sectional studies of random samples of the population in Copenhagen, Denmark.[89] Contact sensitivity to PPD was 0% in 1990 and 0.2% in 1998. In Italy, a rate of 0.5% PPD positive individuals was detected in 593 healthy subjects.[90] More recently, the prevalence of contact sensitisation to PPD in the general population in Germany was reported as 1.5% of 1141 adults.[91] For reference, 40% of the 1141 subjects reacted to at least one of the 25 standard allergens tested with the fragrance mix (15.9%) and nickel (13.1%) having the highest rates. In a consumer complaint-based study, data were obtained by advertising for persons with adverse reactions to hair dyes.[82] Among those responding to the advertisement, 55 cases of severe, acute ACD were identified. Of the 55 subjects 29% (16) were patch tested and all had positive reactions to PPD. From the many published diagnostic patch studies, PPD is the ingredient most often cited as the causative agent of ACD related to hair dye use.

10.4 Summary and Future Directions

Based on all of the animal and human data available it is clear that hair dye ingredients such as PPD and its derivatives have the potential to induce and elicit contact allergy. PPD is by far the most studied ingredient, primarily because it is used as a screening agent for eczema patients in standard patch test trays used by dermatologists around the world. To warn consumers of the risk of allergy, labels have been added to commercial oxidative hair dyes in Europe and in the US. Those labels contain warning phrases such as 'Can cause allergic reactions', 'Contains phenylenediamines' and 'Do not use to dye eyelashes or eye brows'. Consumers are advised also to perform an allergy test prior to the use of the product.[92] Moreover, hairdressers and consumers are cautioned to wear gloves during the dyeing process to limit skin exposures and the potential for the induction/elicitation of ACD.

The relative percentage of PPD-positive patch test individuals in dermatology clinics throughout the world has remained relatively constant over the years (Table 10.3). This is noteworthy because the use of hair dyes by women and men has increased significantly since the 1960s.[93] Interestingly, the NACDG reported a decrease in the 'Significance Prevalence Index Number' (the relative clinical importance of contact allergens in the population) for PPD.[94] The rank for PPD went from three in the 1984 to ten in 1996. Equally important is the fact that the prevalence of confirmed hair dye dermatitis (PPD patch test positive) does not appear to have changed amongst the general population.[89] Thus, an overview of the available patch test data for PPD in Europe, North America and Singapore showed little change in rate of positive reactions in the majority of the clinics against a background of increased exposure to PPD.

There is a strong commitment in industry to reduce the allergenic risk of cosmetic products, including hair dyes. With the advancement of our basic understanding of the chemistry and biology of ACD, progress has been made in the development of new tools (*e.g.* LLNA) which have helped improve the industry's risk assessment capabilities.[95–98] Critical to conducting a sound skin sensitisation risk assessment is having a thorough understanding of the anticipated consumer exposure to the ingredient, as well as knowledge of its allergenic potency and dose response characteristics. In the future, the availability of these tools and approaches will provide the information necessary to improve our risk assessment capabilities for new and existing hair dye chemicals.

10.5 References

1. C.K. Smith and S.A. Hotchkiss, *Allergic Contact Dermatitis: chemical and metabolic mechanisms*, Taylor and Francis, London, 2001.
2. J.D. Bos and M.M.H.M. Meinardi, *Exp. Derm.*, 2000, **9**, 165.

3. M.D. Barratt and D.A. Basketter in *Toxicology of Contact Hypersensitivity*, I. Kimber and T. Maurer (ed), Taylor and Francis, London, 1996, 75.
4. I. Kimber, D.A. Basketter, G.F. Gerberick and R.J. Dearman, *Int. Immunopharm.*, 2002, **2**, 201.
5. J.P. McFadden and D.A. Basketter, *Contact Derm.*, 2000, **42**, 123.
6. I. Kimber, M. Cumberbatch, R.J. Dearman and C.E.M. Griffiths, *Br. J. Derm.*, 2002, **147**, 604.
7. I. Kimber and R.J. Dearman in *Toxicology of Contact Hypersensitivity*, I. Kimber and T. Maurer (ed), Taylor & Francis, London, 1996, 4.
8. J.F. Corbett, *Clin. Derm.*, 1988, **6**, 93.
9. L.J. Wolfram and H.I. Maibach, *Arch. Dermatol. Res.*, 1985, **277**, 235.
10. W. Dressler in *Dermal Absorption and Toxicity Assessment*, M.S. Roberts and K.A. Walters (ed), Marcel Dekker, Inc., 1998, New York, 489.
11. D. Howes and J.G. Black, *Int. J. Cosmetic Sci.*, 1983, **5**, 215.
12. R.L. Bronaugh and E.R. Congdon, *J. Invest. Derm.*, 1984, **83**, 124.
13. R.L. Bronaugh and H.I. Maibach, *J. Invest. Derm.*, 1985, **84**, 180.
14. H. Beck, M. Bracher, C. Faller and H. Hofer, *Cosmet. Toiletries*, 1993, **108**, 76.
15. H. Beck, K. Brain, W. Dressler, R. Grabarz, D. Green, D. Howes, L. Kitchiner, R. Pendlington, M. Python, K. Schröder, R. Sharma, W. Steiling, K. Sugimoto, K. Walters and A. Watkinson in *Perspectives in Percutaneous Penetration*, K.R. Brain and K.A. Walters (ed), STS Publishing, Cardiff, UK, 2000.
16. J.J. Yourick and R.L. Bronaugh, *Tox. Appl. Pharm.*, 2000, **166**, 13.
17. W. Steiling, J. Kreutz and H. Hofer, *Toxic. In Vitro*, 2001, **15**, 565.
18. F. Hueber-Becker, G.J. Nohynek, W.J.A. Meuling, F. Benech-Kieffer and H. Toutain, *Fd. Chem. Toxic.*, 2004, **42**, 1227.
19. E.P. Frenkel and F. Brody, *Arch. Environ. Health*, 1973, **27**, 401.
20. H.I. Maibach and L.J. Wolfram, *J. Soc. Cosmet. Chem.*, 1981, **32**, 223.
21. N. Goetz, P. Lasserre, P. Boré and G. Kalopissis, *Int. J. Cosmetic Sci.*, 1988, **10**, 63.
22. SCCNFP, *Opinion concerning* p-*phenylenediamine (Colipa A 7)*, 27 February 2002.
23. G. Dupuis and C. Benezra, *Allergic Contact Dermatitis to Simple Chemicals: A Molecular Approach*, Marcel Dekker Inc., New York & Basel, 1982.
24. J.-P. Lepoittevin, D.A. Basketter, A. Goossens and A.T. Karlberg, *Allergic Contact Dermatitis: The Molecular Basis*, Springer, Berlin, 1998.
25. S.R. Ahlfors, O. Sterner and C. Hansson, *Skin Pharmacol. Appl. Skin Physiol.*, 2003, **16**, 59.
26. C.K. Smith-Pease, D.A. Basketter and G.Y. Patlewicz, *Clin. Exp. Derm.*, 2003, **2**, 177.
27. D.A. Basketter and B.F.J. Goodwin, *Contact Derm.*, 1988, **19**, 248.
28. P. Lisi and K. Hansel, *Contact Derm.*, 1998, **39**, 304.
29. D.A. Basketter and C. Lidèn, *Contact Derm.*, 1992, **27**, 90.
30. R.J. Schmidt and L.Y. Chung, *Free Rad. Biol. Med.*, 1990, **9 (Suppl. 1)**, 147.
31. M. Picardo, C. Zompetta, M. Grandinetti, F. Ameglio, B. Santucci, A. Faggioni and S. Passi, *Br. J. Dermatol.*, 1996, **134**, 681.

32. M. Krasteva, J-F. Nicolas, G. Chabeau, J-L. Garrigue, H. Bour, J. Thivolet and D. Schmidt, *Int. Arch. Allergy Immunol.*, 1993, **102**, 200.

33. S. Sieben, Y. Kawakubo, T.A. Masaoudi and H.F. Merk, *J. Allergy Clin. Immunol.*, 2002, **109**, 1005.

34. H.F. Merk, J. Baron, M. Hertl, D. Niederau and A. Rübben, *Int. Arch. Allergy Immunol.*, 1997, **113**, 173.

35. Y. Kawakubo, M. Nakamori, E. Schöpf and M. Ohkido, *Derm.*, 1997, **195**, 43.

36. A. Schnuch, G.A. Westphal, M.M. Muller, T.G. Schulz, J. Geier, J. Brasch, H.F. Merk, Y. Kawakubo, G. Richter, T. Fuchs, T. Gutgesell, K. Reich, M. Gebhardt, D. Becker, J. Grabbe, C. Szliska, W. Aberer and E. Hallier, *Contact Derm.*, 1998, **38**, 209.

37. Y. Kawakubo, H.F. Merk, T.A. Masaoudi, S. Sieben and B. Blomeke, *J. Pharm. Exp. Therap.*, 2000, **292**, 150.

38. G.A. Westphal, K. Reich, T.G. Schulz, C. Neumann, E. Hallier and A. Schnuch, *Br. J. Dermatol.*, 2000, **142**, 1121.

39. H.F. Merk, J. Abel, J.M. Baron and J. Krutmann, *Tox. Appl. Pharm.*, 2004, **195**, 267.

40. B. Magnusson and A.M. Kligman, *J. Invest. Dermatol.*, 1969, **52**, 268.

41. B. Magnusson and A.M. Kligman, *Allergic Contact Dermatitis in the Guinea Pig. Identification of Contact Allergens*, Charles C. Thomas, Springfield, IL, 1970.

42. E.V. Buehler, *Arch. Dermatol.*, 1965, **91**, 171.

43. M.K. Robinson, T.L. Nusair, E.R. Fletcher and H.L. Ritz, *Toxicol.*, 1990, **61**, 91.

44. G.F. Gerberick, C.A. Ryan, I. Kimber, R.J. Dearman, L.J. Lea and D.A. Basketter, *Am. J. Contact Derm.*, 2000, **11**, 3.

45. B.F.J. Goodwin, R.W.R. Crevel and A.W. Johnson, *Contact Derm*, 1981, **7**, 248.

46. T. Maurer and R. Hess, *Fd. Chem. Toxic.*, 1989, **27**, 807.

47. D.A. Basketter and G.F. Gerberick, *Contact Derm.*, 1996, **35**, 146.

48. E.V. Warbrick, R.J. Dearman, L.J. Lea, D.A. Basketter and I. Kimber, *J. Appl. Toxic.*, 1999, **19**, 255.

49. R.L. Bronaugh, C.D. Roberts and J.L. McCoy, *Fd. Chem. Toxic.*, 1994, **32**, 113.

50. C. Lidén and A. Boman, *Contact Derm.*, 1988, **19**, 290.

51. Z. Xie, R. Hayakawa, M. Sugiura, H. Kojima, H. Konishi, G. Ichihara and Y. Takeuchi, *Contact Derm.*, 2000, **42**, 270.

52. J. Ashby, D.A. Basketter, D. Paton and I. Kimber, *Toxicol.*, 1995, **103**, 177.

53. D.A. Basketter and E.W. Scholes, *Fd. Chem. Toxic.*, 1992, **30**, 65.

54. Cosmetic Ingredient Review, *J. Am. Col. Toxic.*, 1997, **16(suppl 1)**, 59.

55. Cosmetic Ingredient Review, *J. Am. Col. Toxic.*, 1989, **8**, 749.

56. Cosmetic Ingredient Review, *J. Am. Col. Toxic.*, 1986, **5**, 167.

57. D.A. Basketter, E.W. Scholes and I. Kimber, *Fd. Chem. Toxic.*, 1994, **32**, 543.

58. A.M. Kligman, *J. Invest. Dermatol.*, 1966, **47**, 393.

59. W.L. Epstein and M.K. Taylor, *Acta Derm. Venerol.*, 1979, **85 (Suppl.)**, 55.

60. F.N. Marzulli and H. I. Maibach, *Fd. Chem. Toxic.*, 1974, **12**, 219.
61. D.K.B. Armstrong, A.B. Jones, H.R. Smith, J.S. Ross, I.R. White, R.J.G. Rycroft and J.P. McFadden, *Contact Derm.*, 1999, **41**, 348.
62. E.J. Rudner, W.E. Clendenning, E. Epstein, A.A. Fisher, O.F. Jillson, W.P. Jordan, N. Kanof, W. Larsen, H. Maibach, J.C. Mitchell, S.E. O'Quinn, W.F. Schorr and M.B. Sulzberger, *Contact Derm.*, 1975, **1**, 277.
63. C.W. Lynde, L. Warshawski and J.C. Mitchell, *Contact Derm.*, 1982, **8**, 417.
64. F.J. Storrs, L.R. Rosenthal, R.A. Adams, W.C., E.A. Emmett, A.A. Fisher, W.G. Larsen, H.I. Maibach, R.L. Rietschel, W.F. Schorr and J.S. Taylor, *J. Am. Acad. Dermatol.*, 1989, **20**, 1038.
65. J.G. Marks, D.V. Belsito, V.A. DeLeo, J.F. Fowler, A.F. Fransway, H.I. Maibach, C.G. Mathias, J.R. Nethercott, R.L. Rietschel, L.E. Rosenthal, E.F. Sherertz, F.J. Storrs and J.S. Taylor, *Am. J. Contact Derm.*, 1995, **6**, 160.
66. J.G. Marks, D.V. Belsito, V.A. DeLeo, J.F. Fowler, A.F. Fransway, H.I. Maibach, C.G. Mathias, J.R. Nethercott, R.L. Rietschel, E.F. Sherertz, F.J. Storrs and J.S. Taylor, *J. Am. Acad. Dermatol.*, 1998, **38**, 911.
67. S.A. Devos and P.G.M. van der Valk, *Contact Derm.*, 2001, **44**, 273.
68. A.K. Koopmans and D.P. Bruynzeel, *Contact Derm.*, 2003, **48**, 89.
69. P.G.M. van der Valk, S.A. Devos and P-J. Coenraads, *Contact Derm.*, 2003., **48**, 121.
70. W. Uter, A. Schnuch, J. Geier and P.J. Frosch, *Eur. J. Dermatol.*, 1998, **1**, 36.
71. J. Geier, W. Uter, H. Lessmann and A.Schnuch, *Contact Derm.*, 2003, **48**, 280.
72. N.K. Veien, T. Hattel and G. Laurberg, *Am. J. Contact Derm.*, 1992, **3**, 189.
73. B. Edman and H. Möller, *Contact Derm.*, 1982, **8**, 95.
74. J.E. Wahlberg, M. Tammela, C. Anderson, B. Bjorkner, M. Bruze, T. Fischer, A. Inerot, A-T. Karlberg, C. Liden, M. Lindberg, B. Meding, H. Moller, B. Stenberg and K. Sundberg, *Occup. Environ. Dermatol.*, 2002, **50**, 51.
75. J.T.E. Lim, C.L. Goh, S.K. Ng and W.K. Wong, *Contact Derm.*, 1992, **26**, 321.
76. Y-C. Chan, A-K. Ng and C-L. Goh, *Contact Derm.*, 2001, **45**, 217.
77. R.R. Brancaccio, L.H. Brown, Y.T. Chang, J.P. Fogelman, E.A. Mafong and D.E. Cohe, *Am. J. Contact Derm.*, 2002, **13**, 15.
78. B. Hevre-Bazin, D. Gradiski, P. Duprat, B. Marignac, J. Foussereau, C. Cavelier and P. Bieber, *Contact Derm.*, 1977, **3**, 1.
79. A.Y-J. Goon, N.J. Gilmour, D.A. Basketter, I.R. White, R.J.G. Rycroft and J.P. McFadden, *Contact Derm.*, 2003, **48**, 248.
80. S. Seidenari, L. Mantovani, B.M. Manzini and M. Pignatti, *Contact Derm.*, 1997, **36**, 91.
81. R. Fautz, A. Fuchs, H. Van der Walle, V. Henny and L. Smits, *Contact Derm.*, 2002, **46**, 319.
82. H. Søsted, T. Agner, K.E. Andersen and T. Menné, *Contact Derm.*, 2002, **47**, 299.
83. W. Uter, H. Lessmann, J. Geier and A. Schnuch, *Contact Derm.*, 2003, **49**, 236.
84. H. Dickel, O. Kuss, C.R. Blesius, A. Schmidt and T.L. Diepgen, *Br. J. Derm.*, 2001, **145**, 453.
85. H. Dickel, O. Kuss, C.R. Blesius, A. Schmidt and T.L. Diepgen, *Contact Derm.*, 2001, **44**, 258.

86. P.J. Frosch, D. Burrows, J.G. Camarasa, A. Dooms-Goossens, G. Ducombs, A. Lahti, T. Menné, R.J.G. Rycroft, S. Shaw, I.R. White and J.D. Wilkinson, *Contact Derm.*, 1993, **28**, 180.

87. L. Guerra, A. Tosti, F. Bardazzi, P. Pigatto, P. Lisi, B. Santucci, R. Valsecchi, D. Schena, G. Angelini, A. Sertoli, F. Ayala and F. Kokelj, *Contact Derm.*, 1992, **26**, 101.

88. J.D. Meyer, Y. Chen, D.L. Holt, M.H. Beck and N.M. Cherry, *Occup. Med.*, 2000, **50**, 265.

89. N.H. Nielsen, A. Linneberg, T. Menné, F. Madsen, L. Frølund, A. Dirksen and T. Jørgenesen, *Acta. Derm. Venereol.*, 2001, **81**, 31.

90. S. Seidenari, B.M. Manzini, P. Danese and A. Motolese, *Contact Derm.*, 1990, **23**, 162.

91. T. Schäffer, E. Böhler, S. Ruhdorfer, L. Weigl, D. Wessmer, B. Filipiak, H.E. Wichmann and J. Ring, *Allergy*, 2001, **56**,1192.

92. M. Krasteva, A. Cristaudo, B. Hall, D. Orton, E. Rudzki, B. Santucci, H. Toutain and J. Wilkinson, *Eur. J. Dermatol.*, 2002, **12**, 322.

93. G.J. Nohynek, R. Fautz, F. Benech-Kieffer and H. Toutain, *Food Chem. Toxicol.*, 2004, **42**, 517.

93. M. Maouad, A.B. Fleischer Jr., E.F. Sheretz and S.R. Feldman, *J. Am. Acad. Dermatol.*, 1999, **41**, 573.

94. M.K. Robinson, G.F. Gerberick, C.A. Ryan, P. McNamee, I.R. White and D.A. Basketter, *Contact Derm.*, 2000, **42**, 251.

96. G.F. Gerberick, M.K. Robinson, S.P. Felter, I.R. White and D.A. Basketter, *Contact Derm.*, 2001, **45**, 333.

97. S.P. Felter, M.K. Robinson, D.A. Basketter and G.F. Gerberick, *Contact Derm.*, 2002, **47**, 257.

98. S.P. Felter, C.A. Ryan, D.A. Basketter, N.J. Gilmour and G.F. Gerberick, *Reg. Toxic. Pharm.*, 2003, **37**, 1.

99. Cosmetic Ingredient Review, *J. Am. Col. Toxic.*, 1989, **8**, 569.

100. J.E. Wahlberg and A. Boman in Current Problems in Dermatology: Contact Allergy Predictive Tests in Guinea Pigs, K.E. Andersen and H.I. Maibach (ed), Karger, New York, 1985, **14**, 59.

101. P. Aeby, C. Wyss, H. Beck, P. Griem, H. Scheffler and C. Goebel, *J. Invest. Derm.*, 2004, **122**, 1154.

102. O. Hammershoy, *Contact Derm.*, 1980, **6**, 263.

103. S. Fregert, N. Hjorth, B. Magnusson, H.-J. Bandmann, C.D. Calnan, E. Cronin, K. Malten, C.L. Meneghini, V. Pirilä and D.S. Wilkinson, *Trans. St. John's Hospital Dermatological Soc.*, 1968, **55**, 17.

104. A. Sertoli, S. Francalanci, M.C. Acciai and M. Gola, *Am. J. Contact Derm.*, 1999, **10**, 18.

105. C. Romaguera and F. Grimalt, *Contact Derm.*, 1980, **6**, 309.

106. M.R. Albert, Y. Chang and E. Gonzalez, *Am. J. Contact Derm.*, 1999, **10**, 219.

107. T-S. Hsu, M.D.P. Davis, R. el-Azhary, J.F. Corbett and L.E. Gibson, *J. Am. Acad. Dermatol.*, 2001, **44**, 867.

108. Chemotechnique Diagnostics Patch Test Products catalogue, 2003/2004, 23, www.chemotechnique.se

Hair-Colorant Use and Associated Pathology – Cancer?

TONGZHANG ZHENG, YAWEI ZHANG, YONG ZHU
and LINDSAY MORTON

11.1 Introduction

It was estimated in 1992 that 40% of women and more than 10% of men in the United States (US) and Europe use hair dyes each year.[1] The trend has been constantly increasing according to the number of hairdressers and the volume of hair dyes sold.[2] The results from recent industry surveys have shown that over 70% of women in the US have used hair dyes at least once during their lifetime.[3] Given the considerable prevalence of hair dye use, understanding the role of hair dyes in the development of human cancers is of great public health significance. After the publication by Ames *et al.* in 1975 revealing their findings that a number of hair dye ingredients were mutagenic/carcinogenic, the potential carcinogenic effect from hair dye use has received considerable attention by epidemiologists.[4] The results from epidemiological studies linking hair dye use to human cancer risk, however, have been inconsistent.

Hair dyes can be classified into different types (permanent, semi-permanent and temporary) and different colours (dark, red, brown, blond and others), and their constituents differ from each other.[1] Thus, it is likely that different types of hair dye products may have different impacts on human cancer risk.

In 1979, the US Food and Drug Administration (FDA), prompted by the positive National Cancer Institute (NCI) findings, proposed the requirement of a cancer-warning label on hair dye products containing potential carcinogenic material. 'The resulting concern, under the prevailing opinion that there was no safe dose for a carcinogen, caused manufacturers to reformulate all oxidative dye products during 1978–1982'.[5] This reformulation involved the replacement or elimination of some of the hair dye contents, which had been reported to produce tumors in NCI bioassays, such as 2,4-diaminoanisole and some of the yellow and red direct dyes.[5]

The heterogeneous nature of hair dye products, the changes in hair dye formulation, and the complicated patterns of hair dye use create methodological challenges for epidemiologists studying the relationship between hair dye uses and human cancer risk. The accuracy of exposure assessment is a major issue in explaining epidemiological study results linking hair dye products to human cancer risks.

Occupational exposure to hair dye products as a hairdresser, barber or beautician differs from personal use of hair dyes. For example, occupational exposure to hair dyes is affected by the use of personal protective gloves. The relationship between occupational exposure to hair dyes and human cancer risk is also confounded by occupational exposure to other materials, such as solvents, hair spray, and nail and skin care products.

In this chapter, we briefly summarise the results from major epidemiological studies investigating the association between both personal and occupational exposure to hair dyes and human cancer risk.

11.2 Epidemiological Studies of Hair Dye Use and Human Cancer Risk

A detailed review of epidemiological studies linking both personal and occupational hair dye use to human cancer risk was published in 1993 by the International Agency for Research on Cancer (IARC).[1] That review concluded that occupation as a hairdresser or barber entailed exposures that were probably carcinogenic and that personal use of hair colorants cannot be evaluated as to its carcinogenicity. Since then, many new studies have been carried out to further investigate the relationships between hair dyes and human cancer risks.

11.2.1 Personal Use of Hair Dyes and Human Cancer Risk

11.2.1.1 Prospective Follow-up Studies

Prospective follow-up studies are conducted based on the presence or absence of an exposure of interest without the information regarding disease status. Therefore, prospective follow-up studies are less prone to differential recall bias of the exposure. Furthermore, prospective follow-up studies have the advantage of establishing the temporal relationship between exposure and disease, because the exposure occurred prior to the disease outcome. On the other hand, the losses to follow-up, lack of information on exposure during the follow-up period, and a short follow-up time period may pose an issue for the interpretation of the study results, especially for diseases such as cancers, which have long induction and latency periods and for exposures, such as hair dyes, which may change over time. As pointed out by Zahm *et al.*,[6] cohort studies that ascertain exposure early in life and observe subjects over time without repeated assessment will underestimate lifetime use, since the history of hair dye use is affected by the age at interview.

A retrospective cohort study by Hennekens *et al.*[7] was conducted in the US from 1972 to 1976 among married female nurses in 11 states. They found an increased risk for cancers of cervix uteri and the vagina and vulva for those who reported using permanent hair dyes. An increased risk of breast cancer was also reported for those who reported using permanent hair dyes for more than 20 years. In this study, information on both diseases and exposures was obtained from participants through postal questionnaires; therefore, potential misclassification for both disease and exposure was likely to have occurred in this study. A subgroup of the cohort population without cancer in 1976 was followed up until 1982. However, in the subsequent follow-up study, Green *et al.*[8] found no association between permanent hair dye uses and breast cancer risk.

A prospective follow-up study, the Cancer Prevention Study II (CPS-II), by the American Cancer Society[9] examined the relationship between permanent hair dye use and risk of fatal cancers in US women. This study involved 573,369 American Cancer Society volunteers aged 30 years and older, who were followed between 1982 and 1989. Information on permanent hair dye use, including never/ever use, colour, and frequency and duration of use, was collected upon enrollment using questionnaires. The participants' vital status was determined by personal inquiries in 1984, 1986 and 1988, and subsequently by linkage to the National Death Index. Thun *et al.*[9] reported that women who had ever used permanent hair dyes showed a relative risk (RR) of 0.9 (95% CI: 0.9,1.0) for all fatal cancer combined and an RR of 0.7 (95% CI: 0.5,0.9) for urinary system cancer compared to 'never users', while women who had used black hair dyes for 20 years or more showed an increased risk of fatal non-Hodgkin's lymphoma (RR = 4.4, 95% CI: 1.3, 15.2) and multiple myeloma (RR = 4.4, 95% CI: 1.1, 18.3). However, the increased risks were based on just three cases of non-Hodgkin's lymphoma and on only two cases of multiple myeloma. No association was reported for cancers of the mouth, breast, lung, bladder or cervix. The major strengths of this study include the large size of the initial cohort, prospective study design, information on duration of permanent hair dye use and colour of the products, and information on potential confounders. However, using mortality instead of incidence to define the disease outcome may introduce survival bias. The relatively small number of cases of each type of cancer may limit the study's power to investigate the association. Lack of information on non-permanent hair dye use may pose concern for interpreting the findings of the study.

In 1999, Altekruse *et al.*[10] reported results linking cancer mortality to permanent hair dye use from the CPS-II study through 1994, extending follow-up time from 7 to 12 years. When 'ever users' were compared to 'never users', the study found no association for all cancer combined. An odds ratio (OR) of 1.1 (95% CI: 1.0, 1.2) was found for hematopoietic cancers. The risk, however, increased with increasing duration of use for darker colour products. Women who reported using black dye for ten or more years experienced a significantly increased risk of death from non-Hodgkin's lymphoma (OR = 2.5, 95% CI: 1.1, 5.8) and multiple myeloma (OR = 3.1, 95% CI: 1.0, 9.6). No association was observed for other cancer sites.

In 2001, Henley and Thun[11] examined the relationship between bladder cancer deaths and permanent hair dye use in the CPS-II cohort through 1998, extending the follow-up time from 12 to 16 years. The RR associated with 'ever use' of permanent hair dyes after 7 years of follow-up (84 deaths among 'never users' and 15 deaths among 'ever users') was 0.6 (95% CI: 0.3–1.0),[9] and after 12 years of follow-up (154 deaths among 'never users' and 48 deaths among 'ever users'), was 1.0 (95% CI: 0.7,1.4).[10] After 16 years of follow-up (244 deaths among 'never users' and 92 deaths among 'ever users'), Henley and Thun[11] observed an RR of 1.1 (95% CI: 0.8–1.4) for women who reported ever using permanent hair dyes. After restricting the analysis to those who never smoked, the study[11] also did not observe an increased risk of bladder cancer associated with 'ever use' of permanent hair dye (RR = 0.9, 95% CI: 0.6, 1.5).

The Nurse's Health Study[12] also examined the relationship between permanent hair dye use and risks of incident lymphoma, leukemia and multiple myeloma. A total of 99,067 women aged 30–55 years were followed through 1990. After 8 years of follow-up, 244 incident hematopoietic cancers were identified, including 24 Hodgkin's lymphoma, 144 non-Hodgkin's lymphoma, 44 leukemia and 32 multiple myeloma. No positive associations were observed however, between 'ever use' of permanent hair dyes and risk of all haematopoietic cancers (OR = 0.9, 95% CI: 0.7, 1.2) or specific types, including Hodgkin's disease (OR = 0.9, 95% CI: 0.4, 2.1), non-Hodgkin's lymphoma (OR = 1.1, 95% CI: 0.8, 1.6), multiple myeloma (OR = 0.4, 95% CI: 0.2, 0.9), chronic lymphocytic leukemia (RR = 0.6, 95% CI: 0.3, 1.5), and other leukemias (RR = 0.8, 95% CI: 0.3, 1.9). The study, however, only asked several questions about permanent hair dye use; there was no repeated assessment for personal use of hair dye products during the follow-up period. The study also had no information on semi-permanent or temporary hair dye use. Lack of detailed information on hair dye use would undoubtedly introduce misclassification of exposure. If the misclassification was non-differential, it would cause an underestimation of the association between hair dye use and the risks of the disease of interest, potentially accounting for the null results reported in that study.

11.2.1.2 *Case-control Studies*

Compared to prospective follow-up studies, case-control studies usually have the advantage of including a larger number of cases in a relatively short study period. In population-based case-control studies, controls are randomly selected from the population which produced the cases, and therefore, are more likely to represent the population with regards to the major risk factors. However, if the refusal rate from potentially eligible study subjects is high, selection bias can be introduced into the study. In a hospital-based case-control study, controls are selected from other patients hospitalised in the same hospital where the cases are identified. Hospital-based case-control studies may introduce selection bias if the controls' hospitalisation is associated with the exposure of interest.

Bladder cancer. As reviewed by the IARC,[1] the seven case-control studies conducted before 1992 to investigate the relationship between hair dye use and risk of bladder cancer have produced inconsistent results (Table 11.1). Most of the studies included a small number of cases and controls. One large study, which included 2,982 cases and 5,782 controls, reported an increased risk of bladder cancer associated with ever using black hair dye products for men and women combined (OR = 1.4, 95% CI: 1.0, 1.9).[13]

Since the IARC review, Gago-Dominguez *et al.*[14] reported the results from a population-based case-control study in Los Angeles, California, conducted from 1987 to 1996 to investigate the relationship between hair dye use and bladder

Table 11.1 *Summary of the published literature on the relationship between personal use of hair dyes and risk of bladder cancer*

Authors (year of report)	Study design	Sex	Study population	Disease outcome	Analysed hair dye information	Findings
Henneckens *et al.* (1979)	Retrospective cohort, four years of follow-up USA, 1972–76	F	Cohort: 120,557 Cases: 5	Incidence	Permanent, duration	RR = 0.6 No trend with duration
Thun *et al.* (1994)	Prospective cohort, Seven years of follow-up USA, 1982–89	F	Cohort: 537,369 Cases: not specified	Mortality	Permanent, duration	RR = 0.6 Risk decreases with increasing duration
Altekruse *et al.* (1999)	Extending to 12 years of follow-up, USA, 1982–94	F	Cases: 202	Mortality	Permanent, duration	RR = 1.0 No trend with duration
Henley *et al.* (2001)	Extending to 15 years of follow-up, USA, 1982–98	F	Cases: 336	Mortality	Permanent, duration	RR = 1.1 No trend with duration
Howe *et al.* (1980)	Case-control study (neighbourhood), Canada, 1974–76	M/F	Cases: 480(M) 152(F) Controls: 1:1	Incidence	All hair dyes	F:OR = 0.7 M: no exposed controls
Hartge *et al.* (1982)	Case-control study (population-based), USA, 1977–78	M/F	Cases: 2249(M) 733(F) Controls: 4282(M) 1500(F)	Incidence	All hair dyes, colour, duration, frequency	M: OR = 1.1; F: OR = 0.9 No trend with frequency/ duration M+F: OR = 1.4 For dark hair dye use
Ohno *et al.* (1985)	Case-control study (population-based), Japan, 1976–78	F	Cases: 65 Controls: 143	Incidence	All hair dyes, frequency	RR = 1.7 for use of hair dyes more than once a month
Claude *et al.* (1986)	Case-control study (hospital-based), Germany, 1977–82	M/F	Cases: 340(M) 91(F) Controls: 1:1	Incidence	All hair dyes	M or F: no association

Table 11.1 (*Continued*)

Authors (year of report)	Study design	Sex	Study population	Disease outcome	Analysed hair dye information	Findings
Nomura *et al.* (1989)	Case-control study (population-based), USA, 1977–86	M/F	Cases: 195(M) 66(F) Controls: 2:1	Incidence	All hair dyes, duration	M: OR = 1.3; F: OR = 1.5 No trend with duration for either sex
Gago-Dominguez *et al.* (2001)	Case-control study (population-based), USA, 1987–96	M/F	Cases: 694(M) 203(F) Controls: 1:1	Incidence	All hair dyes type, duration frequency	M: no association F: OR = 1.9* for permanent users and increasing trend with frequency/duration
Gago-Dominguez *et al.* (2001)	Case-control study (population-based), USA, 1987–96	F	Cases: 124 Controls: 122	Incidence	Permanent, duration, frequency	OR = 2.7* for NAT2 slow phenotype, increasing trend frequency/duration
Gago-Dominguez *et al.* (2003)	Case-control study (population-based), USA, 1987–96	F	Cases: 159 Controls: 164	Incidence	Permanent, duration, frequency	OR = 2.9* for NAT2 slow phenotype OR = 2.5* for CYP1A2 slow phenotype OR = 6.8* for non-NAT1*10 genotype Increasing trend with frequency/duration

*95% confidence interval excludes null value
F: female; M: male

cancer risk. A total of 897 incident cases of bladder cancer and equal number of age-, sex-, and ethnicity-matched controls were included in this study. A two-fold increased risk of bladder cancer was found among women who had used permanent hair dyes at least once a month, compared to those who had never used hair dyes. The risk increased to 3.3-fold (95% CI: 1.3,8.4) for women who were regular (at least monthly) users for at least 15 years.

Breast cancer. The seven case-control studies conducted before 1992 had provided inconsistent results linking hair dye use to breast cancer risk.[1] Since then, four more case-control studies[15–18] of breast cancer and hair dye use have been conducted to investigate the relationship (Table 11.2).

Boice *et al.*[15] investigated the relationship between hair dye use and prevalent breast cancer risk among radiologic technologists. Questionnaires, containing information on medical history and hair dye use, were mailed to 105,385 women radiologic technologists certified by the American Registry of Radiologic Technologists for at least two years from 1926 to 1982 and known to be alive at the time of the study. A total of 528 breast cancer cases were identified from respondents who reported a history of breast cancer. A total of 2,628 controls were selected from those who reported no history of breast cancer, and individually matched to case (5 controls:1 case) on gender, date of birth within five years, and year of certification within two years. 'Ever use' of hair dye products was not associated with breast cancer risk (OR = 1.1, 95% CI: 0.9, 1.3) compared to 'never users'. No information was collected on type or colour of hair dyes, or frequency and duration of using the products.

In 1983–90, Cook *et al.*[16] conducted a population-based case-control study in western Washington, involving 844 incident cases and 960 controls, aged 45 years or less. In-person interviews were administered to collect detailed information on hair dye use. Women who had never used hair dye products were used as a reference group for all comparisons. An elevated risk of breast cancer was observed among women who reported using permanent hair dyes (OR = 1.2, 95% CI: 1.0, 1.6) and semi-permanent hair dyes (OR = 1.4, 95% CI: 1.0, 1.8). A 2.5-fold (95% CI: 1.6, 3.9) increased risk was found among women who reported using any hair dye products after bleaching. A significantly increased risk was also observed for women who reported using any rinse (OR = 1.7, 95% CI: 1.2, 2.5) and any frosting/tipping (OR = 1.5, 95% CI: 1.2, 2.0). Among women who reported using two or more types of hair dye products, a 3.1-fold (95% CI: 1.6, 6.1) increased risk of breast cancer was observed for those using hair dyes for 90 or more total episodes during their lifetime. A significantly increased risk was also observed for those with total lifetime exposures of 61 minutes and longer, age at first use less than 30, and time since first use fewer than 20 years.

In 1994–97, we conducted a case-control study[17] of hair dye use and breast cancer risk in Connecticut with 608 cases and 609 controls aged 30–80 years. No increased risk associated with the overall use of hair dye products or exclusive use of permanent or temporary types of hair dye products was found in this study. Among those who reported to have exclusively used semi-permanent types of hair colouring products, some of the ORs were elevated. However, none of the ORs relating to age at first use, duration of use, total number of applications and years

Table 11.2 *Summary of the published literature on the relationship between personal use of hair dyes and risk of breast cancer*

Authors (year of report)	Study design	Sex	Study population	Disease outcome	Analysed hair dye information	Findings
Henneckens et al. (1979)	Retrospective cohort, four years of follow-up, USA, 1972–76	F	Cohort: 120,557 Cases: 270	Incidence	Permanent, duration	RR = 1.1 RR = 1.5 for those used more than 20 years
Green et al. (1987)	Prospective cohort, six years of follow-up, USA, 1976–82	F	Cohort: 118,404 Cases: 353	Mortality	Permanent, duration frequency	RR = 1.1 No trend with frequency/duration
Thun et al. (1994)	Prospective cohort, seven years of follow-up, USA, 1982–89	F	Cohort: 537,369 Cases: not specified	Mortality	Permanent, duration	RR = 1.0 No trend with duration
Altekruse et al. (1999)	Extending to 12 years of follow-up, USA, 1982–94	F	Cases: 2,676	Mortality	Permanent, duration	RR = 0.9 No trend with duration
Kinlen et al. (1977)	Case-control study (hospital-based), UK, 1975–76	F	Cases: 191 Controls: 561	Incidence	Permanent or semi-permanent, duration	OR = 1.0 No trend with duration
Shore et al .(1979)	Case-control study (screening centre) USA, 1964–76	F	Cases: 129 Controls: 193	Incidence	Permanent, colour, frequency, duration	Significant association with cumulative exposure
Stavraky et al. (1979)	Case-control study (hospital-based/ neighbourhood), Canada, 1976–79	F	Cases: 85 Controls: Hospital (100) Neighbourhood (70)	Incidence	Permanent	RR = 1.3 RR = 1.1
Nasca et al. (1980)	Case-control study (population-based), USA, 1975–76	F	Cases: 118 Controls: 233	Incidence	Permanent or semi-permanent, frequency, latency	OR = 1.1 No trend with frequency/latency OR = 3.1* for use of hair dye to change colour

Wynder and Goodman (1983)	Case-control study (hospital-based), USA, 1979–81	F	Cases: 401 Controls: 625	Incidence	All hair dyes, type, frequency, duration	OR = 1.0 No dose-response relationship
Koenig *et al.* (1991)	Case-control study (screening centre) USA, 1977–81	F	Cases: 398 Controls: 790	Incidence	All hair dyes, frequency, duration	OR = 0.8 No trend with increasing use
Boice *et al.* (1995)	Case-control study (nested) USA, 1987–96	F	Cases: 528 Controls: 2,628	Prevalence	All hair dye	OR = 1.1
Cook *et al.* (1999)	Case-control study (population-based) USA, 1983–90	F	Cases: 844 Controls: 960	Incidence	All hair dyes, type, frequency, duration	No association for exclusive use of any type Significant increased risk for use of more than one type of hair dye No trend with frequency/duration
Zheng *et al.* (2002)	Case-control study (hospital and population-based), USA, 1994–97	F	Cases: 608 Controls: 609	Incidence	All hair dyes, type, colour frequency, duration	No association
Petro-Nustas *et al.* (2002)	Case-control study (convenient control), Jordan, 1996	F	Cases: 100 Controls: 1:1	Incidence	All hair dyes	OR = 8.62*

*95% confidence interval excludes null value

F: female

since first use were statistically significant. There was also no increased risk of breast cancer associated with exclusive use of dark or light hair-colouring products, or use of mixed types or colours of hair dye products in this study.

Petro-Nustas *et al.*[18] conducted a case-control study to investigate risk factors associated with breast cancer in Jordan women involving 100 cases and an equal number of controls. They reported that women who had ever used hair dye products experienced a 8.6-fold (95% CI: 3.3, 22.3) increased risk of breast cancer compared to those who had never used hair dye products. However, this study had several limitations, including a convenience sample of controls, lack of detailed information on hair dye use, and small sample size.

Lymphatic and hematopoietic tumors. Four case-control studies relating hair dye use and lymphatic and haematopoietic tumors were published before 1992[1] (Table 11.3). From 1980 to 1983, Cantor *et al.*[19] conducted a population-based case-control study among men in Iowa and Minnesota, USA. This study included a total of 578 leukemia cases, 622 non-Hodgkin's lymphoma cases, and 1,245 controls. They found a significantly increased risk of leukemia (OR = 1.8, 95% CI: 1.1, 2.7) and non-Hodgkin's lymphoma (OR = 2.0, 95% CI: 1.3, 3.0) among men who reported ever using hair dye products compared to those who had never used hair dye products. The low prevalence of hair dye use among men (less than 10%) limited this study's power to further examine the relationship by detailed characteristics of hair dye products and hair dye using pattern, although relatively large samples were involved in this study.

Zahm *et al.*[6] reported a population-based case-control study of hair dye use and the risk of lymphoma multiple myeloma, and chronic lymphocytic leukemia in eastern Nebraska. This study, conducted from 1983 to 1986, recruited a total of 385 non-Hodgkin's lymphoma cases, 70 Hodgkin's disease cases, 72 multiple myeloma cases, 56 chronic lymphocytic leukemia cases and 1,432 gender-and age-matched controls. For non-Hodgkin's lymphoma, a 50% increased risk was observed among women who had ever used any hair-colouring products, and the risk increased significantly with duration of use and younger age at first use of any hair dyes. The risk also varied by type of hair dye and by subtype of non-Hodgkin's lymphoma, with darker and permanent hair dye products showing the highest risk. The authors estimated that use of hair dyes could account for 35% of non-Hodgkin's lymphoma occurring in exposed women, and 20% in all (exposed and non-exposed) women. That study concluded that, if it is causal, hair dye use would explain a larger percentage of NHL cases among women than any other known or suspected risk factor. With very few cases of the other three types of cancers, the study had limited statistical power to investigate the risks by various characteristics of hair dye use.

Brown *et al.*[20] investigated the relationship between hair dye use and risk of multiple myeloma in white men from a population-based case-control study conducted in Iowa from 1981–84. This study included a total of 173 incident cases and 645 controls. They reported a 1.9-fold increased risk of multiple myeloma among 'ever users' based on 14 exposed cases. The greatest risk was observed among those who used hair dyes at least once a month for a year or more (OR = 4.3, 95% CI: 0.9, 19.7) based on just four exposed cases compared to those

Table 11.3 *Summary of the published literature on the relationship between personal use of hair dyes and risk of lymphatic and haematopoietic neoplasms*

Authors (year of report)	Study design	Sex	Study population	Disease outcome	Analyzed hair dye information	Findings
Non-Hodgkin's Lymphoma						
Grodstein *et al.* (1994)	Prospective cohort, eight years of follow-up, USA, 1982–90	F	Cohort: 99,067 Cases: 144	Incidence	Permanent, frequency, duration	RR = 1.1 for NHL overall RR = 1.5 for follicular No trend with frequency/duration
Thun *et al.* (1994)	Prospective cohort, seven years of follow-up, USA, 1982–89	F	Cohort: 537,369 Cases: 350	Mortality	Permanent, colour, duration	RR = 4.4* for using black permanent hair dyes more than 20 years No data for NHL subtype
Altekruse *et al.* (1999)	Extending to 12 years of follow-up, USA, 1982–94	F	Cases: 763	Mortality	Permanent, colour, duration	RR = 2.5* for using black permanent hair dyes between 10 and 19 years; OR = 2.1 for more than 20 years No data on NHL subtype
Cantor *et al.* (1988)	Case-control study (population-based), USA, 1980–83	M	Cases: 622 Controls: 1245	Incidence	All hair dyes, frequency	OR = 2.0* for NHL overall 'ever users' OR = 2.8* for follicular 'ever users'
Zahm *et al.* (1992)	Case-control study (population-based), USA, 1983–86	M/F	Cases: 201(M) 184(F) Controls: 725(M) 707(F)	Incidence	All hair dyes, type, colour, frequency, duration	F: OR = 1.7* for ever use permanent hair dyes OR = 2.0* for follicular ever use permanent OR = 2.5* for follicular ever use dark permanent No trend with frequency/duration M: No association

Table 11.3 (*Continued*)

Authors (year of report)	Study design	Sex	Study population	Disease outcome	Analyzed hair dye information	Findings
Holly *et al.* (1998)	Case-control study (population-based), USA, 1991–95	M/F	Cases: 385(M) 328(F) Controls: 989(M) 615(F)	Incidence	All hair dyes, type, colour, frequency, duration	F: no association M: OR = 2.0* for semi-permanent hair dyes OR = 2.4* for large cell lymphoma semi-permanent hair dye users No trend with frequency/duration
Miligi *et al.* (1999)	Case-control study (population-based), Italy, 1991–93	F	Cases: 611 Controls: 828	Incidence	All hair dyes colour, type	OR = 1.0 for ever use any hair dyes OR = 1.1 for permanent No data for NHL subtype
Schroeder *et al.* (2002)	Case-control study (population-based), USA, 1980–83	M	Cases: 248 Controls: 1245	Incidence	All hair dyes	OR = 1.8 for t(14;18)-positive NHL OR = 2.0* for t(14;18)-negative NHL
Zhang *et al.* (2004)	Case-control study (population-based), USA, 1996–2001	F	Cases: 601 Controls: 717	Incidence	All hair dyes type, colour, frequency, duration	OR = 2.1* for NHL overall ever use dark permanent hair dyes OR = 1.9* for follicular ever use permanent OR = 2.2* for follicular ever use light colour OR = 1.6* for B-cell ever use permanent/dark OR = 1.8* for low grade ever use light colour No clear trend with frequency/duration

Hodgkin's Disease

Grodstein *et al.* (1994)	Prospective cohort, eight years of follow-up, USA, 1982–90	F	Cohort: 99,067 Cases: 24	Incidence	Permanent, frequency, duration	RR = 0.9 for ever use No further analysis for frequency/duration
Thun *et al.* (1994)	Prospective cohort seven years of follow-up, USA, 1982–89	F	Cohort: 537,369 Cases: 31	Mortality	Permanent, duration	RR = 0.6 for ever use No trend with duration
Zahm *et al.* (1992)	Case-control study (population-based), USA, 1983–86	M/F	Cases: 35(M) 35(F) Controls: 725(M) 707(F)	Incidence	All hair dyes, type, colour frequency, duration	F: OR = 1.7 for ever use any hair dyes OR = 3.0* for ever use permanent Trend with decreasing age at first use M: OR = 1.7 or ever use any hair dye
Miligi *et al.* (1999)	Case-control study (population-based) Italy, 1991–93	F	Cases: 165 Controls: 828	Incidence	All hair dyes, colour, type	OR = 0.7 for ever use any hair dyes OR = 0.7 for ever use permanent

Multiple Myeloma

Grodstein *et al.* (1994)	Prospective cohort, eight years of follow-up, USA, 1982–90	F	Cohort: 99,067 Cases: 32	Incidence	Permanent, frequency, duration	RR = 0.4* for ever use No further analysis for frequency/duration
Thun *et al.* (1994)	Prospective cohort, seven years of follow-up, USA, 1982–89	F	Cohort: 537,369 Cases: 195	Mortality	Permanent, duration	RR = 1.1 for ever use RR = 1.4 for more than 20 years of use Risk was increasing with duration
Altekruse *et al.* (1999)	Extending to 12 years of follow-up, USA, 1982–94	F	Cases: 460	Mortality	Permanent, colour, duration	RR = 1.0 for ever use RR = 3.1* for more than 20 years of using dark permanent hair dyes

Table 11.3 *(Continued)*

Authors (year of report)	Study design	Sex	Study population	Disease outcome	Analyzed hair dye information	Findings
Zahm *et al.* (1992)	Case-control study (population-based), USA, 1983–86	M/F	Cases: 32(M) 40(F) Controls: 725(M) 707(F)	Incidence	All hair dyes, type, colour frequency, duration	F: OR = 1.8 for ever use any hair dyes OR = 2.8* for ever use permanent Trends with duration/decreasing age at first use M: OR = 1.8 or ever use any hair dye
Brown *et al.* (1992)	Case-control study (population-based), USA, 1981–84	M	Cases: 173 Controls: 650	Incidence	All hair dyes, frequency	OR = 1.9* for ever use any hair dyes OR = 4.3 for used more than one year and at least once a month
Herrinton *et al.* (1994)	Case-control study (population-based), USA, 1977–81	M/F	Cases: 360(M) 319(F) Controls: 931(M) 744(F)	Incidence	All hair dyes	F: OR = 1.1 for ever use any hair dyes M:OR = 1.3 for ever use any hair dyes
Miligi *et al.* (1999)	Case-control study (population-based), Italy, 1991–93	F	Cases: 134 Controls: 828	Incidence	All hair dyes, colour, type	OR = 0.8 for ever use any hair dyes OR = 1.0 for ever use permanent
Leukemia						
Grodstein *et al.* (1994)	Prospective cohort, eight years of follow-up, USA, 1982–90	F	Cohort: 99,067 Cases: 44 (all types)	Incidence	Permanent, frequency, duration	RR = 0.6 for chronic lymphocytic leukemia ever use RR = 0.8 for acute and chronic myelocytic leukemia and acute lymphocytic leukemia ever use No further analysis for frequency/duration

Thun *et al.* (1994)	Prospective cohort, seven years of follow-up, USA, 1982–89	F	Cohort: 537,369 Cases: 365	Mortality	Permanent, duration	RR = 0.8 for lymphoid leukemia ever use RR = 0.9 for myeloid and monocytic leukemia ever use RR = 1.0 for other leukemia ever use No trend with duration
Altekruse *et al.* (1999)	Extending to 12 years of follow-up, USA, 1982–94	F	Cases: 718	Mortality	Permanent, colour, duration	RR = 1.1 for ever use RR = 1.3 for more than 20 years of use Significant trend with duration RR = 1.5 for more than 20 years of using brown colour
Cantor *et al.* (1988)	Case-control study (population-based), USA, 1980–83	M	Cases: 577 Controls: 1245	Incidence	All hair dyes, frequency	OR = 1.8* for leukemia (all types) ever use OR = 2.9* for dysmyelopietic syndrome 'ever use'
Zahm *et al.* (1992)	Case-control study (population-based), USA, 1983–86	M/F	Cases: 37(M) 19(F) Controls: 725(M) 707(F)	Incidence	All hair dyes, type, colour, frequency, duration	F: OR = 1.0 and 0.8 for chronic lymphocytic leukemia 'ever use' any or permanent hair dyes respectively No further analyses by frequency/ duration M: 1.0 for chronic lymphocytic leukemia ever use any hair dyes No further analyses by frequency/ duration

Table 11.3 *(Continued)*

Authors (year of report)	Study design	Sex	Study population	Disease outcome	Analyzed hair dye information	Findings
Mele *et al.* (1994)	Case-control study (hospital-based), Italy, 1987–90	M/F	Cases: 346(M) 273(F) Controls: 399(M) 762(F)	Incidence	All hair dyes, colour, frequency, duration	F: OR = 1.6 for acute myeloid leukemia use >10 years OR = 1.7 for acute myeloid leukemia using dark colour more than 10 years OR = 2.0 for acute lymphocytic leukemia use >10 years OR = 1.9 for acute lymphocytic leukemia using dark colour more than 10 years M: OR = 2.1 for chronic myeloid leukemia ever use any hair dyes or dark colour dyes
Mele *et al.* (1995)	Case-control study (hospital-based), Italy, 1987–90	M/F	Cases: 18(M) 18(F) Controls: 399(M) 762(F)	Incidence	All hair dyes, colour, frequency duration	F: OR = 1.6 for acute myeloid leukemia use >10 years OR = 1.7 for acute myeloid leukemia using dark colour more than 10 years OR = 2.0 for acute lymphocytic leukemia use >10 years OR = 1.9 for acute lymphocytic leukemia using dark color more than 10 years
Markovic-Denic *et al.* (1995)	Case-control study (hospital-based), Yugoslavia	M/F	Cases: 130 Controls: 130	Incidence	All hair dyes	OR = 1.8* for chronic lymphocytic leukemia ever use any hair dyes

Mele *et al.* (1996)	Case-control study (hospital-based), Italy, 1986–90	M/F	Cases: 14(M) 25(F) Controls: 56(M) 100(F)	Incidence	All hair dyes, colour, frequency, duration	F+M: OR = 5.3* for essential thrombocythemia using dark colour for more than 10 years
Ido *et al.* (1996)	Case-control study (hospital-based), Japan, 1992–93	M/F	Cases: 69(M) 47(F) Controls: 1:1	Incidence	All hair dyes	F: OR = 2.5 for myelodysplastic syndromes ever use any hair dyes M: OR = 1.2 myelodysplastic syndromes ever use any hair dyes
Nagata *et al.* (1999)	Case-control study (hospital-based), Japan, 1995–96	M/F	Cases: 69(M) 42(F) Controls: 459(M) 371(F)	Incidence	All hair dyes, frequency, duration	M+F: OR = 2.0* for myelodysplastic syndromes ever use any hair dyes Significant trend with frequency/duration F: OR = 2.9* for myelodysplastic syndromes ever use any hair dyes Significant trend with frequency/duration
Miligi *et al.* (1999)	Case-control study (population-based), Italy, 1991–93	F	Cases: 260 Controls: 828	Incidence	All hair dyes, colour, type	OR = 0.9 for ever use any hair dyes OR = 2.0 for ever use dark permanent hair dyes
Bjork *et al.* (2001)	Case-control study, Sweden, 1976–93	M/F	Cases: 226 Controls: 251	Incidence	All hair dyes	OR = 0.4* for chronic myeloid leukemia ever use any hair dyes

*95% confidence interval excludes null value
F: female; M: male

who had never used hair dyes. The major limitation involved in this study is the very small number of cases and the rare exposure to hair dyes among men.

Herrinton *et al.*[21] conducted a population-based case-control in four areas of the US from 1977 to 1981 involving 689 incident cases and 1,681 controls. While no association was found for women who were regular hair dye users compared to those who had never used hair dyes, an elevated risk was observed among men who were regular hair dye users compared to those who had never used hair dyes. A limitation of the study is the lack of detailed information on personal hair dye uses.

Mele *et al.*[22] conducted a hospital-based case-control study in Italy between 1986 and 1990 to evaluate the behavioural and environmental risk factors in relation to leukemia and pre-leukemia, acute promyeolocytic leukemia,[23] and essential thrombocyhemia.[21] A total of 619 leukemia and pre-leukemia patients and 1,161 controls were recruited in this study. Among men, dark colour hair dyes were associated with the risk of acute myeloid leukemia (OR = 1.6, 95% CI: 0.4, 5.5), refractory anaemia with excess of blasts (OR = 1.5, 95% CI: 0.3, 8.1), and chronic myeloid leukemia (OR = 2.1, 95% CI: 0.6, 7.2). However, none of the associations achieved statistical significance. The very small number of men reporting hair dye use limited the power to make further analyses by frequency, duration and colour of the products in this group. Among women, elevated risks of acute myeloid leukemia (OR = 1.6, 95% CI: 0.8, 3.0) and acute lymphocytic leukemia (OR = 2.0, 95% CI: 0.7, 5.7) were observed for those who reported using hair dyes for more than ten years compared to those who had never used hair dyes. After stratification by colour of the products used, an increased risk was observed for those using dark colour products but not for light-colour-product users. Acute promyelocytic leukemia was not associated with hair dye product use in a subsequent study by Mele *et al.*[23]

In a hospital-based case-control study involving 39 essential thrombocythemia cases and 156 controls, Mele *et al.*[24] reported an increased risk of essential thrombocythemia for those who reported using dark hair dyes (OR = 2.1, 95% CI: 0.9, 3.8) compared to those who had never used any hair dyes. The risk was found to be the highest for those who reported using dark hair dyes for more than 10 years (OR = 5.3, 95% CI: 1.4, 19.9).

In an another hospital-based case-control study, conducted with 130 cases of chronic lymphocytic leukemia and an equal number of controls in Yugoslavia, Markovic-Denic *et al.*[25] reported an OR of 1.8 (95% CI: 1.0, 3.3) for those who had ever used any hair dyes compared to those who had never used any hair dyes. The study, however, did not investigate the relationship by type or colour of hair dyes, or frequency or duration of using hair dye products.

Ido *et al.*[26] conducted a hospital-based case-control study in Japan from 1992–93 to investigate the risk factors for myelodysplastic syndromes involving 116 cases and 116 controls. Information on hair dye use was derived through self-administered questionnaires. No information on type or colour of hair dyes, frequency or duration of use was collected in this study. A 2.5-fold increased risk (95% CI: 0.97, 6.41) was observed among women who had ever used hair dyes, while no association was observed among men.

Holly *et al.*[27] conducted a population-based case-control study in the San Francisco Bay area counties in 1991–95 involving 747 non-Hodgkin's lymphoma cases and 1,633 controls to investigate the relationship between hair dye use and risk of non-Hodgkin's lymphoma. Detailed information on hair dye use was collected using in-person interviews. No association was found for overall hair dye use or by type or colour of hair dye products used by women. While a two-fold increased risk was observed among men (OR = 2.0, 95% CI: 1.4, 3.8) who reported using semi-permanent hair dyes compared to non-users of any hair dyes, there were no clear trends by duration or frequency of use of the products.

Miligi *et al.*[28] reported results from a population-based case-control study conducted in Italy linking hair dye use to risks of hematolymphopietic malignancies among women. This study recruited a total of 611 cases of non-Hodgkin's lymphoma, 260 leukemia, 134 multiple myeloma, 165 Hodgkin's disease, and 828 controls. There was no increased risk for ever using any hair dyes associated with non-Hodgkin's lymphoma, leukemia, multiple myeloma and Hodgkin's disease. An increased risk of leukemia was found among women who reported using dark permanent hair dyes (OR = 2.0, 95% CI: 1.1, 3.8). A limitation of the study is the lack of information on time and duration of using hair dye products.

A hospital-based case-control study in Japan in 1995–96 investigated the association between hair dye use and risk of myelodysplastic syndromes.[29] A total of 111 cases and 830 controls were included in this study. Ever using hair dyes was significantly associated with the risk of myelodysplastic syndromes in both men and women combined (OR = 2.0, 95% CI: 1.2, 3.4) and in women only (OR = 2.9, 95% CI: 1.4, 6.0). The risks increased with increasing duration and number of hair dye uses. The study, however, provided no information by type or colour of hair dye products used.

A case-control study of chronic myeloid leukemia was conducted in Sweden.[30] The study reported a reduced risk of chronic myeloid leukemia among women who used hair dyes regularly (OR = 0.4, 95% CI: 0.2, 0.7). A major limitation of the study is that information on ever using hair dyes was obtained from interviews of proxies in 81% of cases and 14% of controls. The study also had no information on type, colour, frequency or duration of hair colouring products used.

From 1995 to 2001, a population-based case-control study involving 601 histologically confirmed incident cases and 717 controls was conducted in Connecticut to investigate the relationship between personal hair dye use and non-Hodgkin's lymphoma.[31] In-person interviews were administered to collect detailed information on hair dye use. Comparing to women who had never used hair dye products during their lifetime, women who started using hair dye products before 1980 (see Chapter 13 for changes in regulatory issues) had a slightly increased risk (OR = 1.3, 95% CI: 1.0, 1.8), for permanent product use (OR = 1.4, 95% CI: 1.0, 1.9), and for darker colour product use (OR = 1.4, 95% CI: 1.0, 1.9). Women who started using hair dye products in 1980 or later, however, showed no increased risk. In detailed analyses restricted to women who started using hair dyes before 1980, for both permanent and semi-permanent hair colouring products, users of darker hair-colouring products showed a 50% increased risk of NHL. Among permanent darker hair-colouring product users, risk was the highest among those with the longest duration of use (OR = 2.1, 95% CI 1.0–4.0), and with the largest number of lifetime hair-colouring product applications (OR = 1.7, 95% CI 1.0–2.8). Investigation by NHL subtype showed a significantly increased risk of NHL for follicular lymphoma and B-cell lymphoma among women who used permanent hair-colouring products. A significantly increased risk was also observed for B-cell lymphoma among women who used dark hair-colouring products, and for follicular lymphoma and low grade NHL among women who used light hair-coloring products. In detailed analyses restricted to women who started using hair dyes before 1980, however, no association was found. This is the first study to investigate the relationship between personal hair dye use and risk of non-Hodgkin's lymphoma by time period of use.

Cancers at other sites. As reviewed by the IARC working group in 1993,[1] several case-control studies have evaluated the potential relationship between personal hair dye use and risk of cancers of the cervix,[32] ovary,[32] lung,[32] kidney,[32] brain,[33] salivary gland,[34] and malignant melanoma.[35,36] Too few studies are available on these cancer sites to allow reviewers to make any conclusion as to whether personal hair dye use is associated with the risk of these cancers.

There was only one study published after 1992 linking hair dye use to other cancer sites (Table 11.4). In a hospital-based case-control study, Tzonou *et al.*[37] investigated the relationship between hair dye use and ovarian cancer risk in Athens, Greece. A total of 189 women with histologically confirmed common malignant epithelial tumors of the ovary were compared with 200 hospital controls. A significantly increased risk of ovarian cancer was observed for women who reported using hair dyes greater than four times per year (OR = 2.2, 95% CI: 1.2, 3.9) when compared to 'never users'. The risk was increased with increasing frequency of using hair dye products (P = 0.007).

Childhood cancers. Olshan *et al.*[38] conducted a case-control study to investigate risk factors for Wilms' tumour, a childhood kidney tumour, in the US and Canada (Table 11.5). A total of 200 cases and 233 controls were recruited in this study from 1984 to 1986. A slightly increased risk of Wilms' tumour was associated with maternal use of hair dye products during pregnancy (OR = 1.4, 95% CI: 0.7, 2.9). The study provided no information by type, colour, frequency or duration of hair dye products used.

Bunin *et al.*[39] conducted a case-control study to investigate risk factors for the two most common types of brain tumour (astrocytic glioma and primitive neuroectodermal tumour) in children in the US and Canada from 1986 to 1989. The study included 155 astrocytic glioma cases, 166 primitive neuroectodermal tumour cases and 321 controls, aged 6 years and younger. Information on maternal use of hair dye products during pregnancy was collected through telephone interviews. Maternal use of hair dye products was not associated with the risk of astrocytic glioma or primitive neuroectodermal tumour in this study. Again, this study provided no information by type or colour, duration or frequency of use of hair dye products.

Holly *et al.*[40] conducted a large population-based case-control study of childhood brain tumours on the west coast of the US to investigate risk factors including

Table 11.4 *Summary of the published literature on the relationship between personal use of hair dyes and risk of other cancer sites*

Authors (year of report)	Study design	Sex	Study population	Disease outcome	Analysed hair dye information	Findings
Henneckens et al. (1979)	Retrospective cohort, four years of follow-up, USA, 1972–76	F	Cohort: 120,557 Cases: 773(all)	Incidence	Permanent, duration	No association for all cancer combined RR = 1.4* for cervix uteri RR = 2.6* for vagina and vulva
Thun et al. (1994)	Prospective cohort, seven years of follow-up, USA, 1982–89	F	Cohort: 537,369 Cases: not specified	Mortality	Permanent, duration	No association for all cancers combined No association for cancers of oral cavity and pharynx digestive system, respiratory system, cervix uteri, vagina and vulva, brain and other nervous system
Altekruse et al. (1999)	Extending to 12 years of follow-up, USA, 1982–94	F	Cases: 18599(all) 144 (oral cavity and pharynx) 4754 (digestive) 4100 (respiratory) 134 (cervix uteri) 55 (vagina & vulva) 613 (brain & nervous) 1405 (ovarian)	Mortality	Permanent, duration	No association for all cancers combined No association for cancers of oral cavity and pharynx digestive system, respiratory system, cervix uteri, vagina and vulva, brain and other nervous system No association for ovarian cancer

Table 11.4 *(Continued)*

Authors (year of report)	Study design	Sex	Study population	Disease outcome	Analysed hair dye information	Findings
Stavarky *et al.* (1981)	Case-control study (hospital-based/ neighbourhood), Canada, 1976–79	F	Endometrium: 36 Cervix: 38 Ovary: 58 Lung: 70 Kidney+bladder: 35 Controls: 2:1	Incidence	Permanent, semi-permanent	OR = 1.6 for endometrium cancer OR = 0.7 for cervix cancer OR = 1.6/0.2 for ovarian cancer OR = 0.8/1.7 for lung cancer OR = 1.1 for kidney and bladder
Holman and Armstrong *et al.* (1983)	Case-control study (population-based), Australia, 1980–81	F	Cases: 511 Controls: 1:1	Incidence	All hair dyes	No association between malignant melanoma and permanent hair dyes Significant trend for total applications of semi-permanent and temporary dyes for Hutchinson's melanotic freckle
Osterlind *et al.* (1988)	Case-control study (population-based), Denmark, 1982–85	M/F	Cases: 474 Controls: 926	Incidence	All hair dyes	OR = 0.6* for malignant melanoma
Ahlbom *et al.* (1986)	Case-control study (population-/hospital-based), Sweden, 1980–81	M/F	Cases: 47(M) Controls: population: 56(M) 36(F); hospital: 88(M) 109(F)	Incidence	All hair dyes	OR = 0.8 for astrocytoma (hospital control) OR = 1.5 for astrocytoma (population control)

Burch *et al.* (1987)	Case-control study (hospital-based), Canada, 1977–81	M/F	Cases: 137(M) 78(F) Controls: 1:1	Incidence	All hair dyes	OR = 2.0* for brain cancer ever use hair dye or hair spray
Spitz *et al.* (1990)	Case-control study (hospital-based), USA, 1985–89	M/F	Cases: 37(M) 27(F) Controls: 74(M) 54(F)	Incidence	All hair dyes, duration	F: OR = 4.1* for salivary gland cancer ever use any hair dyes; risk higher for longer duration; no trend with frequency M: no association
Tzonou *et al.* (1993)	Case-control study (hospital-based), Greece, 1989–91	F	Cases: 189 Controls: 200	Incidence	All hair dyes, frequency	OR = 2.2* for ovarian cancer ever use any hair dyes more than four times per year Significant trend with frequency

*95% confidence interval excludes null value
F: female; M: male

Table 11.5 *Summary of the published literature on the relationship between personal use of hair dyes and risk of childhood cancers*

Authors (year of report)	Study design	Sex	Study population	Disease outcome	Analyzed hair dye information	Findings
Kramer *et al.* (1987)	Case-control study (population-based), USA 1970–79	M/F	Cases: 64(M) 40(F) Controls: 52(M) 49(F)	Incidence	All hair dyes (pregnancy)	OR = 3.0* for neuroblastoma
Bunin *et al.* (1987)	Case-control study (population-based), USA, 1970–83	M/F	Cases: 52(M) 36(F) Controls: 1:1	Incidence	All hair dyes (pregnancy)	OR = 3.6* for Wilms' tumour
Kuijten *et al.* (1990)	Case-control study (population-based), USA, 1980–86	M/F	Cases: 91(M) 72(F) Controls: 1:1	Incidence	All hair dyes (pregnancy)	OR = 0.9 for astrocytoma
Olshan *et al.* (1993)	Case-control study (population-based), USA, 1984–86	M/F	Cases: 94(M) 106(F) Controls: 123(M) 110(F)	Incidence	All hair dyes (pregnancy)	OR = 1.4 for Wilms' tumour
Bunin *et al.* (1994)	Case-control study (population-based), USA/Canada, 1986–89	M/F	Cases: glioma: 81(M) 74(F) PNET: 100(M) 66(F)	Incidence	All hair dyes (pregnancy)	OR = 0.7 for astrocytic glioma OR = 1.1 for primitive neuroectodermal tumor (PNET)
Holly *et al.* (2002)	Case-control study (population-based), USA, 1984–91	M/F	Cases: 297(M) 242(F) Controls: 801	Incidence	All hair dyes, type, colour frequency (one month before or pregnancy)	OR = 2.0 for childhood brain tumour for semi-permanent

*95% confidence interval excludes null value
F: female; M: male

maternal use of hair colouring products. Detailed information on type of hair dye products and time of use during pregnancy was collected, while data on colour and frequency of use were not collected. This study did not find an association between maternal use of hair dye products during pregnancy and risk of childhood brain tumours.

11.2.2 Hair Dye Use and Human Cancer Risk by Gene Type

So far, few studies have investigated the relationship between hair dye use and human cancer risk by genetic polymorphisms. Gago-Dominguez *et al.*[14] recently examined the relationship between permanent hair dyes and bladder cancer risk by *N*-acetyltransferase-2 (NAT2) phenotype among females in the Los Angeles Bladder Cancer Study. A total of 124 cases and 122 controls were included in this analysis. Among NAT2 slow acetylators, exclusive use of permanent hair dyes was associated with 2.7-fold increased risk of bladder cancer (95% CI: 1.0, 7.2) and the risk appeared to increase with increasing duration and frequency of use of hair dye products. Hair dye use was not associated with bladder cancer risk among NAT2 fast acetylators (OR = 1.1, 95% CI: 0.4, 2.7).

In 2003, Gago-Dominguez *et al.*[41] reported the effects of a series of potential arylamine-metabolising genotypes/phenotypes (GATM1, GSTT1, GSTP1, NAT1, NAT2, CYP1A2) on the relationship between permanent hair dye uses and bladder cancer risk among female participants in the Los Angeles Bladder Cancer Study. This study included a total of 159 cases and 164 controls. 'Never users' of hair dye products were used as the comparison group for all comparisons. That study again reported a 2.5-fold increased risk of bladder cancer associated with permanent hair dye use among women exhibiting CYP1A2 'slow' phenotype not among women exhibiting CYP1A2 'rapid' phenotype. A significant dose-response relationship for duration and frequency of use of hair dye products was confined to CYP1A2 'slow' phenotype but not to CYP1A2 'rapid' phenotype. As observed in their previous study, they also found that exclusive use of permanent hair dyes was associated with an increased risk of bladder cancer only among NAT2 slow acetylators not among NAT2 fast acetylators. Exclusive permanent hair dye use was also associated with a significant increased risk of bladder cancer (OR = 6.8, 95% CI: 1.7, 27.4) among non-smoking women with the non-NAT1*10 genotype, and the risk showed a significant dose-response by frequency and duration of use of permanent hair dye products (P = 0.01, 0.009 respectively). On the other hand, no such association was found among non-smoking women with the NAT1*10 genotype. No such modifying effects were observed for GSTT1, GSTM1 and GSTP1 genotypes.

In a population-based case-control study conducted in Iowa and Minnesota, Schroeder *et al.*[42] investigated risk factors for NHL subtype defined by the t(14;18) translocation gene. The t(14;18) translocation can cause constant production of bcl-2 protein, a key inhibitor of apoptosis, by joining the immunoglobulin heavy chain gene (IGH) on chromosome 14 with the bcl-2 gene on chromosome 18.[43–46] It was reported that bcl-2 activation may increase the risk of NHL by preventing cell death and allowing subsequent oncogenic mutations to develop and persist.[47,48] A total of

248 cases and 1,245 controls were involved in this analysis. The study found similar association between hair dye uses and NHL risk based on translocation status (t(14:18)-positive: OR = 1.8, 95% CI: 0.9, 3.7; t(14;18)-negative: OR = 2.1, 95% CI: 1.3, 3.4)).

11.2.3 Occupational Exposure to Hair Dyes and Risk of Human Cancer

Occupational exposure to hair-colouring products as hairdressers, barbers or beauticians may differ from personal use of hair colouring products by the general population because hairdressers/barbers/beauticians also may be exposed to greater amounts of many other products such as hair spray, nail and skin care products *etc.*, in addition to hair dye products. The actual exposure by hairdressers/barbers/ beauticians is affected by their use of protective gloves, *etc.* About one-third of tinters wear gloves when applying hair dyes;[49] hence, their dermal exposures may be much less than those of the person whose hair is coloured.[6] Even without gloves, percutaneous absorption of hair dye constituents may be effectively impeded by the horny surface of the hands, and by the lack of sebaceous glands in the palms. This is very different from the case of personal hair dye users where hair dye constituents can readily be absorbed through the scalp which contains numerous sebaceous glands,[50,51] and where penetrations occur largely along hair follicle and sebaceous gland orifices.[52]

Occupational studies usually involve multiple statistical comparisons made by numerous occupations and industries. Some associations would be expected based on chance alone. However, occupational studies usually involve small sample sizes with fewer exposed cases, and therefore, have limited power to detect underlying associations. Furthermore, studies of occupational hair dye exposure usually use the general population as the comparison group. The general population, however, has a high proportion of people who have used hair dyes, and thus have been exposed to the same substances as the hairdressers/barbers/beauticians. Therefore, the observed associations would be anticipated to be smaller than they would be if a pure non-exposed group had been used as the reference.

A brief review of epidemiological studies of occupational hair dye exposure in relation to human cancers is presented below.

11.2.3.1 Bladder Cancer

Twelve cohort studies[2,49,53–62] and eighteen case-control studies[14,56,63–78] have investigated the relationship between occupational exposure to hair colouring products and bladder cancer risk (Table 11.6). Most of the epidemiological studies do not support a positive association between working as a hairdresser, barber or beautician and an increased risk of bladder cancer. For the six cohort studies which reported an increased risk, Guberan *et al.*[56] reported a two-fold increased risk for male hairdressers based on 11 cases from a cohort study conducted in Switzerland. Malker *et al.*[58] observed a 50% increased risk associated with employment as

Table 11.6 *Summary of the published literature on the relationship between occupational exposure to hair dyes and risk of bladder cancer*

Authors (year of report)	Study design	Sex	Study population	Disease outcome	Job title	Findings
Alderson *et al.* (1980)	Follow-up, England and Wales, 1961–78	M	Cohort: 1,831 Cases: 7	Mortality	Hairdressers	RR = 1.3
Kono *et al.* (1983)	Follow-up, Japan, 1953–77	F	Cohort: 7,736 Cases: 0	Mortality	Beauticians	Expected bladder cancer death is 1.0
Teta *et al.* (1984)	Follow-up, USA, 1935–78	M/F	Cohort: 1,805(M) 11,845(F) Cases: 14(F)	Incidence	Cosmetologists	F: SIR = 1.4
Guberan *et al.* (1985)	Follow-up, Switzerland, 1970–80	M/F	Cohort: 703(M) 677(F) Cases: 11(M)/2(F)	Incidence	Hairdressers	F: RR = 1.3 M: RR = 2.1*
Malker *et al.* (1987)	Follow-up, Sweden, 1961–79	M	Cohort: not specified Cases: 52	Incidence	Barbers and beauticians	SIR = 1.5*
Lynge and Thygesen (1988)	Follow-up, Denmark, 1970–80	M/F	Cohort: 4,874(M) 9,497(F) Cases: 41(M)/7(F)	Incidence	Hairdressers	F: RR = 1.8 M: RR = 2.1*
Skov *et al.* (1990)	Follow-up, Norway, 1961–84	M/F	Norway cohort: 2,149(M)/4,356(F) Cases: 23(M)/11(F)	Incidence	Norway: hairdressers, beauticians;	Norway: F: RR = 1.5 M: RR = 1.5
	Sweden, 1961–79		Sweden cohort: 6,522(M)/16,942(F) Cases: 54(M)/6(F)		Sweden: hairdressers, beauticians;	Sweden: F: RR = 0.4 M: RR = 1.5*
	Finland, 1971–80		Finland cohort: 428(M)/9,138(F) Cases: 0(M)/3(F)		Finland: hairdressers, barbers;	Finland: F: RR = 1.7 M: no observed cases

Table 11.6 *(Continued)*

Authors (year of report)	Study design	Sex	Study population	Disease outcome	Job title	Findings
	Denmark, 1970–80		Denmark cohort: 4,874(M)/9,497(F) Cases: 41(M)/7(F)		Denmark: hairdressers, barbers	Denmark: F: RR = 1.8 M: RR = 2.1*
Pukkala *et al.* (1992)	Follow-up, Finland, 1970–87	F	Cohort: 3,637 Cases: 1	Incidence	Hairdressers	RR = 0.4
Dolin and Cook-Mozaffari (1992)	Follow-up, England and Wales, 1965–80	M	Cohort: 2,457 Cases: 6	Mortality	Hairdressers	SMR = 1.2 for most recent employment
Skov *et al.* (1994)	Follow-up, Denmark, 1970–87	M/F	Cohort: 1,777(M) 4,160(F) Cases: 67(M)/12(F)	Incidence	Hairdressers	F: RR = 1.2 M: RR = 1.6*
Lamba *et al.* (2001)	Follow-up, USA, 1984–95	M/F	Hairdresser cohort: 3,035(M)/23,582(F)	Mortality	Hairdressers, barbers	Hairdressers: F: MOR = 1.4* for whites
			Cases: 6(white men) 88(white women) 15(black women)			MOR = 1.2 for blacks M: MOR = 0.6 for whites no observed death among blacks
			Barber cohort: 11,704(M)/400(F)			Barbers: M: MOR = 1.0 for whites
			Cases:75(white men) 9(black men)			MOR = 1.5 for blacks
Czene, *et al.* (2003)	Follow-up, Sweden, 1960–98	M/F	Cohort: 6,824(M) 38,866(F) Cases: 87(M)/51(F)	Incidence	Hairdressers	F: SIR = 1.1 M: SIR = 1.2

Wynder *et al.* (1963)	Case-control (Hospital-based), USA, 1957–61	M	Cases: 300 Controls: 300	Incidence	Hairdressers	4 exposed cases and no exposed controls
Dunham *et al.* (1968)	Case-control (Hospital-based), New Orleans, 1958–64	M	Cases: 265 Controls: 272	Incidence	Barbers	OR = 2.8 for most recent occupation based on 4 exposed cases
Anthony and Thomas, (1970)	Case-control (Hospital-based), (UK, 1959–67	M/F	Cases: 812(M) 218(F) Controls: 340(M) 50(F)	Incidence	Hairdressers	F: no exposed cases M: OR = 4.1 for predominant occupation based on 4 exposed cases
Cole *et al.* (1972)	Case-control (population-based), USA, 1967–68	M/F	Cases: 356(M) 105(F) Controls: 374(M) 111(F)	Incidence	Hairdressers or barbers	F: 1 exposed cases M: OR = 0.6 for ever employment based on exposed cases
Viadana *et al.* (1976)	Case-control (Hospital-based), USA, 1956–65	M	Cases: unspecified Controls: unspecified	Incidence	Barbers	RR = 1.5 for ever employment based on 5 exposed cases RR = 1.8 for 5 or more years employment
Howe *et al.* (1980)	Case-control (Neighborhood-based), Canada, 1974–76	M/F	Cases: 480(M) 152(F) Controls: 1:1	Incidence	Hairdressers or barbers	F: no exposed controls M: no exposed controls
Vineis and Magnani *et al.* (1985)	Case-control (Hospital-based), Italy, 1978–83	M	Cases: 512 Controls: 596	Incidence	Hairdressers, barbers	OR = 0.9 for ever employment based on 9 exposed cases

Table 11.6 *(Continued)*

Authors (year of report)	Study design	Sex	Study population	Disease outcome	Job title	Findings
Morrison *et al.* (1985)	Case-control (Population-based), USA, UK, Japan, 1976–78	M	Cases: 430(USA), 399(UK), 226(Japan) Controls: 397(USA), 493(UK), 443(Japan)	Incidence	Barbers	OR = 1.0 for USA based on 7 exposed cases OR = 1.3 for UK based on 2 exposed cases OR = 1.0 for Japan based on 1 exposed cases
Risch *et al.* (1988)	Case-control (population-based), Canada, 1979–82	M/F	Cases: 826 Controls: 792	Incidence	Hairdressers, barbers	F: OR = 1.0 based on 9 exposed cases M: OR = 0.7 based on 11 exposed cases No trend with duration of employment
Silverman *et al.* (1989)	Case-control (Population-based) USA, 1977–78	M	Cases: 2,100 Controls: 3,874	Incidence	Hairdressers, barbers	OR = 1.3 for ever employment as either based on 28 exposed cases OR = 2.8 for ever employment as hairdressers based on 7 exposed cases
Silverman *et al.* (1990)	Case-control (Population-based), USA, 1977–78	F	Cases: 652 Controls: 1,266	Incidence	Hairdressers	OR = 1.4 for ever employment based on 17 exposed cases

Kunze *et al.* (1992)	Case-control (Hospital-based), Germany, 1977–85	M/F	Cases: 531(M) 144(F) Controls: 1:1	Incidence	Hairdressers	OR = 1.7 for ever employment based on 10 exposed cases
Cordier *et al.* (1993)	Case-control (Hospital-based), France, 1984–87	M/F	Cases: 658(M) 107(F) Controls: 1:1	Incidence	Hairdressers	M: OR = 2.2 for ever employment Based on 5 exposed cases
Siemiatycki *et al.* (1994)	Case-control (population-based), Canada, 1979–86,	M	Cases: 484 Controls: 2,412	Incidence	Hairdressers, barbers	OR = 1.0 for 10 or more years employment Based on 4 exposed cases
Teschke *et al.* (1997)	Case-control (population-based), Canada, 1990–91	M/F	Cases: 88(M)/17(F) Controls: 112(M) 27(F)	Incidence	Hairdressers, barbers	M+F: OR = 2.5 for ever employment based on 1 exposed cases
Sorahan *et al.* (1998)	Case-control, UK, 1991–93	M/F	Cases: 624(M) 179(F) Controls: 2135	Incidence	Hairdressers	M+F: OR = 1.7 for ever employment based on 11 exposed cases
Gago-Dominguez *et al.* (2001)	Case-control study (population-based), USA, 1987–96	M/F	Cases: 1,514 Controls: 1:1	Incidence	Hairdressers, barbers	M+F: OR = 5.1* for 10 or more years employment based on 14 exposed cases
Zheng *et al.* (2002)	Case-control (population-based), USA, 1986–89	M/F	Cases: 1,135(M) 317(F) Controls: 1601(M) 833(F)	Incidence	Barber shops	M: OR = 1.8 for ever employment for at least 5 years based on 5 exposed cases

*95% confidence interval excludes null value

F: female; M: male

barbers and beauticians in a male cohort from Sweden based on 52 cases. Lynge and Thygesen[59] found a two-fold increased risk for male hairdressers and barbers based on 41 cases in Denmark cohort. Skov *et al.*[57] reported a two-fold increased risk for male hairdressers and barbers in a Danish cohort based on 41 cases and a 50% increased risk for male hairdressers and beauticians in a Swedish cohort based on 54 cases. However, the prevalence of smoking was much higher in the Sweden cohort than in the general population based on a previous survey (74% vs. 46%). Due to the lack of data on smoking in this cohort, the observed association was likely to be confounded by smoking. Another cohort study from Denmark also reported an increased risk for male hairdressers based on 67 cases.[62] A recent cohort study from USA reported a 40% increased mortality of bladder cancer among white female hairdressers and barbers.[60] Out of eighteen case-control studies, only one[14] reported a significantly increased risk of bladder cancer associated with ten or more years of employment as a hairdresser or barber.

11.2.3.2 Breast Cancer

Nine cohort or registry-based epidemiological studies[2,49,54–56,60,79–81] have investigated the relationship between occupational exposure to hair dyes and breast cancer risk (Table 11.7). One registry-based study[60] suggested that employment as hairdressers and barbers was associated with 10% increased risk of breast cancer mortality, while another registry-based study[80] reported that hairdressers and beauticians experienced 30% increased risk of breast cancer incidence. Another seven cohort or registry-based studies observed no association between occupational exposure to hair dye products and breast cancer risk. There have been four case-control studies[82–85] that have examined the relationship between occupational exposure to hair dye products and breast cancer risk in female hairdressers.

11.2.3.3 Ovarian Cancer

Nine cohort or registry-based studies[2,49,54–56,60,86–88] have reported the results linking employment as hairdressers, barbers or beauticians to ovarian cancer risk (Table 11.8). Only one study, from Finland, observed a statistically significant association between ovarian cancer and employment as a hairdresser among women[49] based on 21 cases. All other studies showed no association[2,54,55,60,86–88] or a non-significantly reduced risk.[56]

11.2.3.4 Lung Cancer

Eight cohort studies[2,49,53,54,56,57,59,62] have investigated the relationship between working as hairdressers, barbers or beauticians and risk of lung cancer (Table 11.9). Three of them reported a significantly increased risk of lung cancer among hairdressers, barbers or beauticians.[2,57,62] One study reported a non-significantly increased risk among female hairdressers in Finland.[49] Other studies found no association between occupational exposure to hair dye products and lung cancer risk.[2,53,54,56] A case-control study[89] from the USA by Osorio *et al.* reported no association between employment as cosmetologists and lung cancer.

Table 11.7 Summary of the published literature on the relationship between occupational exposure to hair dyes and risk of breast cancer

Authors (year of report)	Study design	Sex	Study population	Disease outcome	Job title	Findings
Kono et al. (1983)	Follow-up, Japan, 1953–77	F	Cohort: 7,736 Cases: 5	Mortality	Beauticians	RR = 0.6
Teta et al. (1984)	Follow-up, USA, 1935–78	F	Cohort: 11,845 Cases: 204	Incidence	Cosmetologists	SIR = 1.0
Guberan et al. (1985)	Follow-up, Switzerland, 1970–80	F	Cohort: 677 Cases: 7	Incidence	Hairdressers	RR = 0.6
Pukkala et al. (1992)	Follow-up, Finland, 1970–87	F	Cohort: 3,637 Cases: 70	Incidence	Hairdressers	RR = 1.2
Morton et al. (1995)	Follow-up, Polland, Vancouver 1963–77	F	Cohort: 2053 Cases: 28	Incidence	Beauticians	Incidence rate not signifcantly different from rate for all women
Calle et al. (1998)	Follow-up, USA, 1982–91	F	Cohort: 46,433 person-years Cases: 16	Mortality	Beauticians	RR = 1.0 for the longest employment
Pollan et al. (1999)	Follow-up, Sweden, 1971–89	F	Cohort: not specify Cases: 199	Incidence	Hairdressers, beauticians	RR = 1.3* for current employment
Lamba et al. (2001)	Follow-up, USA, 1984–95	F	Hairdresser cohort: 23,582 Cases: 1,027 (white women), 153 (black women).	Mortality	Hairdressers	MOR = 1.1 for white women MOR = 1.2 for black women

Table 11.7 *(Continued)*

Authors (year of report)	Study design	Sex	Study population	Disease outcome	Job title	Findings
Czene *et al.* (2003)	Follow-up, Sweden, 1960–98	F	Cohort: 38,866 Cases: 913	Incidence	Hairdressers	SIR = 1.0
Koenig *et al.* (1991)	Case-control (hospital-based) USA, 1977–81	F	Cases: 398 Controls: 790	Incidence	Beauticians	OR = 3.0 for five or more years empolyment based on 12 exposed cases
Habel *et al.* (1995)	Case-control (population-based) USA, 1988–90	F	Cases: 536 Controls: 487	Incidence	Cosmetologists	OR = 1.5 for the longest employment based on seven exposed cases
Coogan *et al.* (1996)	Case-control (population-based) USA, 1988–91	F	Cases: 6,888 Controls: 9,529	Incidence	Hairdressers, cosmetologists	OR = 0.8 for usual occupation based on 72 exposed cases
Band *et al.* (2000)	Case-control (population-based) British Columbia 1988–89	F	Cases: 1018 Controls: 1020	Incidence	Hairdressers, barbers	OR = 5.5* for ever employment based on 14 exposed cases OR = 2.7 for usual employment based on seven exposed cases

*95% confidence interval excludes null value
F: female

Table 11.8 *Summary of the published literature on the relationship between occupational exposure to hair dyes and risk of ovarian cancer*

Authors (year of report)	Study design	Sex	Study population	Disease outcome	Job title	Findings
Kono *et al.* (1983)	Follow-up, Japan, 1953–77	F	Cohort: 7,736 Cases: 5	Mortality	Beauticians	RR = 1.4
Teta *et al.* (1984)	Follow-up, USA, 1935–78	F	Cohort: 11,845 Cases: 48	Incidence	Cosmetologists	SIR = 1.3
Guberan *et al.* (1985)	Follow-up, Switzerland, 1970–80	F	Cohort: 677 Cases: 1	Incidence	Hairdressers	RR = 0.5
Pukkala *et al.* (1992)	Follow-up, Finland, 1970–87	F	Cohort: 3,637 Cases: 21	Incidence	Hairdressers	RR = 1.6*
Boffetta *et al.* (1994)	Follow-up, Sweden, Norway, Finland, Denmark 1971–87	F	Cohort: 29,279 Cases: 127	Incidence	Hairdressers	SIR = 1.2
Vasama Neuvonen *et al.* (1999)	Follow-up, Finland, 1971–95	F	Cohort: not specify Cases: 57(barbers & hairdressers) 3(beauticians)	Incidence	Hairdressers, barbers	SIR = 1.3 for hairdressers and barbers SIR = 1.0 for beauticians
Lamba *et al.* (2001)	Follow-up, USA, 1984–95	F	Hairdresser cohort: 23,582 Cases: 285(white women), 37(black women)	Mortality	Hairdressers	MOR = 1.0 for white women MOR = 1.2 for black women

Table 11.8 *(Continued)*

Authors (year of report)	Study design	Sex	Study population	Disease outcome	Job title	Findings
Shields *et al.* (2002)	Follow-up, Sweden, 1971–89	F	Cohort: not specify Cases: 51	Incidence	Barbers, beauticians	RR = 1.2
Czene *et al.* (2003)	Follow-up, Sweden, 1960–98	F	Cohort: 38,866 Cases: 192	Incidence	Hairdressers	SIR = 1.1

*95% confidence interval excludes null value

F: female

Table 11.9 *Summary of the published literature on the relationship between occupational exposure to hair dyes and risk of lung cancer*

Authors (year of report)	Study design	Sex	Study population	Disease outcome	Job title	Findings
Alderson *et al.* (1980)	Follow-up, England and Wales 1961–78	M	Cohort: 1,831 Cases: 52	Mortaliity	Hairdressers	RR = 1.0
Kono *et al.* (1983)	Follow-up, Japan, 1953–77	F	Cohort: 7,736 Cases: 9	Mortality	Beauticians	SMR = 1.2
Guberan *et al.* (1985)	Follow-up, Switzerland, 1970–80	M/F	Cohort: 703(M) 677(F) Cases: 8(M)/3(F)	Incidence	Hairdressers	SIR not specified but not signifcant
Lynge and Thygesen (1988)	Follow-up, Denmark, 1970–80	M/F	Cohort: 4,881(M) 9,499(F) Cases: 56(M)/12(F)	Incidence	Hairdressers	F: RR = 1.1 M: RR = 1.1
Skov *et al.* (1990)	Follow-up, Norway, 1961–84	M/F	Norway cohort: 2,149(M)/4,356(F) Cases: 47(M)/16(F)	Incidence	Norway: hairdressers, beauticians;	Norway: F: RR = 1.4 M: RR = 1.6*
	Sweden, 1961–79		Sweden cohort: 6,522(M)/16,942(F) Cases: 98(M)/31(F)		Sweden: hairdressers, beauticians;	Sweden: F: RR = 1.6* M: RR = 1.5*
	Finland, 1971–80		Finland cohort: 428(M)/9,138(F) Cases: 3(M)/2(F)		Finland: hairdressers, barbers,	Finland: F: RR = 0.5 M: RR = 1.5
	Denmark, 1970–80		Denmark cohort: 4,874(M)/9,497(F) Cases: 56(M)/12(F)		Denmark: hairdressers, barbers	Denmark: F: RR = 1.1 M: RR = 1.1
Pukkala *et al.* (1992)	Follow-up, Finland, 1970–87	F	Cohort: 3,637 Cases: 13	Incidence	Hairdressers	RR = 1.7

Table 11.9 *(Continued)*

Authors (year of report)	Study design	Sex	Study population	Disease outcome	Job title	Findings
Skov *et al.* (1994)	Follow-up, Denmark, 1970–87	M/F	Cohort: 1,777(M) 4,160(F) Cases: 127(M)/31(F)	Incidence	Hairdressers	M: RR = 1.2* F: RR = 1.0
Czene *et al.* (2003)	Follow-up, Sweden, 1960–98	M/F	Cohort: 6,824(M) 38,866(F) Cases: 141(M) 160(F)	Incidence	Hairdressers	F: SIR = 1.4* M: SIR = 1.4*
Osorio *et al.* (1986)	Case-control (cosmetologists), USA, 1972–82	F	Cases: 81	Incidence	Cosmetologists	PIR = 1.4

*95% confidence interval excludes null value

F: female; M: male

11.2.3.5 Non-Hodgkin's Lymphoma

Two cohort studies[60,62] reported a significantly increased risk of non-hodgkin's lymphoma associated with employment as hairdressers among women (Table 11.10). Another five cohort studies[2,55,86,90,91] did not find a significant association between employment as hairdressers, barbers or beauticians and risk of non-Hodgkin's lymphoma, although most of them[55,86,90,91] reported an evaluated risk. Most of the eight case-control studies[28,92–98] reported an increased risk of non-Hodgkin's lymphoma associated with employment as hairdressers, barbers or beauticians, although only one study[98] achieved statistical significance among black female beauticians.

11.2.3.6 Hodgkin's Disease

Four cohort or registry-based epidemiological studies[2,60,62,99] have investigated the relationship between occupational exposure to hair dyes and risk of Hodgkin's disease (Table 11.10). Robinson *et al.*[90] reported a proportionate cancer mortality ratio of 2.0 (95% CI: 1.3, 2.9) for female hairdressers and cosmetologists. Skov *et al.*[62] also reported a standardised incidence ratio of 2.0 (95% CI: 0.7, 4.4) for male hairdressers. No association was observed from two other studies.[2,60]

Three case-control studies[28,92,94] have investigated the relationship between employment as hairdressers, barbers and beauticians and risk of Hodgkin's disease. None of them have reported a significant association between occupational exposure to hair dye products and risk of Hodgkin's disease.

11.2.3.7 Multiple Myeloma

Cohort or registry-based epidemiological studies[2,49,55,60,100,101] do not support an association between multiple myeloma and employment as hairdressers barbers and beauticians (Table 11.10). Out of ten case-control studies that have investigated the relationship between employment as hairdressers, barbers or beauticians and risk of multiple myeloma,[21,28,94,95,102–107] only three of them reported an increased risk. An Italian case-control study[28] reported a statistically significant increased risk of multiple myeloma among women who self-reported ever being employed as a hairdresser, barber or beautician. A case-control study from the USA[21] observed a six-fold increased risk for multiple myeloma among females who reported working as a hairdresser for 2–5 years, but not for those who reported working as a hairdresser longer than five years. A mortality case-control study[106] reported a 70% increased risk for multiple myeloma associated with occupation as a barber as classified on the death certificate among white males.

11.2.3.8 Leukemia

Nine cohort or registry-based epidemiological studies[2,49,53–55,60,62,90,101] and two case-control studies[28,95] have investigated the relationship between employment as hairdressers, barbers and beauticians and risk of leukemia (Table 11.10). None of

Table 11.10 *Summary of the published literature on the relationship between occupational exposure to hair dyes and risk of lymphatic and haematopoietic neoplasms*

Authors (year of report)	Study design	Sex	Study population	Disease outcome	Job title	Findings
Non-Hodgkin's Lymphoma						
Teta *et al.* (1984)	Follow-up, USA, 1935–78	F	Cohort: 11,845 Cases: 22	Incidence	Cosmetologists	SIR = 1.3
Shibata *et al.* (1989)	Follow-up, Japan, 1976–87	M/F	Cohort: 4,615(M) 3,701(F) Cases: 1(M)/1(F)	Mortality	Barbers	M: RR = 0.4 F: RR = 1.4
Skov *et al.* (1991)	Follow-up, Denmark, 1970–80	M/F	Cases: 6(M)/7(F)	Incidence	Hairdressers	M: RR = 1.3 F: RR = 2.0
Skov *et al.* (1994)	Follow-up, Denmark, 1970–87	M/F	Cohort: 1,777(M) 4,160(F) Cases: 12(M)/16(F)	Incidence	Hairdressers	F: RR = 1.9* M: RR = 1.2
Boffetta *et al.* (1994)	Follow-up, Sweden, Norway, Finland, Denmark 1971–87	F	Cohort: 29,279 Cases: 36	Incidence	Hairdressers	SIR = 1.2

Lamba et al. (2001)	Follow-up, USA, 1984–95	M/F	Hairdresser cohort: 3,035(M)/ 23,582(F) Cases: 43(white men), 1(black men) 227(white women), 17(black women) Barber cohort: 11,704(M)/400(F) Cases:72(white men) 5(black men)	Mortality	Hairdressers, barbers	Hairdressers: F: MOR = 1.2* for whites MOR = 1.1 for blacks M: MOR = 1.5* for whites MOR = 0.5 for blacks Barbers: M: MOR = 0.9 for whites MOR = 0.8 for blacks
Czene et al. (2003)	Follow-up, Sweden, 1960–98	M/F	Cohort: 6,824(M) 38,866(F) Cases: 29(M)/64(F)	Incidence	Hairdressers	F: SIR = 0.9 M: SIR = 0.9
Giles et al. (1984)	Case-control (population-based), Australia, 1972–80	F	Cases: 116 Controls: 1:1	Incidence	Hairdressers	five exposed cases and no exposed controls
Persson et al. (1989)	Case-control (population-based), Sweden, 1964–86	M/F	Cases: 106 Controls: 275	Incidence	Hairdressers	M+F: OR = 2.2 based on one exposed cases
Blair et al. (1993)	Case-control (population-based), USA, 1980–83	M	Cases: 622 Controls: 1245	Incidence	Cosmetologists, barbers	OR = 2.1 for ever employed based on seven exposed cases OR = 2.7 for ever employed in barbershop
Cote et al. (1993)	Case-control (death certificate) USA, 1987–89	M/F	Cases: 2153 Controls: 8612	Mortality	Cosmetologists, beauticians	OR = 0.7 among AIDs patients based on nine exposed cases

Table 11.10 (*Continued*)

Authors (year of report)	Study design	Sex	Study population	Disease outcome	Job title	Findings
Figgs *et al.* (1995)	Case-control (death certificate) USA, 1984–89	M/F	Cases: 11,990(M) 11,900(F) Controls: 59,950(M) 59,500(F)	Mortality	Beauticians	OR = 2.5* for black females based on ten exposed cases
Holly *et al.* (1997)	Case-control (population-based), USA, 1988–95	M	Cases: 312 Controls: 420	Incidence	Hairdressers, cosmetologists	OR = 2.3 for ever employed based on two exposed cases
Miligi *et al.* (1999)	Case-control (population–based), Italy, 1991–93	F	Cases: 611 Controls: 828	Incidence	Hairdressers, beauticians, barbers	OR = 1.9 for ever employed based on nine exposed cases
Costantini *et al.* (2001)	Case-control (population-based), Italy, 1991–93	M	Cases: 811(M) Controls: 828	Incidence	Hairdressers, beauticians, barbers	OR = 0.6 for five or years employment based on five exposed cases
Hodgkin's Disease						
Skov *et al.* (1994)	Follow-up, Denmark, 1970–87	M/F	Cohort: 1,777(M) 4,160(F) Cases: 6(M)/3(F)	Incidence	Hairdressers	F: RR = 0.9 M: RR = 2.0
Robinson *et al.* (1999)	Follow-up, USA, 1984–95	F	Cohort: not specified Cases: 26	Mortality	Hairdressers, cosmetologists	PCMR = 2.0*

Reference	Study design	Sex	Subjects	Type	Occupation	Result
Lamba *et al.* (2001)	Follow-up, USA, 1984–95	M/F	Hairdresser cohort: 3,035(M)/23,582(F) Cases: 6(white men), 1(black men) 23(white women), 4(black women) Barber cohort: 11,704(M)/400(F) Cases: 5(white men) 0(black men)	Mortality	Hairdressers, barbers	Hairdressers: F: MOR = 1.4 for whites MOR = 2.4 for blacks M: MOR = 0.9 for whites MOR = 1.8 for blacks Barbers: M: MOR = 0.9 for whites
Czene *et al.* (2003)	Follow-up, Sweden, 1960–98	M/F	Cohort: 6,824(M) 38,866(F) Cases: 8(M)/11(F)	Incidence	Hairdressers	F: SIR = 0.6 M: SIR = 1.2
Giles *et al.* (1984)	Case-control (population-based), Australia, 1972–80	F	Cases: 32 Controls: 1:1	Incidence	Hairdressers	two exposed cases and no exposed controls
Persson *et al.* (1989)	Case-control (population-based), Sweden, 1964–86	M/F	Cases: 54 Controls: 275	Incidence	Hairdressers	M+F: OR = 2.7 based on one exposed case
Miligi *et al.* (1999)	Case-control (population-based), Italy, 1991–93	F	Cases: 165 Controls: 828	Incidence	Hairdressers, beauticians, barbers	OR = 2.1 for ever employed based on five exposed cases
Multiple Myeloma						
Teta *et al.* (1984)	Follow-up, USA, 1935–78	F	Cohort: 11,845 Cases: 3	Incidence	Cosmetologists	SIR = 0.7
McLaughlin *et al.* (1988)	Follow-up, Sweden, 1961–79	M	Cohort: 3,067 Cases: 11	Incidence	Beauticians	SIR = 1.3

Table 11.10 *(Continued)*

Authors (year of report)	Study design	Sex	Study population	Disease outcome	Job title	Findings
Pukkala *et al.* (1992)	Follow-up, Finland, 1970–87	F	Cohort: 3,637 Cases: 1	Incidence	Hairdressers	RR = 0.4
Linet *et al.* (1994)	Follow-up, Sweden, 1961–79	F	Cohort: not specified Cases: 9	Incidence	Hairdressers, beauticians,	SIR = 1.4
Lamba *et al.* (2001)	Follow-up, USA, 1984–95	M/F	Hairdresser cohort: 3,035(M)/23,582(F) Cases: 7(white men), 3(black men) 102(white women), 30(black women) Barber cohort: 11,704(M)/400(F) Cases: 49(white men) 11(black men)	Mortality	Hairdressers, barbers	Hairdressers: F: MOR = 1.2 for whites MOR = 1.2 for blacks M: MOR = 1.0 for whites MOR = 3.2* for blacks Barbers: M: MOR = 1.2 for whites F: MOR = 1.2 for blacks
Czene *et al.* (2003)	Follow-up Sweden, 1960–98	M/F	Cohort: 6,824(M) 38,866(F) Cases: 18(M)/31(F)	Incidence	Hairdressers	F: SIR = 1.3 M: SIR = 1.2
Giles *et al.* (1984)	Case-control (population-based), Australia, 1972–80	F	Cases: 59 Controls: 1:1	Incidence	Hairdressers	No exposed cases and one exposed control
Flodin *et al.* (1987)	Case-control (population-based) Sweden, 1981–83	M/F	Cases: 75(M)/56(F) Controls: 200(M) 231(F)	Incidence	Hairdressers	F+M: OR = 3.3 for ever employed based on one exposed case

Reference	Study type	Sex	Cases/Controls	Endpoint	Occupation	Result
Boffetta *et al.* (1989)	Case-control (Nested), USA, 1982–86	M/F	Cases: 74(M)/54(F) Controls: 296(M) 216(F)	Incidence	Cosmetologists beauticians, barbers	F+M: no exposed cases and four exposed controls
Eriksson-Karlsson *et al.* (1992)	Case-control (population-based), Sweden, 1982–86	M/F	Cases: 141(M) 134(F) Controls: 1:1	Incidence	Hairdressers, cosmetologists	F+M: OR = 0.7 for ever employed based on two exposed cases
Pottern *et al.* (1992)	Cases-control (population-based), Denmark, 1970–84	F	Cases: 607 Controls: 2596	Incidence	Hairdressers	OR = 0.7 for most recent employment based on one exposed cases
Heineman *et al.* (1992)	Case-control (population-based), Denmark, 1970–84	M	Cases: 835 Controls: 2,979	Incidence	Occupational exposure to hair dyes	OR = 3.6
Herrinton *et al.* (1994)	Case-control (population-based), USA, 1977–81	M/F	Cases: 360(M) 319(F) Controls: 931(M) 744(F)	Incidence	Hairdressers	F: OR = 1.3 for ever employed based on 12 exposed cases; OR = 6.6* for 2–5 years employment based on five exposed cases M: OR = 1.5 for ever employed based on one exposed case
Figgs *et al.* (1994)	Case-control (death certificate), USA, 1984–89	M/F	Cases: 6095(M) 6053(F) Controls: 30,475(M) 30,265(F)	Mortality	Barber	OR = 1.7* of white males based on 26 exposed cases
Miligi *et al.* (1999)	Case-control (population-based), Italy, 1991–93	F	Cases: 134 Controls: 828	Incidence	Hairdressers, beauticians, barbers	OR = 11.1* for ever employed based on three exposed cases
Costantini *et al.* (2001)	Case-control (population-based), Italy, 1991–93	M	Cases: 133(M) Controls: 918(M)	Incidence	Hairdressers, beauticians, barbers	OR = 2.2 for five or years employment based on five exposed cases

Table 11.10 (*Continued*)

Authors (year of report)	Study design	Sex	Study population	Disease outcome	Job title	Findings
Leukemia						
Alderson et al. (1980)	Follow-up, England and Wales, 1961–78	M	Cohort: 1,831 Cases: 3	Mortaliity	Hairdressers	RR = 1.1
Kono et al. (1983)	Follow-up, Japan, 1953–77	F	Cohort: 7,736 Cases: 6	Mortality	Beauticians	RR = 1.4
Teta et al. (1984)	Follow-up, USA, 1935–78	M/F	Cohort: 1,805(M) 11,845(F) Cases: 14(F)	Incidence	Cosmetologists	F: SIR = 1.2
Shibata et al. (1989)	Follow-up, Japan, 1976–87	M/F	Cohort: 4,615(M) 3,701(F) Cases: 3(M)/0(F)	Mortality	Barbers	F: no observed cases M: RR = 1.2
Pukkala et al. (1992)	Follow-up, Finland, 1970–87	F	Cohort: 3,637 Cases: 4	Incidence	Hairdressers	RR = 1.0
Linet et al. (1994)	Follow-up, Sweden, 1961–79	F	Cohort: not specified Cases: 9	Incidence	Hairdressers, beauticians,	SIR = 1.4
Skov et al. (1994)	Follow-up, Denmark, 1970–87	M/F	Cohort: 1,777(M) 4,160(F) Cases: 13(M)/8(F)	Incidence	Hairdressers	F: RR = 0.9 M: RR = 1.0

Lamba *et al.* (2001)	Follow-up, USA, 1984–95	M/F	Hairdresser cohort: 3,035(M)/23,582(F) Cases: 11(white men), 1(black men) 200(white women), 26(black women) Barber cohort: 11,704(M)/400(F) Cases: 97(white men) 10(black men)	Mortality	Hairdressers, barbers	Hairdressers: F: MOR = 1.2 for whites MOR = 1.3 for blacks M: MOR = 0.4 for whites MOR = 0.5 for blacks Barbers: M: MOR = 1.1 for whites F: MOR = 1.1 for blacks
Czene *et al.* (2003)	Follow-up, Sweden, 1960–98	M/F	Cohort: 6,824(M) 38,866(F) Cases: 29(M)/57(F)	Incidence	Hairdressers	F: SIR = 1.0 M: SIR = 1.0
Miligi *et al.* (1999)	Case-control (population-based), Italy, 1991–93	F	Cases: 260 Controls: 828	Incidence	Hairdressers, beauticians, barbers	OR = 2.2 for ever employed based on five exposed cases
Costantini *et al.* (2001)	Case-control (population-based), Italy, 1991–93	M	Cases: 383(M) Controls: 918(M)	Incidence	Hairdressers, beauticians, barbers	OR = 1.0 for five or more years employment based on five exposed cases

*95% confidence interval excludes null value
F: female; M: male

these studies have reported a significantly increased risk associated with occupational exposure to hair dye products, although one case-control study[28] observed an elevated risk of leukemia associated with employment as a hairdresser, barber or beautician. Thus, evidence from these studies does not support an association between occupational exposure to hair dyes and the risk of leukemia.

11.2.3.9 Cancers at Other Sites

Epidemiological studies have also investigated the relationship between occupational exposure to hair dye products and risk of cancers of other sites. For example, a cohort study from Sweden by Czene *et al.*[2] reported a significantly increased risk of cancers of the pancreas, cervix, and skin *in situ* among female hairdressers, and cancers of upper aerodigestive tracts and colorectal cancer among male hairdressers. A death certificate study in the US by Lambda *et al.*[60] observed a significant increased risk of stomach cancer among white male, pharynx cancer among black male, and pancreatic cancer among white female hairdressers. In a population-based case-control study from Sweden, Swanson *et al.*[108] found a greater than two-fold increased risk of salivary gland cancer associated with employment as a hairdresser. Another population-based case-control study from Italy, France, Switzerland and Spain[109] reported a two-fold increased risk of laryngeal/hypopharyngeal cancer associated with employment as a barber or hairdresser.

11.2.3.10 Childhood Cancers

One cohort study[110] reported an evaluated risk of Wilms' tumor associated with paternal occupation as a hairdresser, but not associated with maternal occupation as a hairdresser (Table 11.11). Five case-control studies[111–115] have investigated the relationship between maternal occupation as a hairdresser, barber or beautician and childhood cancer. Only one study[115] observed a statistically significant association between maternal occupation as a hairdresser or barber and childhood neuroblastoma. All other studies either found a non-significantly increased risk[111,112,114] or no association.[113]

11.3 Biological Plausibility

Some permanent and semi-permanent hair-colouring products and their constituents were found to be mutagenic in *Salmonella*,[4,116] *Escherichia coli*[117] and in *Drosophila melanogaster*[118] in early studies. Ames *et al.*[4] showed that 89% of the commercial permanent and semi-permanent hair dye products that contain various aromatic amines or aromatic nitrocompounds are mutagenic in short-term tests. Searle *et al.*[116] also showed strong mutagenicity of two dyes (2-nitro-*p*-phenylenediamine (2-NPPD), 4-nitro-*o*-phenylenediamine (4-NOPD)) in the *Salmonella* test. Blijleven[119] tested mutagenicity of four hair dye constituents given orally to adult male *Drosophila melanogaster*, and found that all four dyes

Table 11.11 Summary of the published literature on the relationship between parental occupational exposure to hair dyes and risk of childhood cancer

Authors (year of report)	Study design	Sex	Study population	Disease outcome	Job title	Findings
Mutanen and Hemminki et al. (2001)	Follow-up, Sweden, 1958–96	M/F	Cohort: not specify Cases: 6 (Wilms' tumor)	Incidence	Hairdressers	SIR = 10.6* for paternal employment SIR = 1.0 for maternal employment
Kuijten et al. (1992)	Case-control (population-based), USA, 1980–86	M/F	Cases: 163 (astrocytoma) Controls: 1:1	Incidence	Hairdressers	OR = 2.5 for maternal employment before conception based on seven exposed cases OR = 1.5 for maternal employment during pregnancy based on five exposed cases OR = 3.0 for maternal employment postnatally based on four exposed cases
MeKean-Cowdin et al. (1998)	Case-control (population-based), USA, 1984–91	M/F	Cases: 298(M)/242(F) (brain tumour) Controls: 448(M) 353(F)	Incidence	Hairdressers	F+M: OR = 1.5 for maternal employment during pregnancy based on seven exposed cases OR = 1.3 for maternal employment before pregnancy base on 11 exposed cases

Table 11.11 *(Continued)*

Authors (year of report)	Study design	Sex	Study population	Disease outcome	Job title	Findings
Smulevich *et al.* (1999)	Case-control (population-based), Russia, 1986–88	M/F	Cases: 593 Controls: 1,181	Incidence	Hairdressers, beauticians	F+M: OR = 2.0 for maternal employment at two month before pregnancy based three exposed cases
Olshan *et al.* (1999)	Case-control (popultion-based), USA and Canada 1992–96	M/F	Cases: 279(M)/225(F) (neuroblastoma) Controls: 251(M) 253(F)	Incidence	Hairdressers, barbers	F+M: OR = 3.3 for paternal employment based on two exposed cases OR = 2.8* for maternal employ ment based on 24 exposed cases
Cordier *et al.* (2001)	Case-control (population-based), Australia, Israel, France, Canada, Italy, Spain, USA 1976–94		Cases: 651(M)/567(F) (brain tumours) Controls: 1,216(M) 1,007(F)	Incidence	Hairdressers	F+M: OR = 1.1 for maternal employment

*95% confidence interval excludes null value
M: male; F: female

tested were mutagenic with a peak mutagenic activity in metabolically active germ cells. In another study, Mohn *et al.*[117] showed mutagenicity of *p*-phenylenediamine, 2,4-diaminoanisole sulfate, 2,4-diaminotoluene (which is no longer in use) and 4-NOPD in *Escherchia coli*.

Hair-colouring products and their constituents were also suggested to be carcinogenic and teratogenic. Some studies show a strong correlation between carcinogenicity and mutagenicity. In a study of more than 300 chemicals, McCann and Ames[120] found that more than 90% of the carcinogens tested were mutagenic. Some animal studies[116,121–128] also support hair colorants or constituents as being carcinogenic. One study showed that hair dye constituents caused a higher incidence of skeletal malformations and other effects in foetuses from treated mice.[129] The concern as to possible mutagenic, carcinogenic, teratogenic and other adverse effects is heightened by studies demonstrating that hair dyes and their constituents can be absorbed through the skin of humans and animals. Marshall and Palmer[130] reported a 45-year-old woman who complained of a dark discoloration of the urine following the application of hair dye at her local beauty salon. Kiese *et al.*[131] examined the urine of five volunteers who had used a simplified hair dye preparation for 40 minutes on the hair and scalp. The authors found that the diacetyl derivative of 2,5-diaminotoluene (the *para* compound) was excreted in the urine. Maibach *et al.*[132] reported the results from a study of eight healthy adult male volunteers who had their scalp hair dyed with two commercially distributed semi-permanent hair dyes. The authors found that the urine of four out of eight individuals contained a purple metabolite of one of the dye components, and this material was excreted at various intervals (one to four days) after application. Marzulli *et al.*[133] studied ten adult male subjects who were given a marketed hair dye containing 2% lead acetate for daily use (once per day) over a period of three months. The study found that the axillary and pubic hair lead levels rose from <6 to 41 ppm at the start and from 27 to 466 ppm at the conclusion of the experiment, which suggests that lead is indeed absorbed into the body when lead-containing hair dyes are applied to scalp hair. It is estimated that a woman undergoing one hair dyeing (with about 4 g of amines) could absorb as much as 40 mg (1%) of hair dye chemicals (precursors, products and side-products) through the scalp.[4]

If hair dyes and their constituents are carcinogens for humans as suggested, but not proven, by some of previous cited studies, then how do they affect human cancer risk? The current literature does not seem to allow us to draw a definitive conclusion. However, we do know that many carcinogens and mutagens are known to be chromosome-breaking agents.[116,120] Kirkland *et al.*[50] found that women who were occupationally exposed to hair dyes and also had their own hair dyed had a statistically significant excess of chromosomal damage in the peripheral-blood lymphocytes (mainly chromatid breaks). Studies[116,120] showed that some hair colorants and constituents induced a considerable number of chromosome and chromatid gaps and breaks in cultured human peripheral blood lymphocytes at concentrations between 50 μg ml^{-1} and 100 μg ml^{-1}. In a study using cultured prostate cells from Chinese hamster, Kirkland *et al.*[51] found that two widely used proprietary hair colorants produced a marked increase in chromosome damage at 25 μg ml^{-1} (22–26% metaphases with aberrations after four day exposures) and that cytotoxicity, manifested by the loss of the ability to form colonies, was also observed at doses as low as 5 μg ml^{-1}. A significant excess of chromosome breaks was also observed for 2-NPPD and 4-NOPD in a study using hamster cell line.[134] Certain aromatic amines, the likely carcinogenic agents in hair dyes, bind to DNA and cause base substitutions and have been associated with mutations in the ras oncogene in experimental studies.[135,136]

Alternatively, hair-colouring products may act by affecting the immune system, thus increasing the risk of NHL.[6] One study has linked the use of hair dyes to connective tissue disorders, such as systemic lupus erythematosus and scleroderma.[137] These acquired disorders of immunity have been associated with an increased risk of NHL.[138] Freni-Titulaer *et al.*[137] hypothesised that the association between connective tissue disease and hair-colouring products may be affected by acetylator phenotype. The aromatic amines in hair-colouring products, which are absorbed through the scalp, are metabolised through acetylation.[6]

11.4 Conclusions

Epidemiological studies so far have not provided convincing evidence to support the hypothesis that personal use of hair dyes or occupational exposure to hair-dye products as a hairdresser, barber or beautician is associated with human cancer risk. However, a possible association between certain types or colours of hair-dye products, such as permanent dark-colour hair dyes, and certain subtypes of cancers, such as bladder cancer and haematopoietic neoplasms, cannot be ruled out. The susceptible subgroup in the population with certain functional polymorphisms in genes involved in arylamine activation or detoxification may also modify the relationship between hair dye use and human cancer risks.

It should be pointed out that, while a relationship between hair dye exposure and human cancer risk is biologically plausible, the results from epidemiological studies linking both personal and professional exposure to hair dyes to human cancer risks have been inconsistent for almost all of the cancer sites investigated. The methodological limitations involved in epidemiological studies create uncertainties in interpreting the results and may account for the conflicting results. The major methodological issue is exposure assessment. The numerous individual chemicals used in hair-dye products have varied over time. The chemicals used by each hair-dye product vary according to the type, colour and brand of hair-dye products. However, in most studies, hair dye was just one of many exposure variables on which information was collected. In a few studies, only a few questions were asked to collect histories regarding lifetime hair-dye uses. This limited scope of exposure assessment diminishes an adequate characterisation of exposure in terms of type or colour of hair dyes, frequency or duration of use, time period of use and age during each period of use. If the patterns of hair-dye use varied over time among users, misclassification of exposures would likely occur and would attenuate the true association resulting from the limited scope of exposure assessment. Other major methodological issues, including potential disease misclassification due to different disease definitions used by different studies or limited statistical power due to small sample size in each individual study, may also account for the conflicting results.

Given the complicated use patterns of hair-dye products, the heterogeneity of many cancer sites, and the potential gene-environmental interaction, future well-designed, large population-based studies are needed to clarify the relationship between hair dyes and human cancer risks. Based on several recent studies, this relationship appears to be affected by specific genetic polymorphisms.

11.5 References

1. IARC working group on the evaluation of carcinogenic risks to humans, *Occupational exposures of hairdressers and barbers and personal use of hair colorants; some hair dyes, cosmetic colorants, industrial dyestuffs and aromatic amines, Proceedings*, Lyon, France, 6–13 October 1992; *IARC Monogr. Eval. Carcinog. Risks Hum.* 1993, **57(7)**, 398.
2. K. Czene, S. Tiikkaja and K. Hemminki, *Int. J. Cancer,* 2003, **105**, 108.
3. A. Correa, A. Mohan, L. Jackson, H. Perry and K. Helzlsouer, *Cancer Invest.,* 2000, **18(4)**, 366.
4. B.N. Ames, H.O. Kammen and E. Yamasaki, *Proc. Nat. Acad. Sciences U. S. A.,* 1975, **72**, 2423.
5. J.F. Corbett, *Dyes and Pigments,* 1999, **41**, 127.
6. S.H. Zahm, D.D. Weisenburger, P.A. Babbitt, R.C. Saal, J.B. Vaught and A. Blair, *Am. J. Pub. Health,* 1992, **82**, 990.
7. C.H. Hennekens, F.E. Speizer, B. Rosner, C.J. Bain, C. Belanger and R. Peto, *Lancet,* 1979, **1**, 1390.
8. A. Green, W.C. Willett, G.A. Colditz *et al., J. Natl. Cancer Inst.,* 1987, **79**, 253.
9. M.J. Thun, S.F. Altekruse, M.M. Namboodiri, E.E. Calle, D.G. Myers and C.W. Heath, Jr., *J. Natl. Cancer Inst.,* 1994, **86**, 210.
10. S.F. Altekruse, S.J. Henley and M.J. Thun, *Cancer Causes Control,* 1999, **10**, 617.
11. S.J. Henley and M.J. Thun. *Int. J. Cancer,* 2001, **94**, 903.
12. F. Grodstein, C.H. Hennekens, G.A. Colditz, D.J. Hunter and M.J. Stampfer, *J. Natl. Cancer Inst.,* 1994, **86**,1466.
13. P. Hartge, R. Hoover and R. Altman, *Cancer Res.,* 1982, **42(11)**, 4784.
14. M. Gago-Dominguez, J.E. Castelao, J.M. Yuan, M.C. Yu and R.K. Ross, *Int. J. Cancer,* 2001, **91**, 575.
15. J.D. Boice, Jr., J.S. Mandel and M.M. Doody, *J.A.M.A.,* 1995, **274**, 394.
16. L.S. Cook, K.E. Malone, J.R. Daling, L.F. Voigt and N.S. Weiss, *Cancer Causes Control,* 1999, **10**, 551.
17. T. Zheng, T.R. Holford, S.T. Mayne *et al., Eur. J. Cancer,* 2002, **38**, 1647.
18. W. Petro-Nustas, M.E. Norton, I. al-Masarweh, *J. Nurs. Scholarsh.,* 2002, **34**, 19.
19. K.P. Cantor, A. Blair, G. Everett *et al. Am. J. Public Health,* 1988, **78**, 570.
20. L.M. Brown, G.D. Everett, L.F. Burmeister and A. Blair, *Am. J. Public Health,* 1992, **82**, 1673.

21. L.J. Herrinton, N.S. Weiss and T.D. Koepsell *et al. Am. J. Pub. Health*, 1994, **84**, 1142.
22. A. Mele, M. Szklo, G. Visani *et al., Am. J. Epidemiol.*, 1994, **139**, 609.
23. A. Mele, M.A. Stazi, A. Pulsoni *et al., Haematologica*, 1995, **80**, 405.
24. A. Mele, G. Visani, A. Pulsoni *et al.*, Italian Leukemia Study Group, *Cancer*, 1996, **77**, 2157.
25. L. Markovic-Denic, S. Jankovic, J. Marinkovic and Z. Radovanovic, *Neoplasma*, 1995, **42**, 79.
26. M. Ido, C. Nagata, N. Kawakami *et al., Leuk. Res.*, 1996, **20**, 727.
27. E.A. Holly, C. Lele and P.M. Bracci, *Am. J. Pub. Health*, 1998, **88**, 1767.
28. L. Miligi, A. Seniori Costantini, P. Crosignani *et al., Am. J. Ind. Med.*, 1999, **36**, 60.
29. C. Nagata, H. Shimizu, K. Hirashima *et al., Leuk. Res,*. 1999, **23**, 57.
30. J. Bjork, M. Albin, H. Welinder *et al., Occup. Environ. Med.*, 2001, **58**, 722.
31. Y. Zhang, T.R. Holford, B. Leaderer *et al., Am. J. Epidemiol.*, 2004, **159**, 148.
32. K.M. Stavraky, E.A. Clarke, and A. Donner, *Br. J. Cancer,* 1981, **43**, 236.
33. J.D. Burch, K.J. Craib, B.C. Choi, A.B. Miller, H.A. Risch and G.R. Howe, *J. Nat. Can. Instit.*, 1987, **78**, 601.
34. M.R. Spitz, J.J. Fueger, H. Goepfert and G.R. Newell, *Archives of Otolaryngology – Head & Neck Surgery,* 1990, **116**, 1163.
35. C.D. Holman and B.K. Armstrong, *Br. J. Cancer.*, 1983, **48**, 599.
36. A. Osterlind, M.A. Tucker, B.J. Stone and O.M. Jensen, *Int. J. Cancer,* 1988, **42**, 825.
37. A. Tzonou, A. Polychronopoulou, C.C. Hsieh, A. Rebelakos, A. Karakatsani and D. Trichopoulos, *Int. J. Cancer,* 1993, **55**, 408.
38. A.F. Olshan, N.E. Breslow, J.M. Falletta *et al., Cancer,* 1993, **72**, 938.
39. G.R. Bunin, J.D. Buckley, C.P. Boesel, L.B. Rorke and A.T. Meadows, *Cancer Epidemiol., Biomarkers & Preven.*, 1994, **3**, 197.
40. E.A. Holly, P.M. Bracci, M.K. Hong, B.A. Mueller and S. Preston-Martin, *Paediatr. Perinat. Epidemiol.* 2002, **16**, 226.
41. M. Gago-Dominguez, D.A. Bell, M.A. Watson *et al., Carcinogenesis,* 2003, **24**, 483.
42. J.C. Schroeder, A.F. Olshan, R.B. Dent *et al., Cancer Causes Control,* 2002, **13**, 159.
43. M.L. Cleary, S.D. Smith and J. Sklar, *Cell,* 1986, **47**, 19.
44. A. Bakhshi, J.J. Wright, W. Graninger *et al., Proc. Nat. Acad. Sciences U.S.A.*, 1987, **84**, 2396.
45. T.J. McDonnell, N. Deane, F.M. Platt *et al., Cell,* 1989, **57**, 79.
46. Y. Tsujimoto and C.M. Croce, *Proc. Nat. Acad. Sciences U.S.A.*, 1986, **83**, 5214.
47. I. Magrath, *Cancer Research,* 1992, **52**, 5529s.
48. M. Potter, *Cancer Research,* 1992, **52**, 5522s.
49. E. Pukkala, P. Nokso-Koivisto and P. Roponen, *Int. Arc. Occ. Environ. Health,* 1992, **64**, 39.
50. D.J. Kirkland, S.D. Lawler and S. Venitt, *Lancet,* 1978, **2**, 124.
51. D.J.V. Kirkland, *S. Mutation Research,* 1976, **40**, 47.

52. E.P. Frenkel and F. Brody, *Arc. Environ. Health*, 1973, **27**, 401.
53. M. Alderson, *J. Epidemiol. Comm. Health*, 1980, **34**, 182.
54. S. Kono, S. Tokudome, M. Ikeda, T. Yoshimura, M. Kuratsune, *J. Nat. Cancer Inst.*, 1983, **70**, 443.
55. M.J. Teta, J. Walrath, J.W. Meigs and J.T. Flannery, *J. Nat. Cancer Inst.*, 1984, **72**, 1051.
56. E. Guberan, L. Raymond and P.M. Sweetnam, *Int. J. Epidemiol.*, 1985, **14**, 549.
57. T. Skov, A. Andersen, H. Malker, E. Pukkala, J. Weiner and E. Lynge, *Am. J. Ind. Med.*, 1990, **17**, 217.
58. H.S. Malker, J.K. McLaughlin, D.T. Silverman *et al.*, *Cancer Res.*, 1987, **47**, 6763.
59. E. Lynge and L. Thygesen, *Int. J. Epidemiol.*, 1988, **17**, 493.
60. A.B. Lamba, M.H. Ward, J.L. Weeks and M. Dosemeci, *J. Occ. Environ. Med.*, 2001, **43**, 250.
61. P.J. Dolin and P. Cook-Mozaffari, *Br. J. Cancer*, 1992, **66**, 568.
62. T. Skov and E. Lynge, *Skin Pharmacol.*, 1994, **7**, 94.
63. L.J. Dunham, A.S. Rabson, H.L. Stewart, A.S. Frank and J.L. Young., *J. Nat. Cancer Inst.*, 1968, **41**, 683.
64. H.M. Anthony, G.M. Thomas, P. Cole and R. Hoover, *J. Nat. Cancer Inst.*, 1971, **46**, 1111.
65. P. Cole, R. Hoover and G.H. Friedell, *Cancer*, 1972, **29**, 1250.
66. E. Viadana, I.D. Bross and L. Houten, *J. Occ. Med.*, 1976, **18**, 787.
67. G.R. Howe, J.D. Burch, A.B. Miller *et al.*, *J. Nat. Cancer Inst.*, 1980, **64**, 701.
68. P. Vineis and C. Magnani, *Int. J. Cancer*, 1985, **35**, 599.
69. A.S. Morrison, A. Ahlbom, W.G. Verhoek *et al.*, *J. Epidemiol. Comm. Health*, 1985, **39**, 294.
70. H.A. Risch, J.D. Burch, A.B. Miller, G.B. Hill, R. Steele and G.R. Howe, *Br. J. Ind. Med.*, 1988, **45**, 361.
71. D.T. Silverman, L.I. Levin, R.N. Hoover and P. Hartge, *J. Nat. Cancer Inst.*, 1989, **81**, 1472.
72. D.T. Silverman, L.I. Levin and R.N. Hoover, *Am. J. Epidemiol.*, 1990, **132**, 453.
73. K. Teschke, M.S. Morgan, H. Checkoway *et al.*, *Occ. Environ. Med.*, 1997, **54**, 443.
74. T. Zheng, K.P. Cantor, Y. Zhang and C.F. Lynch, *J. Occ. Environ. Med.*, 2002, **44**, 685.
75. E. Kunze, J. Chang-Claude and R. Frentzel-Beyme, *Cancer*, 1992, **69**, 1776.
76. J. Siemiatycki, R. Dewar, L. Nadon and M. Gerin, *J. Epidemiol.*, 1994, **140**, 1061.
77. T. Sorahan, L. Hamilton, D.M. Wallace, S. Bathers, K. Gardiner and J.M. Harrington, *Br. J. Urol.*, 1998, **82**, 25.
78. S. Cordier, J. Clavel, J.C. Limasset *et al.*, *Int. J. Epidemiol.*, 1993, **22**, 403.
79. W.E. Morton, *J. Occ. Environ. Med.*, 1995, **37**, 328.
80. M. Pollan and P. Gustavsson, *Am. J. Pub. Health*, 1999, **89**, 875.
81. E.E. Calle, T.K. Murphy, C. Rodriguez, M.J. Thun, C.W. Heath, Jr., *Am. J. Epidemiol.*, 1998, **148**, 191.

82. K.L. Koenig, B.S. Pasternack, R.E. Shore and P. Strax. *Am. J. Epidemiol.*, 1991, **133**, 985.
83. L.A. Habel, J.L. Stanford, T.L. Vaughan *et al.*, *J. Occ. Environ. Med.*, 1995, **37**, 349.
84. P.F. Coogan, R.W. Clapp, P.A. Newcomb *et al.*, *Am. J. Ind. Med.*, 1996, **30**, 430.
85. P.R. Band, N.D. Le, R. Fang, M. Deschamps, R.P. Gallagher and P. Yang, *J. Occ. Environ. Med.*, 2000, **42**, 284.
86. P. Boffetta, A. Andersen, E. Lynge, L. Barlow and E. Pukkala, *J. Occ. Med.*, 1994, **36**, 61.
87. K. Vasama-Neuvonen, E. Pukkala, H. Paakkulainen *et al.*, *Am. J. Ind. Med.*, 1999, **36**, 83.
88. T. Shields, G. Gridley, T. Moradi, J. Adami, N. Plato and M. Dosemeci, *Am. J. Ind. Med.*, 2002, **42**, 200.
89. A.M. Osorio, L. Bernstein, D.H. Garabrant and J.M. Peters, *J. Occ. Med.*, 1986, **28**, 291.
90. A. Shibata, R. Sasaki, N. Hamajima and K. Aoki, *Nippon Ketsueki Gakkai Zasshi*, 1990, **53**, 116.
91. T. Skov and E. Lynge, *Scand. J. Soc. Med.*, 1991, **19**, 162.
92. B. Persson, A.M. Dahlander, M. Fredriksson, H.N. Brage, C.G. Ohlson and O. Axelson, *Br. J. Ind. Med.*, 1989, **46**, 516.
93. A. Blair, A. Linos, P.A. Stewart *et al.*, *Am. J. Ind. Med.*, 1993, **23**, 301.
94. G.G. Giles, J.N. Lickiss, M.J. Baikie, R.M. Lowenthal and J. Panton, *J. Nat. Cancer Inst.*, 1984, **72**, 1233.
95. A.S. Costantini, L. Miligi, D. Kriebel *et al.*, *Epidemiol.*, 2001, **12**, 78.
96. T.R. Cote, M. Dosemeci, N. Rothman, R.B. Banks and R.J. Biggar, *Am. J. Pub. Health.*, 1993, **83**, 598.
97. E.A. Holly, C. Lele and P. Bracci, *J. Acq. Imm. Def. Syndr. Human Retrovirol.*, 1997, **15**, 223.
98. L.W. Figgs, M. Dosemeci and A. Blair, *Am. J. Ind. Med.*, 1995, **27**, 817.
99. C.F. Robinson and J.T. Walker, *Am. J. Ind. Med.*, 1999, **36**, 186.
100. J.K. McLaughlin, H.S. Malker, M.S. Linet *et al.*, *Arc. Environ. Health*, 1988, **43**, 7.
101. M.S. Linet, J.K. McLaughlin, H.S. Malker *et al.*, *J. Occ. Med.*, 1994, **36**, 1187.
102. U. Flodin, M. Fredriksson and B. Persson, *Am. J. Ind. Med.*, 1987, **12**, 519.
103. P. Boffetta, S.F. Stellman and L. Garfinkel, *Int. J. Cancer*, 1989, **43**, 554.
104. M. Eriksson and M. Karlsson, *Br. J. Ind. Med.*, 1992, **49**, 95.
105. L.M. Pottern, E.F. Heineman, J.H. Olsen, E. Raffn and A. Blair, *Cancer Causes & Control*, 1992, **3**, 427.
106. L.W. Figgs, M. Dosemeci and A. Blair, *J. Occ. Med.*, 1994, **36**, 1210.
107. E.F. Heineman, J.H. Olsen, L.M. Pottern, M. Gomez, E. Raffn and A. Blair, *Cancer Causes & Control*, 1992, **3**, 555.
108. G.M. Swanson and P.B. Burns, *Ann. Epidemiol.*, 1997, **7**, 369.
109. P. Boffetta, L. Richiardi, F. Berrino *et al.*, *Cancer Causes & Control*, 2003, **14**, 203.

110. P. Mutanen and K. Hemminki, *J. Occ. Environ. Med.*, 2001, **43**, 952.

111. R.R. Kuijten, G.R. Bunin, C.C. Nass and A.T. Meadows, *Cancer Res.*, 1992, **52**, 782.

112. R. McKean-Cowdin, S. Preston-Martin, J.M. Pogoda, E.A. Holly, B.A. Mueller and R.L. Davis, *J. Occ. Environ. Med.*, 1998, **40**, 332.

113. S. Cordier, L. Mandereau, S. Preston-Martin *et al.*, *Cancer Causes & Control*, 2001, **12**, 865.

114. V.B. Smulevich, L.G. Solionova and S.V. Belyakova, *Int. J. Cancer*, 1999, **83**, 718.

115. A.F. Olshan, A.J. De Roos, K. Teschke *et al.*, *Cancer Causes & Control*, 1999, **10**, 539.

116. C.E. Searle, D.G. Harnden, S. Venitt and O.H. Gyde, *Nature*, 1975, **255**, 506.

117. G.R.S. Mohn, F.J. de Serrers, *Mut. Res.*, 1976, **38**, 116.

118. M.J. Fahmy and O.G. Fahmy, *Mut. Res.*, 1977, **56**, 31.

119. W.G. Blijleven, *Mut. Res.*, 1977, **48**, 181.

120. J. McCann, and B.N. Ames, *Proc. Nat. Acad. Sciences U.S.A.*, 1976, **73**, 950.

121. N. Ito, Y. Hiasa, Y. Konishi and M. Marugami, *Cancer Res.*, 1969, **29**, 1137.

122. J.M. Sontag, *J. Nat. Cancer Inst.*, 1981, **66**, 591.

123. National Cancer Institute, *NCI Tech. Rep. Ser.*, 1978, **94**, 1.

124. National Cancer Institute, *NCI Tech. Rep. Ser.*, 1978, **84**, 1.

125. National Cancer Institute, *NCI Tech. Rep. Ser.*, 1978, **162**, 1.

126. National Cancer Institute, *NCI Tech. Rep. Ser.*, 1979, **169**, 1.

127. C.E. Searle and E.L. Jones, *Br. J. Cancer*, 1977, **36**, 467.

128. M.M. Jacobs, C.M. Burnett, A.J. Penicnak *et al.*, *Drug Chem. Toxicol.*, 1984, **7**, 573.

129. M. Inouye and U. Murakami *Food & Cosme. Toxicol.*, 1977, **15**, 447.

130. S. Marshall and W.S. Palmer, *JAMA*, 1973, **226**, 1010.

131. M. Kiese and E. Rauscher, *Toxicol. Appl. Pharmacol.*, 1968, **13**, 325.

132. H.I. Maibach, M.A. Leaffer and W.A. Skinner, *Arc. Dermatol.*, 1975, **111**, 1444.

133. F.N. Marzulli, P.M. Watlington and H.I. Maibach, *Curr. Prob. Dermatol.*, 1978, **7**, 196.

134. W.F. Benedict, *Nature*, 1976, **260**, 368.

135. W. Lutz and B. Krajewska, *Medycyna Pracy*, 1991, **42**, 477.

136. F.P. Perera, K. Hemminki, E. Gryzbowska *et al.*, *Nature*, 1992, **360**, 256.

137. L.W. Freni-Titulaer, D.B. Kelley, A.G. Grow, T.W. McKinley, F.C. Arnett and M.C. Hochberg, *Am. J. Epidemiol.*, 1989, **130**, 404.

138. M.H. Greene, in *Cancer Epidemiology and Prevention*, D.D. Schottenfeld and J.F. Fraumeni, Jr., (ed), Saunders, Philadelphia, 1983.

The Chemistry of Hair Care Products: Potential Toxicological Issues for Shampoos, Hair Conditioners, Fixatives, Permanent Waves, Relaxers and Depilatories

J. JACHOWICZ

12.1 Introduction

Hair care is a growing segment of the personal care product industry, which is valued in sales at about $42.5 billion worldwide according to Euromonitor. The growth is due to a constant 3–5% expansion of the developed markets in the United States, Western Europe and Japan, in addition to the dramatic increase in sales in Asia, South America and Eastern Europe. There is a tremendous diversity of hair care formulations because of different preferences in various countries for consumer perceivable benefits. Cosmetic companies, which market hair care products, have at their disposal thousands of ingredients of both synthetic and natural origin. As in any market-driven field, the innovation is through new formulations, new claims, the introduction of new raw materials, or the discovery of new physical phenomena. There is a significant multidisciplinary R&D effort in major cosmetic and specialty chemical companies, which results in introduction of new raw materials including surfactants, polymers, oils, emollients, sunscreens, photo-absorbers, antioxidants, free-radical scavengers, preservatives, dyes, fragrances, botanicals *etc*. All new and old materials undergo continuous performance evaluation and toxicological assessment. It should be pointed out that, while hair care products are designed to affect the properties of hair, they actually come into contact with skin and since the products are used frequently over a long period of time the absorption of surfactants and other actives through skin cannot be excluded. Thus, the performance analysis includes the study of raw material interactions with both hair and skin. Toxicological assessment encompasses skin compatibility (dermal irritation), mucous membrane compatibility (eye irritation), sensitisation and percutaneous absorption. Systemic effects are assessed by measuring acute toxicity, chronic toxicity, mutagenicity, carcinogenicity and teratogenicity. In addition, there is an emerging trend to include biodegradability and aquatic toxicity in the overall toxicological and safety assessment of cosmetic products and raw materials.

Another important aspect of cosmetic safety is the production and storage of cosmetic products, which have to be carefully controlled in order to avoid microbial growth. The use of preservatives is common, with manufacturers frequently using biocide blends in order to achieve broad-spectrum antimicrobial protection against gram-positive and gram-negative bacteria as well as against yeasts and molds. Packaging has to be designed in such a way that makes the transfer of contamination during usage difficult. So far, there have been few reported incidents of microbial contamination of cosmetics in the United States and Europe.

This chapter will review the latest information on hair care product (non-colourant) chemistry (see Chapter 9 for discussion on colourant chemistry) as currently used, with a focus on potential associated toxicological issues.

12.2 Safety Evaluations

Toxicological safety studies are carried out on new and established raw materials in order to find out about the potential health risks for the user of the final product. An important goal is to prevent and eliminate any conceivable adverse effects including those which may result from product misuse. The tests are carried out according to various protocols described in a number of publications.[1-5] The regulatory status of special classes of cosmetic raw materials, such as sunscreens and preservatives, are covered in several review articles.[6-12] Detailed descriptions of the experimental procedures for testing chemicals are collected in the official publication of the Organisation for Economic Cooperation and Development (OECD).[1] The publication consists of four sections: Section 1 – physical/chemical properties, Section 2 – effects on biotic systems, Section 3 – degradation and accumulation and Section 4 – health effects. Section 2 deals with several aspects of aquatic toxicity such as growth inhibition test of alga, acute toxicity test of fish, prolonged toxicity test of fish, growth test of terrestrial plants, acute oral toxicity test of honeybees *etc.* Section 3 describes the procedures employed in the evaluation of the biodegradation of chemicals including ready biodegradability, inherent biodegradability, biodegradability in sewage, inherent biodegradability in soil and biodegradability in seawater. Section 4 reviews the methodologies used in the studies of acute oral toxicity, acute dermal toxicity, acute inhalation toxicity, acute dermal irritation/corrosion, acute eye irritation/corrosion, skin sensitisation, toxicity in rodents and non-rodents, teratogenicity, toxicokinetics, neurotoxicity and carcinogenicity. It also includes information on genetic toxicology testing.

A brief description of basic types of toxicological tests employed in the evaluation of cosmetic actives is given below.

12.2.1 Animal Safety Tests

12.2.1.1 Acute Toxicity[13]

Acute toxicity is measured by introducing large amounts of an investigated substance into an animal by the oral route, inhalation or skin penetration. For oral ingestion, the frequently used parameter is LD_{50} which provides the dose of the investigated chemical, in mg kg^{-1} of bodyweight, which results in the death of 50% of the population of tested animals. For example, for cosmetic surfactants LD_{50} values range from several hundred to several thousand mg kg^{-1} of bodyweight.[2]

Acute dermal toxicity is assessed by judging adverse effects occurring within a short time of the dermal application of a single dose of a test substance to about 10% of the animal's body surface. The test is typically performed with neat liquids, concentrated solutions or solid powders. The results can be reported in terms of median lethal dose, LD_{50}.

Similarly, acute inhalation toxicity is performed by animal exposure to high concentrations of dusts or aerosols containing surfactants or polymers. The test, whose objective is to detect disturbances of lung function, is designated primarily for materials employed in the manufacturing of hair sprays.

12.2.1.2 Acute Dermal Irritation/Corrosion[14]

The test is performed by applying a substance in a single dose to a small skin area of an experimental animal (albino rabbit). The degree of irritation/corrosion (erythema and edema) is evaluated and scored at specific time intervals. According to the rating scale, a substance can be classified as corrosive, irritating, or non-corrosive/non-irritating.

12.2.1.3 Acute Eye Irritation[15]

In measurements of acute eye irritation the analysed substance is applied in a single dose to one of the eyes of an animal while the untreated eye is used as a control. Adult albino rabbits are used in these experiments. The eyes are examined at 1, 24, 48 and 72 hours for signs of irritation lesions in conjunctivae, cornea and iris.

This, and the test described above are based on the procedures originally developed by Draize *et al.*[3] to study skin corrosivity and irritation by using rabbit skin and eye.

12.2.1.4 Skin Sensitisation[16,17]

The guinea pig maximisation test (GPMT) of Magnusson and Kligman is performed to assess the sensitisation potential of a chemical substance.[16] According to the test procedure the animals are initially treated with the studied material by intradermal injection and/or topical epidermal application. After a rest period of 6–8 days, during which an immune response may develop, the animals are subjected to an induction, topical application of a challenge dose of the tested material or a skin irritant such as sodium lauryl sulfate. After 21 days, the treated sites are subjected to a challenge-topical application with the test sample, which is kept in contact with the skin by an occlusive dressing for 24 hours. After removing the patch, the treated area is cleaned, examined, and scored with the following ratings: no change, discrete or patchy erythema, moderate or confluent erythema, or intense erythema and swelling.

Recently, a different test referred to as the localised lymph node assay (LLNA) was introduced.[17] The basic principle underlying the LLNA is that sensitisers induce a primary proliferation of lymphocytes in the lymph node draining the site of chemical application. This proliferation is proportional to the dose applied (and to the potency of the allergen) and provides a simple means of obtaining an objective, quantitative measurement of sensitisation. Radioactive labelling with ^{3}H-methyl thymidine is employed to quantify cell proliferation. Unlike guinea pig tests the LLNA does not require that challenged-induced dermal hypersensitivity reactions be elicited.

12.2.1.5 90-Day Inhalation[18]

This test provides information on health hazards associated with exposure by the inhalation route over a period of 90 days. According to experimental protocol, the animals are exposed to an atmosphere containing the tested substance under controlled conditions of exposure concentration, temperature, humidity *etc*. At the conclusion of the test the animals are sacrificed, necropsied and their organs are subjected to histopathologic examination.

Inhalation tests are usually applicable when there is a potential for inhalation exposure, for example, in the case of hairspray resins.

12.2.1.6 Chronic Toxicity – Repeated Insult Patch Test

Chronic toxicity is evaluated by a repeated treatment with small amounts of a cosmetic raw material over an extended period of time. For example, oral intake of surfactants by animals has been investigated for all important classes of these materials indicating no observable effects for dosages at concentrations of several thousand parts per million (ppm).[2]

Chronic dermal toxicity can be performed by a variety of different protocols,[2,3] which may extend over a period of 28 days or 90 days. The results of the tests based on repeated insult patch test, which may result in skin sensitisation or transient skin irritation, have been described in the literature for several classes of materials.[19]

12.2.1.7 Mutagenicity and Carcinogenicity[20]

Mutagenicity is the ability of a given chemical to effect irreversible changes in the genetic material, while carcinogenicity is the ability to produce cancer.

The Ames test is frequently used to screen cosmetic chemicals for mutagenicity. The background assumption is that any substance that is mutagenic may also prove to be carcinogenic. The test involves the use of a special strain of bacteria, which can be mutated by an investigated chemical. The test can be adapted to use eukaryotic yeast or mammalian cells, which may be a better model of human body.

12.2.1.8 Teratogenicity[21]

Teratogenicity is the ability of a chemical substance to cause permanent structural or functional abnormalities during the period of embryonic and foetal development. The test protocol of teratogenicity calls for graduated dose administration of the material under investigation to pregnant experimental animals (rat or rabbit). After sacrifice or death, the uterus is removed and its content is examined for embryonic or foetal deaths and the number of live foetuses.

12.2.1.9 Photo-toxicity and Photo-allergenicity

Photo-toxic reactions can be produced by systemic or topical application of certain materials, which are not toxic except when exposed to UV light. In their presence and in response to UVA or UVB irradiation, skin may produce redness or blisters. Chronic reaction may involve skin hyper-pigmentation or thickening. Photo-allergy is a delayed hypersensitivity reaction requiring prior sensitisation. Induction and elicitation of a photo-toxic reaction can result from a topical exposure to a photo-allergen and is referred to as photo-contact dermatitis. The photo-toxic or photo-allergenic raw materials, which could be potentially employed in the formulation of cosmetics, include anti-microbials, fragrances, sunscreens or photo-absorbers. New methods to assess photo-toxicity of cosmetic actives were recently discussed in the literature.[6,19,22,23]

12.2.2 Non-animal Safety Testing

There has been a growing trend in the cosmetics industry to avoid and eliminate animal testing in the evaluation of the toxicological potential of raw materials. Several companies have actually pledged not to employ chemicals which were tested on animals. In connection with this, there is a significant research effort to develop and validate alternatives to the use of animals (vertebrates). The literature in this field has been recently compiled as a bibliography with abstracts by the National Library of Science/National Institute of Health and is available on-line.[24] Two examples of *in vitro* tests employed in the cosmetics industry are shown below.

12.2.2.1 Cell Toxicity Test[25]

This test measures cell culture growth inhibition. For example, it can use as cells normal human epidermal keratinocytes in a serum-free medium. Cell quantity analysis is performed by neutral red (NR) assay. The concentration causing a 50% growth reduction NR50 (in ppm) is used as a parameter, which ranges from 3.4 and 4.1 for highly irritating sodium lauryl ether sulfate and sodium lauryl sulfate, respectively, to 35.8 and 306 for less irritating sodium *N*-lauroyl *N*-methyltaurate and sodium lauroyl glutamate, respectively.

12.2.2.2 *Primary Eye Irritation*[26]

This *in vitro* test (Eytex™) evaluates of the permeability of a surfactant in the form of a 5% aqueous solution through a membrane (eye mucosa model) and its subsequent reaction with the protein. The Eytex™ score varies from 8.5 for sodium cocoyl glutamate (low irritation) to 38.9 for sodium lauryl sulfate (high irritation), respectively.

12.2.3 Biodegradability

Biodegradability of cosmetic formulations is of growing interest to cosmetic scientists. In recent years several hair care product lines were commercialised which claim compositions containing 'natural' and 'biodegradable' ingredients.

The objective of aerobic and anaerobic biodegradability tests is to predict how long it takes to eliminate the tested material from the environment under aerobic and anaerobic conditions. The biodegradability of surfactants and polymers can be evaluated by a number of methods.[27,28] In the OECD method, a biodegradation score (BOD) is presented in the form of a plot of %BOD as a function of time in days. For example sodium lauryl ether sulfate and sodium *N*-lauroyl *N*-methyltaurate showed 75% and 92% BOD after 28 days indicating higher rate of biodegradation for the latter surfactant.

12.3 Cosmetic Ingredient Review

In the United States, the cosmetic ingredient review was established by the Cosmetic, Toiletry, and Fragrance Association (CTFA) to review and assess the safety of ingredients used in cosmetics in an unbiased and expert manner. Ingredient safety data are studied by an expert panel and the conclusions are publicly announced without any interference by the cosmetics industry. The review process, which includes a scientific literature review and a series of public discussions, leads to the publication of a tentative report. It classifies the materials as: 1) safe as used, 2) safe with certain qualifications, 3) unsafe, and 4) insufficient data to support safety. After reviewing additional public comments, the final report is issued for publication in the literature.

The information related to specific ingredients and presented in this paper is drawn from the 2004 CIR Compendium,[29] which provides a compilation of information taken from final reports.

12.4 Review of the Chemistry of the Main Hair Care Product Categories

12.4.1 Hair Shampoos

The main function of hair shampoos is to clean hair by removing dirt and dust, accumulated sebum, sweat, cosmetic treatment residues and other contaminants. They should rinse off easily with water and produce copious amounts of foam.

The shampoos should be designed in such a way that they do not excessively strip hair of natural lipids. Secondary functions of a shampoo include deposition of modifiers to improve hair shine, manageability, body and conditioning. The shampoos can be classified according to their physical appearance (*e.g.* clear, opaque), performance properties (*e.g.* conditioning, non-conditioning cleansing, styling), content of special ingredients (*e.g.* anti-dandruff, colouring, herbal) and properties of the employed surfactant system (*e.g.* mild, baby shampoo). The mode of usage of a shampoo includes a dilution of a product with a small amount of water, application into the hair followed by lathering, then rinsing with water. The time of contact for a dilute shampoo, with a total concentration of actives estimated to be in the range from 1% to 5%, with hair, scalp and skin is typically on the order of minutes.

Shampoos are complex mixtures of surfactants, oils, polymers and preservatives. There are a lot of commercial formulations and their ingredients are chosen from hundreds of available raw materials such as surfactants, lather boosters, conditioning agents, thickeners, opacifiers, preservatives, sunscreens, photo-absorbers, antioxidants, dyes, and other functional raw materials. Table 12.1 lists the main classes of cosmetic raw materials used to formulate shampoos.[30] The most important are primary and secondary surfactants and their concentrations, which determine the cleansing power of a shampoo. Viscosity modifiers and foam boosters as well as thickeners are used to adjust the rheological properties (*i.e.* flow) of a shampoo formulation as well as the density of the lather. An important component of a shampoo is a conditioning system that usually contains cationic surfactants, cationic polymers and hydrocarbon and/or silicone oils. All these materials interact with each other in a shampoo composition, forming a variety of colloidal structures such as micelles, mixed micelles, polymer-surfactant complexes, surfactant-surfactant complexes, emulsions, lamellar phases, vesicles *etc.* During lathering and rinsing the shampoo cleans or conditions by interacting with the grime on the fibre as well as with the fibre itself. Additional processes include adsorption/desorption of surfactants, polymers and complexes as well as solubilisation or deposition of oils. All of these processes have been characterised qualitatively and quantitatively on hair by physicochemical methods.[31,32] The deleterious reactions accompanying shampooing include excessive stripping of natural hair lipids or excessive deposition of conditioning actives, which may permanently modify the properties of hair. For example, high molecular cationic polymers may be irreversibly adsorbed on the hair surface affecting its natural affinity to other cosmetic actives, as well as its texture and feel.

The potential for skin irritation and skin damage by shampoos also has to be taken into consideration. It is usually connected with the surfactant penetrating into skin and inducing irritation. The deleterious effects of surfactants may manifest themselves by stratum corneum swelling,[33] appearance of redness,[34] denaturation of epidermal keratin,[35] and damage to the barrier properties of skin leading to permeation of the stratum corneum.[36] There have been several studies comparing the irritation potential of various individual surfactants as well as surfactant mixtures. Single component surfactant solutions have been investigated or reviewed by Lang and Spenglet,[37] Rhein,[5] Rhein *et al.*,[38] Tavss *et al.*,[39] and Gabbianelli *et al.*[40] The most intensively studied surfactant is sodium lauryl sulfate

Table 12.1 *Typical shampoo compositions*[30]

Shampoo constituents	Raw materials	Concentration [%], active
Primary surfactant	Sodium lauryl sulfate Ammonium lauryl sulfate TEA lauryl sulfate Sodium lauryl ether sulfate Ammonium lauryl ether sulfate α-Olefin sulfonate Sulfosuccinates: disodium lauryl ether sulfosuccinate Sarcosinates Taurates Acylglutamates Alkyl polyglucosides: decyl polyglucose	9–5
Secondary surfactant	Cocoamphocarboxyglycinate Propionates Sultaines Crypto-anionics: sodium trideceth-7-carboxylate Non-ionics Alkyl betaines Alkyl aminoxide Alkyl amidoaminoxide	0–5
Viscosity/foam stabiliser	Lauryl diethanolamide	1–5
Rheology modifier	$NaCl$, NH_4Cl, PEG-120 methyl glucose Dioleate, PEG-150 distearate, Cellulose ethers (hydroxyethylcellulose, hydroxyethyl methylcellulose, hydroxypropyl methylcellulose, hydroxypropyl cellulose, methylcellulose) Carbomer Fatty alcohols	0–2
pH Adjustment agent	Citric acid	0–1
Preservative	Methyl paraben Quaternium-15	0–1
Fragrance		0–2
Conditioning agent	Polyquaternium-10 Polyquaternium-11 Cationic guar gum	0–1
Antioxidant	Tocopherol	0–0.5
Sunscreen/photo-absorber	Benzophenone-4 Lauryl pabamidopropylammonium chloride	0.25–2
Other speciality additives	Anti-dandruff: zinc pyrithione, coal tar	0–2

(SLS), which was recently reviewed by the European Society of Contact Dermititis.[41–43] In addition to clinical studies, several instrumental techniques including transepidermal water loss (TEWL), scanning laser-doppler velocimetry, ultrasound imaging, conductance, squamometry and corneosurfametry and fluorescent dye staining[5,34,44–46] have been employed to detect subclinical reactions to anionic surfactants.

Tavss *et al.*[39] correlated Duhring chamber measurements of pH on mixing surfactants with bovine serum albumin solutions with clinical observations of degree of irritation caused by many surfactants to human forearms at a 10% concentration, in neutral pH and during a 5-day application period. The test was earlier shown to produce similar results, in terms of irritancy ranking of surfactants, as realistic washing tests. The test ranking for skin irritation of 11 surfactants showed: 1) severe reaction within 24 hours for sodium lauryl sulfate and linear alkylbenzene sulfonate; 2) intense redness by the fourth day for sodium laureth sulfate (3EO) and triethanol ammonium lauryl sulfate; 3) mild-to-moderate redness after five days for ammonium laureth sulfate (3EO), ammonium laureth sulfate (6EO), and Igepon TC-42 (non-ionic surfactant); and 4) no irritation after five days exposure for sodium laureth sulfate (6EO), sodium laureth sulfate (9EO), ammonium laureth sulfate (9EO), and ammonium laureth sulfate (12EO).

Gabianelli *et al.*[40] evaluated the mildness/irritation potential of a series of widely used surfactants by employing clinical patch testing. A 48 hour patch test was used to assess irritation resulting from a single protracted exposure while 14 or 21 day cumulative patch testing was employed to determine the effect of repeated and extended treatment. Subjective evaluation by professional clinicians was complemented by corneometer and chromameter measurements to quantify surfactant induced dryness and reddening (erythema). Based on a 14 day cumulative irritation test of 0.5% surfactant solutions, the following ranking for pure surfactants was obtained: SLS (mean irritation score $= 40.3$) $>$ SLES (30.9) $>$ sodium cocoamphoacetate (SCA, 11.8) $>$ potassium monoalkyl (C12) phosphate (MAPS) (4.48).

Additional results on a variety of anionic surfactants are reviewed by Rhein[5] and include the work carried out by Lang and Spengler[37] and other authors.

For surfactant mixtures it has been observed that they lead to lowering of the surfactant monomer concentration and result in reduced skin irritation.[47–53] Rhein *et al.*[38] studied the mixtures SLS-AEOS-3EO (laureth-3 sulfate), SLS-AEOS-6EO (laureth-6 sulfate) and SLS-cocamidopropylbetaine. They reported significant reductions in stratum corneum swelling for 1:1 and 1:0.5 surfactant mixtures. A similar pattern was confirmed in an *in vivo* patch test for SLS and AEOS-6EO.[38] Comparable results were also reported by Dominguez *et al.*[47] for mixtures of cocamidopropyl betaine with SLS, by Dillarstone and Paye[48] for SLS-AEOS-2EO, SLS-cocamide DEA, LAS-AEOS-2EO, and LAS-C15–8EO. More recently, this observation was also made by studying the systems SLES-SCA,[49] and SLS-dimethicone copolyol phosphate.[50]

Another factor which could affect the swelling and irritation of skin on contact with surfactant solutions is pH. Lower pH was reported to reduce swelling of stratum corneum but the irritation data were not conclusive.[5]

Thus, mild, non-irritating shampoos can be formulated by using alkyl ether sulfates with a high ethylene oxide number, anionic surfactants with triethanolamine as a counter-ion, sulfosuccinates,[54] as well as nonionic or amphoteric surfactants.[29] Such products may also contain anti-irritant agents such as polysorbate-20, and they are usually formulated at low viscosities.[29]

Conditioning agents are frequently used in shampoo formulations (see below).[32,55] They smooth the cuticle surface and lower wet or dry frictional coefficients, which eases combing and improves the tactile properties of hair. The most frequently used ingredients for this purpose are cationic polymers (*e.g.* polyquaternium-10 or cationic guar gum), silicones (including high molecular weight dimethicone), and cationic surfactants. Cationic polymers readily deposit on hair in the amount of less than 1 mg of polymer per g of hair, which is enough have a distinct effect on the tactile perception of wet hair and improve combing, especially for chemically processed hair.

A special class of shampoos is designed to limit the formation of dandruff, which manifests itself by a characteristic flaking and scaling of the scalp.[56–58] Several shampoo ingredients are typically used to ease the condition in over the counter (OTC) products such as zinc pyrithione, coal tar, selenium disulfides, polidocanol and salicylates. Other anti-mycotics used are azoles (climbazole, cloritrimazole, ketoconazole, oxiconazole), polyenes (amphotericin B) and hydroxypyridones (ciclopirox, octopirox).

It has been recently reported that the repetitive use of a shampoo containing 1% zinc pyrithione can increase the scalp barrier by increasing the level of lipids (fatty acids, cholesterol, triglycerides and ceramides) in the scalp stratum corneum.[58] The techniques employed in this study included rating by an experienced trichologist, quantitative high performance thin-layer chromatography (HPTLC) analysis of tape-stripped sebum, stratum corneum protein quantification, corneosurfametry and squammometry.

12.4.2 Hair Conditioners

Hair conditioners are designed to improve combing, ease manageability, reduce static charges, and facilitate styling by surface modification of hair.[59] In its basic form a conditioner should not be visually detectable because it should form a very thin layer on the surface of the hair. There are several different types of hair conditioners used in present cosmetic practice:

(1) Rinse-off conditioner sometimes referred to as a cream rinse. It is typically formulated as a thick cream, which is applied to wet hair in the amount of a few grams, worked in (in order to provide uniform coverage), left on the hair for one to five minutes, and then rinsed off. The rinse-off conditioners are typically used after shampooing or after chemical hair treatments such as dyeing, waving, relaxing, bleaching *etc.*

(2) Leave-in conditioner. It is in the form of a water-thin aqueous solution of actives. It is applied to hair, spread uniformly with a comb or brush and

then left to dry. Leave-in conditioners can sometimes be applied to hair by a pump spray.

(3) 'Intensive hair treatment' conditioner. It is usually a very thick paste, which is rubbed into wet hair and left to react or penetrate with the hair shaft for about 30 minutes, and then rinsed. The formulations of intensive hair treatments typically contain high concentrations of proteins and cationic surfactants.

(4) 'hot oil treatment' conditioner. It is usually designed as a water-based liquid formulation heated to 30–40°C prior to application to hair for an extended period of time (10–30 minutes). Elevated temperature increases the penetration of surfactant into the hair, resulting in a more durable conditioning effect that can survive several shampooings. Hot oil treatments are sometimes used as an after-treatment for semi-permanent hair dyes to increase the durability of the dyeing effect.

(5) 'Shine treatment' conditioner. It is oil-based, frequently a solution of high molecular weight silicone in cyclomethicone. After cyclomethicone evaporates, a thin layer of high molecular weight film is left on the surface of hair producing an increase in fibre lustre. The product can be also formulated based on hydrocarbon raw materials.

Examples of typical raw materials used for the formulation of hair conditioners are listed in Table 12.2. They are generally characterised by good safety profiles, which are published in cosmetic ingredient reviews. One safety aspect of their usage, a potential for skin irritation by cationics, was discussed in several studies.[50,65–67] Early work was based on the Draize rabbit skin test and involved concentrated cationic surfactant solutions. They established the high irritation potential of quaternary compounds such as C_{12-16} alkyl dimethyl benzyl ammonium chloride or didecyl dimethyl ammonium chloride.[67] Subsequently, a number of different techniques including TEWL, histopathology, colorimetry and clinical scoring were used to confirm and expand earlier data. The recent trend has been to eliminate animal tests through the use of human volunteers, because there is little justification to judge human skin irritation based on the Draize test in rabbits. General conclusions about structure-irritation property relationships of diverse types of surfactants were drawn by Rieger,[67] who proposed that a compound's molecular characteristics and usage determine the severity of skin reaction. According to this work lower skin irritancy is more likely for water-soluble surfactants characterised by high molecular weight. Lower irritancy is also more plausible in the case of rinse-off products, where the application conditions result in little or no skin permeation. In contrast to this, higher skin irritancy can result from the use of lipid-soluble low molecular weight surfactants, applied as leave-on products, under conditions favouring high skin permeation.

Toxicological profiles relevant to cosmetic applications for the most frequently used cationic surfactants and polymers are published in the cosmetic ingredient review. A few examples of such data are given in Table 12.3.

An additional way of reducing the irritation potential of quaternary materials is to employ them in mixtures with other surfactants. For example, combinations of

Table 12.2 *Typical conditioner composition*

Conditioner constituents	Raw materials	Concentration (%), active
Emulsifier/thickener (used in rinse-off or intensive care conditioners for thickening)	Cetearyl alcohol Glyceryl stearate Ceteareth-20 Glycol monostearate PEG-100 stearate	2–5
Cationic surfactant[60]	Behentrimonium methosulfate (and) cetearyl alcohol Stearalkonium chloride Cetrimonium chloride Steramidopropyl dimethylamine	1–5
Cationic polymer[61]	Polyquaternium-10 Polyquaternium-7 Polyquaternium-11	0.25–2
Oils and emollients	Mineral oil Hydrogenated polyisobutylene	0.5–5
Proteins[62]	Hydrolysed collagen Hydrolysed keratin Hydrolysed soya and wheat protein Quaternised protein hydrolysates	0.5–3
Silicones[63,64]	Dimethicone Phenyl trimethicone Amodimethicone Dimethicone copolyol Dimethicone copolyol stearate Silicone phosphate esters	0.5–2
Sunscreens or photo-absorbers	Octyl methoxy cinnamate Benzophenone-4	0–5
Antioxidants, moisturisers, etc.	Tocopherol Panthenol	0.1–2

quaternaries and carboxysilicones such as cetrimonium chloride–dimethicone copolyol phthalate and stearalkonium chloride–dimethicone copolyol phthalate were found to have significantly lower irritation scores than the corresponding pure cationics.[50]

Another aspect of safety for surfactants in general, is their chemical purity and the absence in their composition of significant amounts of reactive species in the form of alkylating or ethoxylating agents, as well as residual monomers, solvents and catalysts. The most toxic contaminants have to be identified and concentration limits are imposed on raw materials containing them. For example, for acrylamide-based polymers and co-polymers, the maximum allowed concentration is 10 ppm.

Table 12.3 *Safety data for cationic conditioning agents used in shampoos and conditioners*[29]

Conditioning agent	Concentration in product (%)	Toxicological status
Polyquaternium-7	0.1–5.0	A (acrylamide < 10 ppm) Ocular exposure – mild irritation, repeated insult patch test – mild irritant, non-sensitiser
Polyquaternium-10	0.1–5.0	A Mild irritant to eye or skin; inhalation, dermal and ocular data – low toxicity
Polyquaternium-11	0.1–5.0	A No skin or eye irritation; no sensitisation
Stearalkonium chloride	0.5–4.0	A Slight eye irritation. at 1.25%
Cetrimonium chloride	<0.25% for leave-in no limit for rinse off, typical range 0.5%–4%	A Mild eye Irritation at 0.5%, 0.25% limit for leave-on products

A: Determined to be safe as a cosmetic ingredient in the present practices of use

Strict concentration limits are also established for ethoxylated surfactants and polymers in terms of the concentration of residual 1,4-dioxane, ethylene oxide and epichlorohydrin.

12.4.3 Hair Waving and Depilating Products

Hair waving, depilating, and relaxing involve the use of disulfide bond-reducing agents. The objective of waving is to impart to hair a durable shape that is different from its native configuration. The first step in the waving process is to soften hair keratin by breaking disulfide cross-links in keratin protein, then moulding hair into a desired configuration, and finally fixing the newly imparted geometry by re-oxidation of thiol groups with hydrogen peroxide.

The markets are dominated today by formulations based on thioglycolic acid (TGA) and its derivatives. The popularity of TGA stems from a number of factors. Its long history of use has built an impressive evidence of adequate medical safety.[68,69] The incidence of injury has been extremely low and so has been the frequency of sensitisation. High adaptability of TGA to various formulation types, which provide markedly different end benefits, coupled with the performance reliability and low price, have all contributed to its success. The unpleasant odour of TGA has remained its most perceptible drawback. Although some progress has been made in fragrancing of TGA-based lotions, the results so far are at best mediocre.

Conventional waving lotions contain 0.5–0.8M TGA adjusted to pH 9.1–9.5. The neutralising base can be ammonia, alkanol amines, sodium carbonate or a mixture thereof. Ammonia appears more effective than the other bases in facilitating diffusion of the TGA through hair. It is also preferred over non-volatile amines because it escapes during processing and the resultant drop in pH reduces the activity of the lotion with time and thus minimises the danger of over processing.

A TGA derivative, glyceryl monothioglycolate (GMTG) has some practical importance and is used in so-called 'acid waves'. In terms of waving performance, GMTG works better than TGA at low pH. However, the resulting wave lacks the crispness and durability of the conventional alkaline TGA wave. This is somewhat compensated by lower hair damage. To increase the efficacy of GMTG, the waving process is often carried out with the aid of heat. There have also been reports of associated skin sensitisation, which has limited the use of GMTG to beauty shop (salon) applications.[69,70]

It is also worthwhile mentioning TGA-based formulations referred to as 'self-timing' or 'self-heating' waves. The self-timing wave contains dithiodiglycolic acid (DTDGA), the oxidation product of TGA, to prevent hair over-processing without negatively affecting the waving performance. The exothermic wave products contain a small vial of aqueous H_2O_2 (separate from the neutraliser!), which is added to the waving lotion just prior to its use. Oxidation of TGA, which in this case is in excess of concentrations required for waving, generates some heat as well as small quantities of DTDGA.

Other reducing agents used in waving formulations include cysteine and sodium bisulfite. Cysteine is claimed to provide a 'natural' and non-odorous alternative to TGA and to wave the hair without damage. However, the waving efficacy of cysteine is low, although it can be significantly increased by incorporation of high concentrations of urea (2–3 M). Sodium bisulfite offers consumers perceptible attributes such as lack of odour and low hair damage although it does not achieve the waving efficacy nor durability of TGA systems. Also, while the CIR expert panel concluded that sodium bisulfite is safe as used in cosmetic formulations, the toxicological assessment of sodium bisulfite showed positive results in mutagenicity and genotoxicity tests.[71]

The aspect of hair damage most readily perceived by consumers is the deterioration in the hair shaft's tactile characteristics and in hair combability. It is probably caused by etching of the fibre surface, and resulting loss of protein and surface lipids. Reformation of the cleaved disulfides in the neutralisation step is also not fully complete, even in TGA-waved hair, due to steric hindrance and irreversible formation of cysteic acid.

All of these factors mentioned above have a small but measurable effect on the properties of hair. Waving makes hair slightly weaker and more extensible when wet although little changes are seen in dry fibres. The porosity of hair is increased and this is readily perceived and negatively assessed by consumers who also use hair colourants. The hair is also more sensitive to weathering and lightens much faster in the sun.

To mask and/or limit the reducing agent-associated damage, one has to use conditioning shampoos and conditioners after waving. Alternatively, one can employ waving formulations that contain conditioning agents. Cationic polymers or silicones are particularly effective in depositing a lubricating layer on the surface of hair during the waving procedure. Finally, the use of sunscreens or photo-absorbers in shampoos, conditioners and fixatives could provide an additional measure of damage protection.

Chemical depilation of hair is carried out with actives capable of weakening the mechanical properties of hair to the point that they may be easily detached or pulled off the surface of skin. Sodium and calcium thioglycolates are widely used in combination with calcium hydroxide, raising the pH of the depilatory to about 11.5. The high pH of such products increases their irritating potential, which can be partially alleviated by the use of mild emulsifiers and oils in the preparation of the product. Also, the timing of the treatment has to be strictly limited to 5–10 minutes in order to avoid undesirable skin reactions.

12.4.4 Hair Relaxing Products

Hair relaxing is the process that aims to straighten very curly African hair. This can be accomplished in several ways, such as by employing chemical processes or by physical transformation using hot irons typically operating in the temperature range 120–175°C. The temperature range of some curling irons may even extend to 220–240°C where the melting (denaturation) of the hair crystalline phase occurs. The damage to hair accompanying high temperature treatment may be extensive and includes surface damage (reflected by an increase in combing forces), colour changes, tryptophan degradation, decrease in fibre mechanical strength and secondary grooming damage.[72] If hot iron treatment is carried out at lower temperatures (120–175°C), the straightening effect is reversible and hair fibres will curl up again when exposed to higher humidity. More durable effects can be achieved by using chemical processes, although they also differ in degree of permanency. The use of TGA or bisulfite/urea systems, similar to those employed in waving formulations, results in only semi-permanent effects. Truly permanent straightening can be achieved by using alkali such as sodium hydroxide, lithium hydroxide and calcium hydroxide at a pH of 12.[73] The products based on these materials are referred to as lye relaxers (Table 12.4). An effective straightening system can also be formulated by a two-component composition, one part containing guanidine carbonate and the other calcium hydroxide. After mixing, guanidine hydroxide is formed, which acts as an active species in the process of hair straightening. Such products are referred to as no-lye relaxers (Table 12.4).[74]

The process of relaxing can cause severe damage to hair because of irreversible disulfide bond breakage with the formation of lanthionine and lysinealanine. In addition the hair may undergo surface damage related to hydrolysis of covalently bound lipids (18-methyleicosanoic acid) and removal of loosely bound fatty acids. Thirdly, hair may experience a phenomenon known as supercontraction, which is due to transformations in the α-helical crystalline phase in the hair. The formation of sulfur in the reaction of keratin disulfide with thiolate in may result in yellowing of hair, an effect evident especially in gray hair. Application of a conditioner amends the combing forces, however, the repair effect is only temporary and a single shampooing restores elevated combing forces in the damaged hair.

Table 12.4 *Lye and no-lye relaxer compositions*[75]

Constituents	Content (%)
Lye relaxer	
Petrolatum and mineral oil	30.0
Fatty alcohol	1.0
Emulsifying wax	11.0
Emulsifiers	2.5
Water	51.3
Propylene glycol	2.0
Sodium hydroxide	2.2
No-lye relaxer	
A:	
Petrolatum and mineral oil	40.0
Fatty alcohol	7.0
Emulsifiers	2.5
Water	43.0
Propylene glycol	2.0
Anhydrous calcium hydroxide	5.5
B:	
Water	75.0
Guanidine carbonate	25.0

Apart from hair damage, the application of relaxers may cause stinging, itching or burning to the scalp, with irritation localised to the site of product contact with skin. It has been suggested that the irritation is non-immunological and in most instances disappears after rinsing the relaxer. It has also been demonstrated that no-lye relaxers are less irritating than sodium hydroxide-based products.[74] While lye relaxers were shown to cause severe, moderate, mild and minor irritation in 1.33%, 6.67%, 7.2% and 5.1% of users, respectively, for no-lye straighters the corresponding numbers were 0.83%, 0%, 2.1% and 2.5%.

12.4.5 Hair Styling Products

The function of fixative products is to keep hair in place after styling. They can be divided into several sub-categories:

(1) Hair sprays are formulated as non-viscous, alcohol, aqueous or ethanol–aqueous solutions of a polymer (1–8% w/w), with a low proportion of other ingredients such as surfactants, neutralising agents, plasticisers,

anti-corrosion agents, photo-filters, anti-oxidants, fragrances and preservatives. They can be dispensed from a can with a special pump nozzle as a coarse aerosol or as fine particle aerosols from pressurised cans loaded with a product concentrate and a propellant.

(2) Styling gels are prepared by using a styling polymer at a concentration of 1–4%, a gelling polymer at a concentration of 0.25–2%, and other ingredients such as surfactants, neutralising agents, plasticisers, anti-corrosion agents, photo-filters, anti-oxidants and preservatives.

(3) Mousses are formulated by using a styling polymer at 1–4%, a surfactant system to produce foam, propellant and other ingredients similar to those mentioned above. The mousse is dispensed from an aerosol can in the form of a foam, which is then worked into the hair to provide a fixative effect after drying.

(4) Styling lotions or creams are typically prepared with a fixative polymer as the main ingredient and additional polymers, surfactants, oils and emulsifiers to adjust viscosity, texture and appearance. Other ingredients, as pointed out for other types of products, may also be included in the final product.

The main components of hair styling products are synthetic or natural polymers. They are subjected to a battery of toxicological tests prior their use in cosmetic formulations. High molecular weight materials typically do not penetrate skin, thus, irritation and sensitisation potentials are low. Table 12.5 presents few examples of styling polymers and their toxicological characterisation based on CIR Compendium data. Typically, the main point of concern is the presence of low molecular weight contaminants such as residual monomers, solvents and catalysts. The manufacturers of raw materials usually provide information about the polymer composition as well as a description of the methods employed to characterise a given raw material. Considering this information and the material's safety profile the formulator can make a judgment about the health hazard posed to the user of the final product.

12.5 Preservatives

Preservatives are widely used to eliminate microbial contamination of cosmetics, which may be a health hazard and may also lead to deleterious changes in the product's efficacy, odour, colour and viscosity.[9] The preservatives are usually very effective but the process of their selection is complex and a number of factors such as effectiveness in relation to various types of microbes, type of formulation, safety, stability, regulatory status, shelf life, exposure conditions, packaging of the consumer product, anticipated use and misuse of the product have to be taken into consideration. Preservation is required for aqueous systems in the pH range 4–9. On the other hand, preservative-free products can be manufactured under strictly sterile conditions and distributed in self-closing packaging, which prevents the penetration of microbes into the formulation during application. A multi-lab study by the Cosmetics, Toiletry, and Fragrance Association (CTFA) demonstrated,[75] by using a method employing ten species of challenge organisms (such as bacteria, yeasts and moulds) that adequately preserved formulations, including shampoos, conditioners,

Table 12.5 *Selected styling polymers used in hair care products[29]*

Styling polymer	Concentration in product (%)	Toxicological status
Ethyl and butyl esters of PVM/MA copolymer	0–7	Non-toxic in oral toxicity, non-toxic in subchronic inhalation, non-irritant, non-sensitiser, and non-photosensitiser
Acrylate copolymers, styrene/acrylate copolymers, ethylene/acrylic acid/VA copolymers, vinyl caprolactam/PVP/DMEMA copolymers *etc.*	0–8	Polymers are typically non-toxic and non-irritating or with low irritation potential. Residual monomers may be toxic and irritating and should be kept at the lowest possible level. For example, residual acrylic acid and alkyl acrylates <100 ppm in 20%–50% polymer concentrate.
Polyvinylpyrrolidone (PVP)	0–5	Non-toxic, non-irritating, non-sensitising. Short-term inhalation produced mild lymphoid hyperplasia and fibroplasia in animals, but no inflammatory response
PVP/VA copolymers	0–5	Non-toxic, mild irritant at high concentrations, non-sensitising. Favourable epidemiological surveys of cosmetologists exposed to inhalation of the copolymer
Carbomer	0.1–2.0	Low acute toxicity, low potential for skin irritation and sensitisation.

w/o emulsions and o/w emulsions were, according to United States pharmacopeia (USP) criteria,[76] effective in preservation.

The evaluation of the safety of preservatives consists of the tests for acute toxicity, ocular irritation, dermal irritation, sensitisation, photo-toxicity, mutagenicity and embryological (or developmental) toxicity mentioned above. The key objective is to determine skin response to a biocide and level of preservative that can elicit a negative response under ordinary use and foreseeable misuse exposure levels.[9]

Table 12.6 collects most important preservatives used in cosmetic products.[77,78] Parabens are most frequently used and are particularly effective against fungi and gram-positive bacteria. They produce low rates of sensitisation especially when used in leave-in, topically-applied products. Formaldehyde is currently being replaced by formaldehyde-releasing agents, due to its relatively high rate of irritation and sensitisation. They are primarily effective against yeasts, moulds and bacteria and they are frequently combined with parabens for increased protection against yeast and fungi. Methylchloroisothiazolinone/methylisothiazolinone are now strictly regulated because of their relatively high sensitisation potential. The use of methyldibromoglutaronitrile/phenoxyethanol, which were introduced in the 1980s, is also restricted because of accumulating evidence about high allergenicity of this material.[77] Iodopropynyl butylcarbamate is an effective organo-iodine fungicide, bactericide, and parasiticide with a low number of reported sensitisation cases.

Table 12.6 *Preservatives used in cosmetic formulations*[77,78]

Preservative	Concentration in product (%)	Toxicological status
Parabens		
Methyl paraben	0.1–0.8	Excellent safety as stable, effective and non-irritating. Low rate of sensitisation
Ethyl paraben		
Propyl paraben		
Butyl paraben		
Formaldehyde and formaldehyde releasers		
Formaldehyde	0.1–0.2	Irritant and sensitiser; <0.2% of free formaldehyde is recommended by CTFA >0.05% product labeled as sensitiser
Diazolidinyl urea	0.1–0.5	Weak sensitiser
Imidazolidinyl urea	0.03–0.2	Low allergic sensitisation
DMDM hydantoin	0.1–1.0	Weak sensitiser
Quaternium-15	0.02–0.3	Sensitiser
Bronopol		Irritant and sensitiser
Other		
Methylchloroisothiazolinone/ methylisothiazolinone	<15 ppm	Irritant and allergen <7.5 ppm in leave-on and 15 ppm in rinse-off products
Methyldibromoglutaronitrile/ phenoxyethanol	0.0075–0.06	Weak sensitiser
Iodopropynyl butylcarbamate	<0.1	Weak sensitiser
Benzoic acid	0.1–0.5	Effective in food and cosmetic products
Sodium benzoate		
Sorbic acid	0.1–0.3	Primarily active against moulds and yeasts
Potassium sorbate		
Dehydroacetic acid		Bacteriostat and fungistat
Sodium dehydroacetate		
Chloroxylenol	0.2–0.8	Antibacterial and antifungal
Phenoxyethanol	0.5–2	Active against gram-negative bacteria
Ethanol	15–20	Antimicrobial activity against bacteria and fungi
Benzalkonium chloride	0.1–0.3	Activity against gram-positive and some gram-negative bacteria

Table 12.6 *(Continued)*

Preservative	Concentration in product (%)	Toxicological status
Glutaral	0.01–0.1	Broad spectrum preservative
Salicylic acid	0.1–0.5	Bacteriostat and fungistat

In summary, currently used cosmetic preservatives are highly efficacious in preventing microbial spoilage of products. Those that are more effective may display irritating and/or sensitising activity, which is usually documented in clinical research. The risk for the public can be minimised by compulsory labelling and disclosure, strict adherence to the guidelines of safe use and through search of alternative non-sensitizing and non-irritating molecules.

12.6 Ingredients with Restricted Use

A number of shampoo and conditioner ingredients should be employed at limited concentrations because of their potential for *N*-nitrosamine formation, skin irritation or sensitisation, skin absorption or due to the presence of certain impurities (see Table 12.7). A complete list of raw materials with restricted use is given in reference.[29]

12.7 Conclusions

Hair care products are used daily by hundreds of millions of people around the world. There is a tremendous variety of formulations to address every conceivable problem of hair washing, conditioning, waving, relaxing, dyeing, bleaching and styling. Special products are formulated for hair of people with various racial origins such as African, Caucasian, Chinese, Japanese and Indian descent. Thousands of different raw materials are used in all these formulations. Safety of these ingredients is thus a significant public health issue, which is addressed by specialised regulatory bodies in many countries including the Cosmetics, Toiletry, and Fragrance Association (CTFA) in the United States, The European

Table 12.7 *Selected ingredients for shampoos and conditioners with restricted use*

Ingredient	Conditions for usage
Potential *N*-nitrosamine formation	
Secondary and tertiary amines	
Diethanolamine, diisopropanolamine, diisopropylamine, triethanolamine, triisopropanolamine	Should not be used in the presence of *N*-nitrosating agents

Table 12.7 *(Continued)*

Ingredient	Conditions for usage
Alkyl amides	
Cocamide DEA, cocamide MEA, isostearamide DEA and MEA, lauramide DEA, linoleamide DEA, myristamide DEA and MEA, oleamide DEA	Should not be used in the presence of N-nitrosating agents or in products in which N-nitroso compounds may be formed
Sarcosinates	
Ammonium cocoyl sarcosinate, ammonium lauroyl sarcosinate, sodium cocoyl sarcosinate, sodium lauroyl sarcosinate, sodium myristoyl sarcosinate	≤5% in leave-in products; should not be used in the presence of N-nitrosating agents or in products in which N-nitroso compounds may be formed
Sarcosines	
Cocoyl sarcosine, myristoyl sarcosine, oleoyl sarcosine, searoyl sarcosine	≤5% in leave-on products; should not be used in the presence of N-nitrosating agents or in products in which N-nitroso compounds may be formed
Ethoxylated Alcohols Ceteareth-2, -3, -4,…,-100	Should not be used on damage skin or in products in which N-nitroso compounds may be formed
Lecithin and hydrogenated lecithin	≤15% in leave-in products; should not be used in the presence of N-nitrosating agents or in products in which N-nitroso compounds may be formed.
Leave-on vs. rinse-off distinctions	
Sarcosinates and sarcosines	As above
Lecithin and hydrogenated lecithin	As above
Alkyl amides	≤10%-40% in leave-in products
Alkyl Sulfates	
Ammonium lauryl sulfate, sodium lauryl sulfate,	≤1% in leave-in products because of potential irritation
Alkenyl Sulfonates	
Sodium C12–14 olefin sulfonate	≤2% in leave-in. Sultone impurity limits specified in the range from 0.1 ppm to 10 ppm depending on the type of sultone.
Alkylamineoxides	
Lauramine oxide,	≤3.7% in leave-in; irritation
stearamine oxide	≤5.0% in leave-in; irritation
Betaine	
Cocamidopropyl betaine	≤3% in leave-in; skin irritation

Table 12.7 *(Continued)*

Ingredient	Conditions for usage
Ethoxylated alcohols	
Nonoxynol-1, . . .,-8, octoxynol-1,. . ., 8	≤5 in leave-in; potential for penetration; limits on residual ethylene oxide and 1,4-dioxane.[80]
Coal tar	0.5–5%

Cosmetic, Toiletry, and Perfumery Association (COLIPA) in Europe, and the Japanese Cosmetic Industry Association in Japan. Publicly- and privately-funded research institutions investigate safety profiles of established and new raw materials from synthetic and natural sources. The data are published in open literature and there is also a substantial amount of proprietary safety data in private companies.

The analysis of the information pertaining to hair care products, including shampoos, conditioners, waves, relaxers and styling formulations, suggests that they are in general safe to use and do not present a foreseeable public health hazard. Occasionally, ingredients with questionable safety profiles are identified and the limitations of their use in cosmetic products have to be delineated. The examples of such materials, creating controversies and under investigation and discussion in recent years, include ethoxylated surfactants containing 1,4-dioxane and ethylene oxide as by-products from synthesis, formaldehyde-releasing preservatives, parabens, nitrosamine-forming amines and amides, coal tar as an active in anti-dandruff shampoos, mutagenicity and carcinogenicity of dyes and dye precursors (See Chapter 9 for discussion on the latter), sensitisation and photo-sensitisation potential of fragrance ingredients, heavy metal ions in certain raw materials, inhalation data for polymers, as well as polymer and surfactant biodegradability and aquatic toxicity. The use of animals in safety testing is also criticised and several countries have introduced legislation to restrict or ban animal experiments. As new raw materials are put forward on a continuous basis, the industry has to come up with reliable alternatives for a non-animal assessment of their safety.

12.8 Acknowledgements

The author would like to thank Trish Pranke of ISP's Toxicological Department for providing information about methods employed in the evaluation of cosmetic raw materials and referring to literature sources with toxicological data. My thanks also go to John Merianos, Joe Albanese, Ray Rigoletto, Linda Foltis, and Leszek Wolfram for sharing their knowledge and literature on preservatives, polymers and hair waving. Finally, I would like to express my gratitude to Roger McMullen for critical reading and correction of the manuscript.

12.9 References

1. OECD, *Guidelines for the Testing of Chemicals*, Paris, 2003.
2. W. Sterzel in *Surfactants in Cosmetics*, M.M. Rieger and L.D. Rhein (ed), Surfactant Science Series, Vol. 68, Marcel Dekker, New York, 1997.
3. J.H. Draize, G. Woodard and H.O. Galvey, *J. Pharmacol. Exp. Ther.*, 1944, **82**, 377.
4. O.H. Mills and A.M. Kligman, *Arch. Dermatol.*, 1982, **118**, 903.
5. L.D. Rhein, *J. Soc. Cosmet. Chem.*, 1997, **48**, 253.
6. W.L. Billhimer in *Clinical Safety and Efficacy Testing of Cosmetics*, W.C. Waggoner (ed), Marcel Dekker, New York, 1990.
7. E.G. Murphy in *Surfactants in Cosmetics*, M.M. Rieger and L.D. Rhein (ed), Surfactant Science Series, Vol. 68, Marcel Dekker, New York, 1997.
8. A. Janousek, in *Surfactants in Cosmetics*, M.M. Rieger and L.D. Rhein (ed), Surfactant Science Series, Vol. 68, Marcel Dekker, New York, 1997.
9. D.K. Brannan, *J. Soc. Cosmet. Chem.*, 1995, **46**, 199.
10. D.C. Steinberg, *Cosmetics & Toiletries*, 2004, **119(1)**, 55.
11. D.C. Steinberg, *Cosmetics & Toiletries*, 2002, **117(6)**, 18.
12. D.C. Steinberg, *Cosmetics & Toiletries*, 2002, **117(8)**, 24.
13. OECD, *Guidelines for the Testing of Chemicals*, Paris, 2003. Test OECD 401 and 402.
14. OECD, *Guidelines for the Testing of Chemicals*, Paris, 2003. Test OECD 404.
15. OECD, *Guidelines for the Testing of Chemicals*, Paris, 2003. Test OECD 405.
16. OECD, *Guidelines for the Testing of Chemicals*, Paris, 2003. Test OECD 406.
17. OECD, *Guidelines for the Testing of Chemicals*, Paris, 2003. Draft New Guideline 429.
18. OECD, *Guidelines for the Testing of Chemicals*, Paris, 2003. Test OECD 413.
19. D.J. Arquette, E.M. Bailyn, J. Palenske, D. DeVorn Bergman and L. Rheins, *J. Cosmet. Sci.*, 1998, **49**, 377.
20. OECD, *Guidelines for the Testing of Chemicals*, Paris, 2003. Test OECD 409.
21. OECD, *Guidelines for the Testing of Chemicals*, Paris, 2003. Tests OECD 421 and 422.
22. S.R. Rachui, T. Boufaissal, E.A. Newcombe and R.J. Allen, *J. Soc. Cosmet. Chem.*, 1996, **47**, 315.
23. E. Selvaag, H. Anholt, J. Moan and P. Thune, *J. Soc. Cosmet. Chem.*, 1996, **47**, 167.
24. http://sis.nlm.nih.gov/animalalt/aa1998_No_1.htm
25. Cell toxicity test, Microbiological Associates, Inc., 1998.
26. Primary Eye Irritation, Eytex™, In-Vitro International, 1998.
27. ASTM D 5209 – 92 – Standard test method for determining the aerobic biodegradation of plastic materials in the presence of municipal sewage sludge, *Annual Book of ASTM Standards*, 2001, **8.03**, 417; *ibid.* ASTM D 5210–92, 421.
28. OECD, *Guidelines for the Testing of Chemicals*, Paris, 2003. Test OECD MITI.
29. *Cosmetic Ingredient Review Compendium*, 2004, published by CTFA.
30. K. Klein, *Formulating shampoos*, Cosmetech Laboratories, Inc., 1998.

31. E.D. Goddard and K.P. Ananthapadmanabhan in *Interactions of Surfactants with Polymers and Proteins*, CRC Press, 1993.

32. J. Jachowicz and C. Williams, *J. Soc. Cosmet. Chem.*, 1994, **5**, 309.

33. G.J. Putterman, N.F. Wolejsza, M.A. Wolfram and K. Laden, *J. Soc. Cosmet. Chem.*, 1977, **28**, 521.

34. K.P. Wilhelm, C. Surber and H.I. Maibach, *Arch. Dermatol. Res.*, 1989, **281**, 293.

35. S.P. Harrold, *J. Invest. Dermatol.*, 1959, **32**, 581.

36. R. Scheuplein and L. Ross, *J. Soc. Cosmet. Chem.*, 1970, **21**, 853.

37. C. Lang and J. Spengler, *Proceedings of the 14th IFSCC Congress*, 1986, 25.

38. L.D. Rhein, C.R. Robbins, K. Fernee and R. Cantore, *J. Soc. Cosmet. Chem.*, 1986, **37**, 125.

39. E.A. Tavss, E. Eigen and A. Kligman, *J. Soc. Cosmet. Chem.*, 1988, **39**, 267.

40. A. Gabbianelli, R. Hillermeier, M. Tate and M. Prendergast, *J. Cosmet. Sci.*, Preprints of the 1998 Annual Scientific Meeting, 1998, **49**, 194.

41. R.A. Tupker, C. Willis, E. Berardesca, C.H. Lee, M. Fartasch, T. Agner and J. Serup, *Contact Dermititis*, 1997, **37**, 53.

42. T. Agner and J. Serup, *J. Invest. Dermatol.*, 1990, **95**, 543.

43. K.-P. Wilhelm, A.B. Cua, H.H. Wolff and H.I. Maibach, *J. Invest. Dermatol.*, 1993, **101**, 310.

44. V. Goffin, G.E. Pierard and C. Pierard-Franchimont, *J. Soc. Cosmet. Chem.*, 1994, **45**, 269.

45. A. Pagnoni, A.M. Kligman and T. Stoudemayer, *J. Cosmet. Sci.*, 1998, **49**, 33.

46. M. Misra, *J. Soc. Cosmet. Chem.*, 1997, **48**, 219.

47. J.G. Dominiguez, F. Balaguer, J.L. Parra and C.M. Pelejero, *Int. J. Cosmet. Sci.*, 1981, **3**, 52.

48. A. Dillarstone and M. Paye, *Contact Dermatitis*, 1993, **28**, 198.

49. P.N. Moore, A. Schiloach, S. Puvvada and D. Blankenschtein, *J. Cosmet. Sci.*, 2003, **54**, 143.

50. A.J. O'Lenick, *J. Cosmet. Sci.*, 1998, **49**, 137.

51. T.J. Hall-Manning, G.H. Holland, G. Rennie, P. Revell, J. Hines, M.D. Barratt and D. Basketter, *Fd. Chem. Toxic.*, 1998, **36**, 233.

52. L.D. Rhein, F.A. Simion, R.L. Hill, R.H. Cagan, J. Mattai and H.I. Maibach, *Dermatologica*, 1990, **180**, 18.

53. L. Rigano, T. Cavalletti, S. Benetti and S. Traniello, *Int. J. Cosmet. Sci.*, 1995, **17**, 27.

54. A. Allardice, *J. Cosmet. Sci.*, 1998, **49**, 44.

55. R.Y. Lochhead, *Cosmetics & Toiletries*, 2001, **116(11)**, 55.

56. V. Goffin, C. Pierard-Franchimont and G.E. Pierard, *J. Dermatol. Treatment*, 1996, **7**, 215.

57. P. Mayser, H. Argembeaux and F. Rippke, *J. Cosmet. Sci.*, 2003, **54**, 263.

58. H. Meldrum, C.R. Harding, J.S. Rogers, A.M. Moore, C.J. Little, P.L. Bailey, C. Arrowsmith and K. Darling, *IFSCC Magazine*, 2003, **6(1)**, 3.

59. M. Westman in *Conditioning Agents for Hair and Skin*, R. Schueller and P. Romanowski (ed), Marcel Dekker, Inc., New York and Basel, 1999, 281.

60. M.F. Jurczyk, D.T. Floyd, B.H. Grüning in *Conditioning Agents for Hair and Skin*, R. Schueller and P. Romanowski (ed), Marcel Dekker, Inc., New York and Basel, 1999, 223.

61. B. Idson in *Conditioning Agents for Hair and Skin*, R. Schueller and P. Romanowski (ed), Marcel Dekker, Inc., New York and Basel, 1999, 251.

62. G.A. Neudahl in *Conditioning Agents for Hair and Skin*, R. Schueller and P. Romanowski (ed), Marcel Dekker, Inc., New York and Basel, 1999, 139.

63. E.S. Abrutyn in *Conditioning Agents for Hair and Skin*, R. Schueller and P. Romanowski (ed), Marcel Dekker, Inc., New York and Basel, 1999, 167.

64. A.J. O'Lenick, Jr. in *Conditioning Agents for Hair and Skin*, R. Schueller and P. Romanowski (ed), Marcel Dekker, Inc., New York and Basel, 1999, 201.

65. R.A. Cutler and H.P. Drobeck in *Cationic Surfactants*, E. Jungerman (ed), Marcel Dekker, New York, 1970, 527.

66. H.P. Drobeck in *Cationic Surfactants*, J. Cross and E.J. Singer (ed), Marcel Dekker, New York, 1994.

67. M.M. Rieger, *J. Soc. Cosmet. Chem.*, 1997, **48**, 307.

68. C.R. Robbins, *Chemical and Physical Behavior of Human Hair*, 3rd edn, Springer Verlag, 1994, 126.

69. *Cosmetic Ingredient Review Compendium* 2004, 18.

70. N.O. Wesley and H.I. Maibach, *Cosmetics & Toiletries*, 2004, **119(1)**, 26.

71. *Cosmetic Ingredient Review Compendium*, 2004, 250.

72. R. McMullen and J. Jachowicz, *J. Soc. Cosmet. Chem.*, 1998, **49(4)**, 223.

73. M. Wong, G. Wis-Surel and J. Epps, *J. Soc. Cosmet. Chem.*, 1994, **45**, 347.

74. A.N. Syed and A.R. Naqvi, *Cosmetics & Toiletries*, 2000, **115(2)**, 47.

75. G. Fishler, *J. Soc. Cosmet. Chemists*, 1997, **48**, 62.

76. Antimicrobial preservatives–effectiveness, *The United States Pharmacopeia*, **XXII**, 1990.

77. D. Sasseville, *Dermatologic Therapy*, 2004, **17**, 251.

78. J. Mufti, D. Cernasov and R. Macchio in *Preserving Personal Care and Household Products*, Happi, 2001, 69.

79. D. Song, S. Zhang, W. Zhang and K. Kohlhof, *J. Soc. Cosmet. Chem.*, 1996, **47**, 177.

Hair Care Products – Regulatory Issues

P. RANIERO DE STASIO

13.1 Introduction

A chapter on regulatory issues of hair products may seem almost too simplistic for the academic scope of this book as it comes after several others focused on the safety and toxicology of hair products. These are the drivers of all hair products regulations virtually everywhere in the world, ensuring that users do not get hurt when they use hair products. Hair products, particularly those using active chemistry (for example hair colorants and permanent waves) do present important challenges to safety assessors and regulators world-wide. This chapter attempts to explain why products using complex chemistry can be safely used by millions of consumers world-wide with a fairly simple regulatory approach based on robust safety assessments.

Hair products are regulated in most of the world as 'cosmetics'. It is important to say that the word 'cosmetic' is often used in many languages to denote something trivial. This is not the case in regulatory terms. Regulation of cosmetics is taken very seriously by health authorities world-wide. Cosmetics are the most important category of consumer products after foods both in terms of quantity and time of exposure to humans. Regulations in this area do not generally require strict pre-marketing approval controls in most developed countries; but the level of responsibility and liability that cosmetic safety assessments involve is very similar to food and drug assessments world-wide.

Another consideration to quantify the magnitude and importance of cosmetics globally is that this market was estimated in 2003 to be worth $200 billion (US). This is approximately a fifth of the packaged food industry in terms of global business volume (Euromonitor data[2]). Over a quarter of this market (*i.e.* $50 billion) is estimated to be hair products.

The definition of 'cosmetic product' used in this chapter is taken from the European Cosmetics Directive 76/768/EEC[1]: 'A "cosmetic product" shall mean any substance or preparation intended to be placed in contact with the various external parts of the human body (epidermis, hair system, nails, lips and external genital organs) or with the teeth and the mucous membranes of the oral cavity with a view exclusively or mainly to cleaning them, perfuming them, changing their appearance and/or correcting body odours and/or protecting them or keeping them in good condition.'

13.2 Hair Product Regulations

13.2.1 European Regulations

13.2.1.1 The EU Cosmetics Directive

Hair products in the European Union and the European Economic Area are regulated by the Cosmetics Directive 76/768/EEC. The central pillar of this directive is Article 2 which states that cosmetics must not cause any harm to the user under normal or foreseeable conditions of use taking into account the conditions of use of the product. This concept is the key to the directive as it defines the need for a safety assessment of cosmetics based on exposure rather than the mere summation of the hazards of the single ingredients. Article 2 states: 'A cosmetic product put on the market within the Community must not cause damage to human health when applied under normal or reasonably foreseeable conditions of use, taking account, in particular, of the product's presentation, its labelling, any instructions for its use and disposal as well as any other indication or information provided by the manufacturer or his authorized agent or by any other person responsible for placing the product on the Community market ...'

Article 2 therefore lays out the responsibility and liability for the manufacturer to produce proof of safety. Subsequent articles of the directive state that cosmetics in the EU can be introduced onto the market without any checks or pre-authorisation by the authorities. Checks on products are done post-marketing and a complex and exhaustive product information dossier, including full formulation details, manufacturing procedures, stability data, adverse events data, and proof of label claims must be kept on file by the manufacturer ready for inspection by the authorities.

Ingredients are specifically regulated in the EU. There is an EU Inventory of Ingredients.[3] The inventory is not meant to be an exhaustive or 'positive list' of all ingredients used in cosmetics but is established mainly to define a nomenclature that does not require translation in all EU languages: the international nomenclature of cosmetic ingredients (INCI). Cosmetics are the only category of consumer products that have such an inventory defining an international nomenclature.

The Annexes to the Cosmetics Directive are used to control particular classes of substances that may or may not be used in cosmetics:

Annex II is the list of forbidden substances in cosmetics and it includes pharmaceutical ingredients that must not be used in cosmetics, *e.g.* minoxidil, lignocaine and benzoyl peroxide or particular chemical impurities such as nitrosamines. Annex II includes today over 1,100 compounds and the list is constantly updated.

Annex III is a list of substances permitted under particular conditions or concentrations. Considering all recent additions Annex III includes more than 150 entries including many hair colorant precursors and couplers, *e.g.* paraphenylenediamine, amino-phenols and resorcinol. Other reactive chemical ingredients, for example thioglycolate salts used for permanent hair waves and depilatory products are controlled in the EU by virtue of Annex III. Annex II and III are the two key regulatory tools used by authorities and developers alike to ensure ingredient control in cosmetics.

Annex IV contains the list of colorants allowed in cosmetics including decorative cosmetics but excluding hair colorants. Annex IV includes 149 colours allowed in cosmetics for different uses.

Annex V is designed to list substances excluded from the scope of the directive (different from Annex II). This annex has never been used by regulators and is effectively empty.

Annex VI is the list of preservatives and anti-microbials allowed in cosmetics products in the EU. Substances with antimicrobial activity other than preservation, *e.g.* anti-fungal or anti-plaque are also listed in Annex VI. Annex VI includes 56 substances.

Annex VII lists the UV filters allowed in the EU. Annex VII includes 27 substances.

Adaptations of the Cosmetics Directive. All Annexes to the Cosmetics Directive are kept up to date by the ATP procedure (Adaptation to Technological Progress), which through a relatively rapid system maintains ingredient regulations. The EU Commission group that is responsible for recommending changes *via* the ATP procedure is the Scientific Committee for Cosmetics and Non-Food Products (SCCNFP). The SCCNFP as of July 2004 was superseded by the Scientific Committee for Consumer Products (SCCP)[4] a new committee with a similar mandate to the SCCNFP to advise the EU Commission on all consumer products that are non-food and non-drug products.

Amendments to the Cosmetics Directive. A more lengthy procedure for changes that require a modification of the text or articles of the directive is the amendment procedure. There have been seven amendments to the directive since 1976. The latest (7th Amendment)[5] could be the subject of a whole chapter and it is therefore not discussed in detail here. The amendments have not modified the spirit of the Cosmetics Directive, *i.e.* the central role of the safety evaluation based on calculation of exposure-based risk. The EU regulatory process is a 'self-registration' mechanism that has proven very effective to protect EU cosmetics users since the 1980s when the legislation was introduced. To date, there have not been any major cases of public health issues related to cosmetics reported, unlike other consumer categories that have experienced issues in the EU, for example transmissible spongiform encephalopathy (TSE) and salmonella in the food industry. Less sensational issues like dioxin contamination have not affected cosmetics directly. Cosmetics in the EU (and the rest of the world) are considered safe consumer products that are effectively regulated by 'self-registration'.

13.2.1.2 Specific Considerations on the EU Cosmetic Directive

As said before hair products are regulated by applying the Cosmetics Directive. This consists of: i) complying with all limitations imposed by the various annexes; and ii) preparing and maintaining the technical dossier (including the safety assessment) for the authorities' inspection. For example, anti-dandruff products have to comply with Annex III and Annex VI listings for the various anti-fungal ingredients that manufacturers use to combat the scalp yeast that causes dandruff. Similarly, hair colorants have to comply with Annex III listings for precursors, couplers and hydrogen peroxide and a set of specific labelling provisions especially designed for hair colorants. These labelling provisions address the allergenic potential of some ingredients such as *para*-phenylenediamine (see Chapter 10 elsewhere in this volume). Specifically Annex III requires the following labelling for *para*-phenylenediamine:

- Can cause an allergic reaction.
- Contains *para*-phenylenediamine.
- Do not use to dye eyelashes or eyebrows.

There has been talk for many years about the potential introduction of a separate annex to list all hair colorant ingredients allowed in the EU. Some consumers groups advocate a 'positive list' (*i.e.* an annex) for hair colorants ingredients. This process is happening already but without the need to create a new annex. A regulatory revision process is currently taking place whereby the EU Commission has compiled a list of potential hair colorant ingredients and requested the industry to provide state of the art safety data to support each ingredient by September 2004.[6] Industry is providing data for all those ingredients that are actually used, but not for those that are of no commercial interest. All ingredients that receive a favourable review by the Commission will be listed in Annex III (expected by end 2005). All other ingredients for which there is no commercial interest and no data submitted will be automatically prohibited by listing in Annex II. One may ask: 'why a regulatory revision on hair colorants and why now? Is there a particular safety concern which the EU Commission is trying to tackle by performing a thorough review of all hair colorant ingredients?' The answer is that there is a general need for clarity on which ingredients are actually used by manufacturers (and therefore supported by safety data) and which ones could be simply ignored (or prohibited by listing in Annex II). Additionally, there has been raised awareness by the popular press to cancer and allergy allegations with the use of hair colorants. The latter considerations regarding the popular press are tackled in further detail in Section 13.3 below and elsewhere in this volume.

13.2.2 US Regulations

The Food, Drugs and Cosmetics Act designates the FDA as the regulatory body responsible for cosmetics safety.[7] As in the EU, cosmetics may be marketed freely without any pre-marketing checks. 'Self-registration' in the US is again the rule of the game and even though a product information dossier is not specifically required, manufacturers are still responsible and liable for user safety. The definition of a cosmetic product is in part similar to the EU: 'Articles to be rubbed, poured, sprinkled, or sprayed on or introduced into, or otherwise applied to the human body or any part thereof for cleansing, beautifying promoting attractiveness, or altering the appearance, and article intended for use as a component of any such articles . . .'

Some products that are cosmetics in the EU are OTC drugs in the US and thus regulated by the FDA OTC monograph system, *i.e.* active ingredient(s) must comply with the OTC monograph requirements and specifications published by the US FDA.[8] In hair care, the only relevant example is anti-dandruff products. Other non-hair care products in this category are sunscreens, anti-perspirants, anti-caries toothpaste and medicated skin lotions and protectants that are not the subject of this chapter.

Cosmetic colour ingredients that are used for colouring hair are regulated in one of two ways. Firstly, there is a list of approved colours that can be used for colouring hair, and other cosmetics, *e.g.* eye-shadows. Only 36 FDA 'certified colours' (and some lakes of those that are soluble) and 23 'permitted colours' are on the approved colour list for use in cosmetics.[9] Secondly, hair dyes that are derived from coal tar can be used to colour hair as long as the product is labelled with the following statement: 'Caution – This product contains ingredients which may cause skin irritation on certain individuals and a preliminary test according to the accompanying directions should first be made. This product must not be used for dyeing the eyelashes or eyebrows; to do so may cause blindness'. The accompanying package must provide directions for carrying out the sensitivity test.

The FDA cosmetic regulations do provide a list of forbidden substances (the equivalent in Europe is 'Annex II') but this includes only a handful of ingredients. The CIR (Cosmetic Ingredient Review)[10] process is a mechanism set up by all interested parties to review cosmetic ingredients. The panel includes six independent scientists who are voting members and three non-voting members one each from the FDA, the Cosmetics Industry Association (CTFA), and the Consumer Association.

The CIR has reviewed to date many hundreds of ingredients and of those they found two-thirds to be considered safe, almost another third to be considered safe under specific conditions of use and less than 10% with insufficient data. Only a very few have been declared unsafe for use. The CIR panel in the US serves a similar function to the SCCNFP in the EU although the regulatory systems to execute CIR and SCCNFP recommendations are different in the US vs. the EU. There is no direct regulatory action following CIR reviews however CIR deliberations are regarded very highly and all reputable manufacturers comply with them as if they were mandatory. CIR reviews may also influence other regulatory bodies – it is known for example that the CIR and the SCCNFP often compare their deliberations and conclusions on ingredients.

13.2.3 Japanese Regulations

All cosmetic ingredients in Japan have to be listed in a 'positive list': the CLS (Comprehensive Licensing Standards of Cosmetics by Category).[11] Until 2002 all cosmetics also had to be authorised by the Ministry of Health and Welfare (MHW) *before* marketing. After 2002 the MHW granted a de-regulation for cosmetics and now only listing in the CLS is sufficient to market cosmetics. However, many product categories considered cosmetics in the US and the EU are considered quasi-drug in Japan. These are: hair colorants, permanent waves, products to combat bad breath, bath preparations, talcum powder, depilatories, shaving lotions and skin products. These still require pre-registration with the MHW. Japan is the only country in the developed world still working with a partial pre-registration system for cosmetics.

13.2.4 Some Regulatory Considerations for Other Countries

Many countries around the world are looking at the EU and US regulations as a model. For example, in Latin America many countries have adopted a de-regulated system that is expected to embrace more and more the self-registration mechanism. China recently relaxed its cosmetics importation rules in line with European legislation, although hair colorants still require a cumbersome registration and ingredients have to be on a positive list. Even the recent Japanese de-regulation is based on the EU 'self-registration' mechanism aided by ingredients lists. This de-regulation trend will probably continue on the basis that cosmetics regulations in the EU and the US provide a strong level of consumer protection without the need for pre-market registration.

13.3 Hair Colorants – Special Considerations

Why have colorants attracted regulatory and media attention? The complexity of hair colorant chemistry and their toxicology has been described elsewhere in this volume. From a regulatory viewpoint hair colorants attract more attention than other cosmetic categories for two main reasons: i) the reactive chemical nature of oxidative hair colorants and the fact that many dye precursors are arylamines although hair colorants have been consistently and repeatedly cleared from any association with cancer over the years; ii) the contact allergy potential of some hair colorant ingredients such as *para*-phenylenediamine – allergy potential is a well known trait of hair colorants and continues to be carefully monitored by dermatologists.

13.3.1 Cancer Claims

Attention to the carcinogenic potential of hair colorants is periodically drawn by sensationalised and inaccurately interpreted scientific data in the media. Historically, this started in the 1970s when the Ames test[12] was introduced to screen compounds for their carcinogenic potential. Some hair colorants were initially found positive in this test and this immediately triggered a flurry of attention in the media. The Ames test is a basic mutagenicity assay using bacterial cells and it is useful for many classes of chemicals but not for arylamines. Subsequently, more sophisticated tests based on a tier battery approach including many types of different system (*e.g* mammalian cells instead of bacterial) have improved our knowledge in this area and provided clearer data in support of the safe use of hair colorants.

Later some epidemiology studies reported an association between hair colourant use and various types of cancer. In spite of the many isolated case-reports the overwhelming weight of evidence from large-scale epidemiology studies demonstrates no association with any type of cancer, even with life-time use of hair colorants. A key example of a study that triggered much debate in the popular press is the 2001 study investigating an alleged association between bladder cancer and hair colorant use.[13] Following this, the SCCNFP requested more data[14] on hair colorants and cancer – this is perhaps one of the reasons (but not the only reason) that precipitated the current regulatory review by the EU Commission.

In the US, the CIR decided to undertake a major review of all the epidemiology data available on hair colorants to date. A report analysing 83 epidemiology studies on hair colorants was submitted to the CIR Epidemiology Review Panel in August 2003.[15] The CIR panel states on this '... The authors found insufficient evidence to support a causal association between personal hair dye use and a variety of tumors and cancers ... The authors concluded that the available evidence is insufficient to conclude a causal association between personal hair dye use and bladder cancer, non-Hodgkin's lymphoma, and multiple myeloma. With respect to other cancers, including leukemia, breast cancer, or childhood cancers, and autoimmune disease or adverse developmental/reproductive effects, the authors concluded that the evidence also did not demonstrate a causal association with hair dye use...'

In conclusion, the CIR review is very useful in clarifying the allegations of links between the use of hair colorants and cancer and it is hoped that the CIR statement will also be adopted in the EU and across the world.

13.3.2 Allergy

Regulatory authorities and the industry have developed very thorough labelling requirements for hair colorants. These products in most countries require a warning on the label stating 'can cause an allergic reaction'. Manufacturers often add voluntary labelling that goes beyond regulatory requirements. As an example, in Europe many manufacturers recommend users to perform a self-patch test 48 hours prior to each use of a hair colorant. Therefore, allergy is a known trait of hair colorants that is effectively managed by regulations and manufacturers with very clear instructions to users. The result is that hair colorants are probably the best labelled category among all consumer products on the market. Recent evidence suggests that allergy incidence to hair colorants has been steady in the past few years or even declining if one considers the generalised increase in use of these products.[16]

13.4 Conclusions

All regulations reviewed here can be considered less complicated when compared with pharmaceuticals. However, the post-market controls by authorities and the regulatory responsibilities imposed on the manufacturers and their safety assessors are rigorous and are sufficient to ensure that hair products are strictly controlled. No major concern around users' safety exists and the key fundamentals of the 'self-registration' regulatory models employed in the US and the EU are being adopted more and more by other countries world-wide. Key recent examples in Japan, China and Latin America lead us to believe that cosmetics regulations are moving towards a common direction. There is a good precedent in this area for global harmonisation: the INCI nomenclature is a virtually global language for cosmetic ingredients – no other consumer products can boast such an historic regulatory achievement. On this basis one can be optimistic and expect that eventually cosmetics will be self-regulated globally, perhaps with a set of harmonised rules that will allow total global exchange and consumer-safe circulation of these products.

13.5 References

1. *European Commission Directive 76/768/EEC*. All updates to the directive can be found on the 'Eudra' website at: http://dg3.eudra.org/F3/cosmetic/CosmLex Updates.htm
2. Euromonitor website: http://www.euromonitor.com/
3. *European Commission Decision 96/335/EC*.
4. SCCNFP and SCCP website: http://europa.eu.int/comm/health/ph_risk/committees/sccp/sccp_en.htm
5. *European Directive 2003/15/EC*.
6. European Commission's communications with Colipa – the European Cosmetics, Perfumery and Toiletry Industry Association.
7. The FDA cosmetics regulations website: http://vm.cfsan.fda.gov/~dms/costoc.html
8. US FDA homepage: http://vm.cfsan.fda.gov/list.html
9. FDA cosmetic colours regulations: http://www.fda.gov/opacom/laws/fdcact/fdcact7b.htm
10. CIR website: http://www.cir-safety.org
11. Japanese Ministry of Health and Welfare homepage: http://www.mhlw.go.jp/english/index.html; Cosmetic regulations: http://www1.mhlw.go.jp/english/wp_5/vol2/p2c5.html
12. B. Ames *et al.*, *Mutation Research*, 1975, **31**, 347.
13. M. Gago-Dominguez *et al.*, *Int. J. Cancer*, 2001, **91**, 575.
14. http://europa.eu.int/comm/health/ph_risk/committees/sccp/docshtml/sccp_out-143_en.htm
15. CIR agenda and results – CIR homepage: http://www.cir-safety.org
16. G. Nohynek *et al.*, *Food and Chem. Toxicol.*, 2004, **42**, 517.

Part 4
Hair in Archaeology

Hair as a Bioresource in Archaeological Study

ANDREW S. WILSON

14.1 The Value of Hair in Bioarchaeology – An Introduction

The robust structural morphology and chemistry of the hair shaft is responsible for its survival in the archaeological record.[1] Sometimes it is found to be better preserved than archaeological tissue remains such as bone collagen that are in more common usage.[2] As a consequence archaeological scientists have sought to use this material for a variety of different purposes. Early studies focussed on morphological variation of human hair,[3] whereas modern bioarchaeological investigations have concentrated on the analysis of diet[4] and drug use[5,6] with areas such as dating, genetics, disease and environmental toxicology also considered.[7–11] The following chapter will provide a critical overview of these techniques as applied to archaeological hair.

The particular value of hair for archaeological investigation stems from the biology of hair formation and growth. Since hair grows at a relatively constant rate (1 cm will roughly equal 1 month's growth in Caucasoid scalp hair), although this is known to be mildly affected by racial characteristics,[12] age, sex and endocrine function,[13,14] information from the hair shaft is of good chronological resolution. Furthermore, once the hair shaft has undergone keratinisation it does not remodel further as do other tissues used in archaeology, such as bone or muscle. Turnover rates for bone, which remodels throughout life, are poorly documented, but complete replacement may take 10–30 years in adults, producing an averaged biogenic signal over that timeframe.[15] As such, the hair shaft captures a snapshot of information that will reflect the final months or years of an individual's life depending on the hair length. This 'recent life history' provided by hair can again be contrasted with information derived from tooth enamel, which although it does not remodel reflects the mineralisation of teeth during formation and so only covers the period of childhood to early adulthood depending on the tooth in question.

As a direct result of technological developments, analyses can now be conducted using small sample sizes and indeed can often use a portion of an individual fibre as in the reconstruction of isotopic histories;[16,17] or bundled segments for drug analysis.[18] Destructive analysis of single fibres, particularly from intact mummy bundles, is far less controversial than the invasive and destructive sampling of long bones or teeth (Figure 14.1). These are crucial developments for bioarchaeology

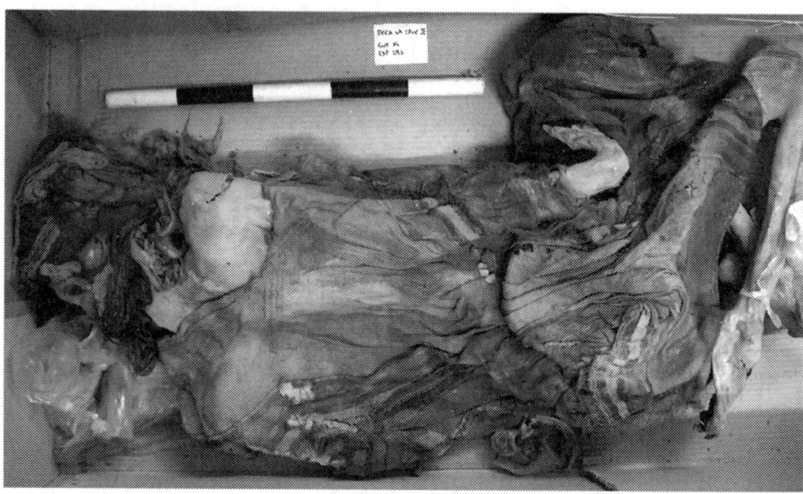

Figure 14.1   *Well-preserved pre-Inca mummy bundle excavated in 2003 from a cemetery site in the lower reaches of the Osmore Valley, Peru dated to roughly AD1000*

where sampling size is critical to conservators, curators and collections managers, as these have a responsibility to safeguard collections over the longer term, overseeing the best interests of what is ultimately a finite resource.

The investigation of hair highlights important parallels between forensic science, archaeology and conservation science. Identification and characterisation of fibres using morphological criteria has long been established in archaeology and anthropology,[19–24] the museum world[25,26] and forensic science.[27] Hair, along with other fibres, is regarded as important trace evidence in forensic science,[28] however recognition of the increasing potential of hair has resulted in renewed interest from within medico-legal investigation.[29] Despite the current importance of largely 'fresh' shed hairs as forensic trace evidence, hair may also be recovered with corpses – whether from missing persons/accidental death, homicide victims or exhumed cemetery remains. Depending on the prevailing environmental conditions and the post-mortem interval, the soft tissue remains may have undergone some degree of putrefaction and hair may have started to detach from the cranium as part of 'skin-slippage'. Yet, even in these circumstances when soft tissues are gone or decomposed, hair can still survive and may be of vital importance in determining systemic poisoning, drug abuse or as part of individualisation.

Thus, human remains, whether ancient or modern, have the potential to tell a variety of stories – for example, what sort of diet a person was eating, where they came from, whom they were related to, what pollutants they were exposed to in life, whether they used drugs, how they died *etc*. Different body tissues provide different threads to this story. Hair provides unique and detailed information relating to the months or years immediately prior to death. This is extremely useful for archaeological populations, because in the case of nutrition, for instance, dietary intake was likely to be far more seasonally-defined in the past than in modern populations.

14.2 The Early Use of Hair as a Resource for Physical Anthropology

Lockets of hair have long been collected as curios.[30] There are numerous samples attributed to prominent individuals in historical collections the world over, much of them without a fully documented curation history[31] (Figure 14.2). However, it's not until the end of the 19th century that we begin to see the systematic collection and study of hair. At the World's Columbian Exposition held in Chicago in 1893, for instance, hair samples were collected from different population groups for comparative study. These early anthropological studies focussed on the morphological

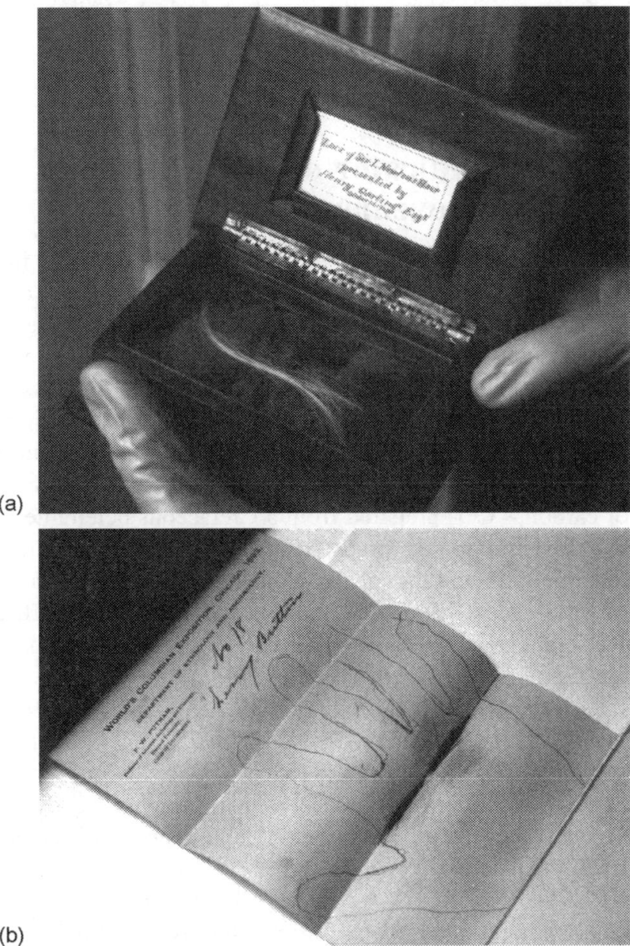

(a)

(b)

Figure 14.2 *Collections of hair include:* (a) *high profile hair samples attributed to famous individuals such as Sir Isaac Newton curated at the Royal Society in London and* (b) *samples from Native Americans collected at the World's Columbian Exposition held in Chicago in 1893*

variation of human hair and established basic classifications for different racial hair types.[32,33]

The earliest detailed study of archaeological hair examined the physical traits of hair from mummy bundles discovered on the Paracas peninsula on Peru's south central coastline.[3] Similar studies have also been conducted for material from Egypt and Sudanese Nubia in attempts to assign racial origin.[34–36] Often it is hair colour that helps to fuel questions over the racial origin of archaeological samples,[37] although there are distinct chemical changes that are known to affect the colour of degraded samples.[8] The more widespread scientific study of ancient hair began with the work of Brothwell and Spearman when they surveyed the types of archaeological environment in which hair has survived[38] and began to consider the factors responsible for its preservation. This work was followed by a histological study of South American mummified and American Indian remains, which examined the hair follicle in scalp tissue for evidence of condition and the presence of palaeopathological lesions.[39]

14.3 Hair as an Indicator of Past Diet and Population Movement

One of the major current uses of archaeological hair lies in the area of palaeodietary analysis. The principle that 'you are what you eat'[4] can be exploited by the use of stable isotopes as chemical signatures used to reconstruct broad categories of diet such the contribution of a marine vs. terrestrial component to the diet, or the consumption of C3 (*e.g.* wheat) versus C4 plants (*e.g.* maize) – categories defined by the photosynthetic pathway that the plant uses to derive carbon from CO_2. Stable isotopes (in hair – carbon, nitrogen, sulfur, hydrogen and oxygen) cannot be measured directly and so are expressed relative to international standards (for carbon – CO_2 prepared from a Cretaceous belemnite found in South Carolina[40] and for nitrogen – atmospheric N_2) so that the relative difference between the sample ratio and that of the standard is expressed as a δ value in parts per mil (‰). Researchers have used modern individuals (see Chapter 8 in this volume) with documented dietary information to validate the use of stable isotopes to characterise diet from hair and breath (as a bone proxy).[41,42] They have also compared archaeological bone and hair samples from an 18th–19th century crypt population excavated from Christ Church, Spitalfields, London and Romano-British samples from Poundbury, Dorset with modern samples. The combined results of this work on modern and archaeological samples suggest that carbon and nitrogen isotopic values from hair 'keratin' and bone collagen are related but cannot be directly equated.[43–45]

Isotopic palaeodietary analysis was first applied to archaeological hair by White and co-workers looking at mummy hair from naturally desiccated Sudanese Nubian mummies, representing X-group (AD 350–550) and Christian periods (AD 550–1300) from the Wadi Halfa area of present day Sudan.[46] Similar work is also underway for samples from South American mummies.[47,48] Analysis of $\delta^{13}C$ values from the Nubian hair provided the first direct archaeological evidence that the modern crop rotation system in N. Sudan has its origins in ancient times. Extrapolating seasonality data on crop cycling from serial measurements along Nubian hair it was found that overall seasonal mortality patterns (defined by the proximal hair root segments) corresponded well with the hot dry Nubian summer that would be demanding of both plant and human physiology.[46,49] Recent re-evaluation of this data reinforces the opinion that the populace were consuming crops soon after they were harvested with only limited use of stored grains and that this pattern persisted for more than 1000 years.[50] This seasonal isotopic variation in hair excavated from the Nilotic sites of the Wadi Halfa group, was contrasted with hair from individuals from the Kharga Oasis, Egypt (AD 400–700). The individuals from the Kharga Oasis showed seasonal stability of carbon isotopic data and consequently a seasonally uniform diet reflecting the stable ecology of the oasis environment.[51]

Isotopic palaeodietary analyses have also been used to show population movement. Analyses of eleven naturally mummified individuals (early Alto Ramirez highland migrants settling in the coastal valleys *ca.* 1000 BC) recovered from a beach burial site at Pisagua, N. Chile were conducted using surviving hair and muscle tissue. Sulfur isotopes in hair (which reflect local geological values at inland sites and marine values at coastal sites), were used for the first time in addition to carbon and nitrogen isotopes to confirm the adaptation of these highland migrants to a predominantly marine diet.[2] The diverse ecology of South America has also enabled conclusions to be drawn as to the geographic origin of Inca capacocha sacrificial victims recovered from high elevation shrines sited on mountains in the high Andes. Alone, stable carbon and nitrogen isotopic analyses of bone collagen from the mummy from Mount Aconcagua, Argentina could not provide definitive information about food sources – indicating either a continental diet with a high percentage of maize, or one including a certain amount of marine products, or a combination of both. Yet the combined analysis of carbon, nitrogen and in particular sulfur isotopes from the hair provided conclusive evidence for a clear terrestrial, in other words non-marine dietary source.[52,53] A further published study on another capacocha show that different behaviour patterns pertained for the young girl known as the Chuscha mummy.[54] Other recent studies on the Aconcagua hair and hair from a woolly mammoth from Western Siberia show that it is now possible to refine isotopic studies of palaeoenviron-ment (inclusive of elevation) and palaeodietary conditions using carbon, nitrogen and hydrogen isotopes.[55,56]

Despite the valuable potential of isotopic data it is important not to push the interpretation beyond reasonable boundaries. In 1991 a well-preserved body was discovered by climbers just below the Hauslabjoch in the Ötztal Alps. Radiocarbon dates eventually proved that the body was more than 5000 years old.[57] Despite the limited survival of hair on the 'Iceman's' body,[58] hair fibres from the Iceman were submitted for isotopic analysis[59] and the data contrasted with hair from a range of sites and time periods derived from modern and archaeological samples.[59] The data was used to suggest that the Iceman had a diet consistent with that observed for modern vegans.[59] This finding was strongly criticised by others[60] who concluded that other evidence indicated that the Iceman had an omnivorous diet. They pointed to the presence of muscle fibre, putative connective tissue, and a hair of animal origin amongst other items in the gut contents of the Iceman. In addition, a critical appraisal of the isotopic data highlighted problems with the reporting of 'purity data' and the fact that, as with all isotopic analyses, δ^{15}N values are not absolute, but rather can be affected by climate and geology.[60]

Work is currently underway to further reinforce the reliability of isotopic data from hair. Here it must be noted that researchers using isotopic data have in general assumed hair to be pure keratin, without considering the additional keratin-associated proteins, melanin and lipids that comprise the hair shaft. Early studies lacked the advantages of modern instrument sensitivity, requiring bundles of on average 15 strands of hair, sequentially cut into 2 cm sections, and also experienced difficulty in undertaking measurements of δ^{15}N. Instrumental developments in the field of isotope geochemistry now make it possible to perform measurements for both carbon and nitrogen along portions of the same fibre and thus provide high-resolution data that enables us to reconstruct, seasonality-related information from small samples.[16,17,48,61] (Figure 14.3)

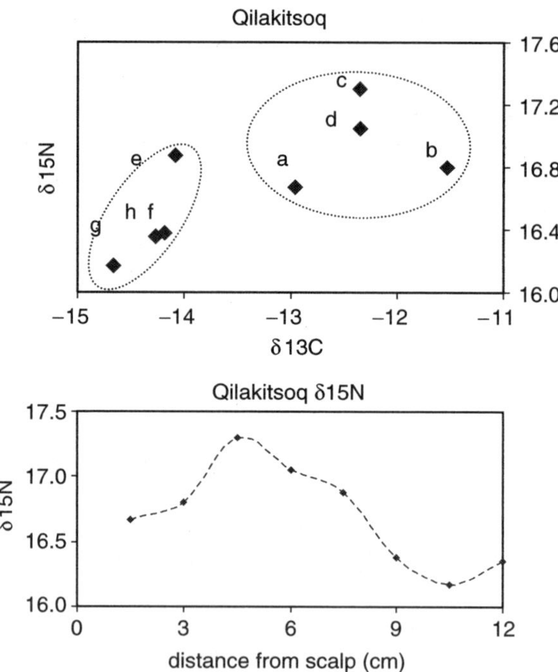

Figure 14.3 *Seasonal variation in the diet of Mummy 3 from Qilakitsoq (W. Greenland) obtained from serial segments along a single hair fibre shows two clear groupings illustrating that their diet had marked seasonal influences. The high values for δ^{15}N show this to be largely due to variation in the availability of marine foodsources*

14.4 Trace Elements from Archaeological Hair and Their Controversial Use in Dietary Analysis and Disease Recognition

Whilst stable isotopes are now the most widely used indicator of past human diet in hair, it is worth noting that the earliest published work used trace elements to yield presumptive dietary information. However, the validity of this approach continues to be questioned for modern[62–64] as well as ancient samples.[65] Importantly, it is the risk of contamination from the burial environment that may pose the greatest threat to gaining reliable data from archaeological samples. Furthermore, the use of different analytical techniques to determine trace element concentrations makes some comparisons between studies difficult because of the relative sensitivity/ resolution of the wide range of different analytical techniques that are used.

There are numerous commercial laboratories that offer to assess nutritional status using trace elements from hair. Such commercial testing has long had its critics even for modern hair samples.[66] When one fellow hair researcher submitted an ancient hair sample from an adult male Maitas Chiribaya mummy from the Azapa valley, Chile (*ca.* 1000 years old) to one such commercial lab for 'nutritional analysis' the data supplied was surprising. The health report supplied suggested the need to take nutritional supplements, including tablets for stress and fibre depletion [67] – not that they would have had much effect on the archaeological individual!

Trace element analysis of archaeological hair has been adapted from biomedical applications such as nutritional and environmental monitoring and has been reviewed by various researchers.[68,69] The first large-scale study using trace elements from archaeological hair was published in 1983,[70] but is also discussed elsewhere.[69,71,72] The results of analyses (using inductively-coupled plasma emission spectrometry) on hair from 168 individuals recovered from two Christian-era Nubian cemeteries at Kulubnarti in Sudan were assessed alongside palaeopathological data. The high incidence of 'porotic hyperostosis' in the Nubian sample (pathological lesions to the upper margins of the orbits frequently attributed to iron-deficiency anaemia), was related to individuals with low concentrations of iron and magnesium as determined from their hair. Whilst the possibility for exogenous contamination was discussed the authors state that this should 'not obviate biocultural considerations of nutrition and disease',[70] despite the fact that iron and magnesium are very mobile elements within the burial environment.

The earliest archaeological study of trace elements from hair used atomic absorption and spectrophotometric analysis to study hair from coastal archaeological sites in Peru.[73] Whilst discussing the potential for future investigations using trace elements to examine seasonal differences in hair, the researchers indicated that the reported analyses were inconclusive since they were uncertain as to 'whether the differences in mineral composition were due to the adjacent soil composition or reflected real differences in mineral composition of the original hair'.[73] Similarly when a small number of samples (5th–7th century AD) from the Antinoe necropolis, south of Cairo were investigated using X-ray fluorescence spectrometry, the researchers again cautioned that soil elemental data would be required to determine whether the data was influenced by contaminants from the burial environment.[74]

Some researchers have recognised that their results from ancient hair are so unrelated to modern samples that the possibility of uptake from the burial environment must be considered to be the cause of such anomalous results.[75] In the case of the Iceman's hair, metals that were elevated were all considered to be derived from the sedimentary rocks of the glacier bed.[76] Strong correlations between the trace element content of hair and adherent soil were identified in samples of hair from a medieval cemetery in Schleswig, N. Germany (11th/12th century) and from an 18th century site in Braunschweig, Germany. The researchers concluded that despite being able to detect differences between groups of individuals coming from different environmental settings, valid interpretations could not be made and that further work into metal transfer from soil to hair was required.[77] Some work has now demonstrated the importance of water and soil percolation, correspondent with burial depth, in the mobilisation of elements such as iron[78] and calcium[79] from soil into hair. Other researchers have stressed that elements such as sodium, potassium, calcium, magnesium and iron which are particularly mobile should not be used in archaeological hair investigations.[9] It is now possible to study this uptake in hair and textile fibres by tracking the relative abundance of trace elements along a single fibre using laser ablation inductively coupled plasma mass spectrometry (LA-ICP-MS)[8] or using elemental mapping techniques and time-of-flight secondary ion mass spectrometry.[80]

In addition to contamination by trace elements derived from the depositional environment it is also possible that other trace elements may be derived from the decomposition of the buried body itself. Re-evaluation of trace element data for the Kulubnarti samples using multivariate statistics found that soil samples from the site at Kulubnarti were of little use in interpreting the presence of exogenous contaminants.[81] Other researchers who examined 46 hair samples from 5 different Inca sites in Peru have stressed that not all hair trace elements are affected in the same way by post-mortem processes.[82] Despite the numerous caveats concerned with post-depositional migration of trace elements in the burial environment, researchers recently investigating trace elements in hair from ten Archaic/Formative period Chinchorro mummies (dating to more than 3500 years old) from coastal N. Chile, failed to discuss this as a potential problem in their interpretation.[83]

14.5 Hair as an Indicator of Exposure to Pollutants and as a Record of the Micro-environment

Set against the same potential risks as described in Section 14.4 above, determining the presence of toxic pollutants in hair samples has enormous importance in environmental monitoring and medicine, particularly since uptake in hair is very tightly time-resolved[84] (see also Chapter 6 and Chapter 5 in this volume). Various archaeological and anthropological studies have investigated the presence of pollutants in hair. For example, the presence of copper and arsenic in the hair of the Iceman[85] was used to suggest long-term and direct involvement with copper working, and high levels of mercury were found in hair from a Roman burial in Dorchester.[86] A 1971 study compared the lead content of human hair from urban and rural populations in the United States with historic samples collected between 1871 and 1923. The results indicated that the lead content of human hair had markedly decreased in the past 50 years despite a general increase in atmospheric lead concentrations.[87] This highlighted the potential for widespread exposure to lead in the past resulting primarily from the ingestion of lead that was used in the collection, storage and piping of drinking water, production of glazed earthenware, leaded paints and cosmetics.[88] Similar work, comparing mercury levels in archaeological hair with exposure levels in modern populations post-industrialisation, has been conducted at the Karluk site in Kodiak, Alaska;[89] Barrow, Alaska[9] and Nunguvik, North Baffin Island;[90] and between 15th century Inuit hair from Qilakitsoq, Greenland and modern Inuit samples.[91,92] Based on the hair data, the researchers suggest that levels of exposure to certain toxic metals such as mercury and lead have increased over time, as increased pollution and divergence from traditional subsistence lifestyles have occurred. However, post-depositional alteration at the Karluk site could not be discounted.[89]

There are many famous instances of possible cases of poisoning discussed within the literature. These include arsenic in the hair and bones of Rudolf Brun, first mayor of Zurich,[93] suspicions of arsenic in the hair of US President Zachary Taylor,[94] mercury in the hair of Robert Burns,[95] US President Andrew Jackson's exposure to mercury and lead[96] and Anastasia Romanova and grand duchess Yelena Galinskaya, respectively the wife and mother of Russian Tsar Ivan IV ('Ivan the Terrible'), who were said to have been killed by mercury poisoning.[97,98]

'was Ivan poisoned at the women? –if the latter was should be ever'

Again there are caveats associated with the use of archaeological or long-curated samples for studies involving exposure to pollutants. For example, chronic arsenic poisoning has long been implicated in the death of Napoleon Bonaparte.[88,99] Segmental analysis of serial sections (~3 mm each in length) along an individual hair attributed to Napoleon Bonaparte analysed by neutron activation analysis was used to suggest that Napoleon Bonaparte had ingested arsenic regularly for several months, corresponding with reports of Napoleon's state of health during 1816.[88] Yet controversy still surrounds these findings. A recent theory is more sceptical, suggesting that the high levels of arsenic result from contamination during the long uncertain curation histories of the hair samples that have been analysed, since arsenic was frequently used both as a pesticide,[100] and in cosmetic materials[88] during the 19th century.

Hair has the potential to provide a very good indication of micro-environment. Since hair is imbued with natural oils *in vivo* it is hardly surprising that both forensic science and archaeology have exploited the potential for both organic and inorganic residues to become naturally trapped in hair. An important current use is with the recovery of explosives residues derived from the firing of handguns[101] or exposure to other explosive traces.[102] However, hair can equally be exploited for pollen, plant or microbial remains when considering victims of homicide.[103] Similarly, archaeological hair samples have been used in so-called 'ecological profiling' to build up a picture of season of deposition and body disposal patterns.[104,105]

14.6 A Record of Drug Use in Archaeological Hair Samples

Humans have exploited the hallucinogenic properties of certain plants for thousands of years. The Inca, for instance, controlled the use of coca[106] and its association with ritual is clear from the carbonised fragments of coca leaves that have been identified from ritual vessels[107] and that have been found with sacrificial capacocha victims in high elevation shrines[108,109] and with the coca metabolites that have been found to be present in their hair.[106] Coca leaves have also been found in funerary contexts that predate the Inca, as in the Osmore valley, Peru.

Drug analysis in hair has developed extensively in the last two decades with the forensic assessment of drugs of abuse,[110–112] therapeutic drug monitoring[113,114] and use of sports doping controls[115–118] (see also Chapter 4 in this volume). Despite the now widespread use of hair in drug analysis there remain issues concerning the differential binding of drugs to hair, as a function of pigmentation differences in hair.[119] Heavily pigmented Negroid type hair is considered more susceptible to uptake of drugs than bleached or blond Caucasoid type hair.[120] Similarly, individuals with greying hair have differential uptake of drugs between pigmented and grey hair fibres.[121] However, hair pigmentation is just one variable that may affect drug incorporation, with different drugs expressing different affinities for melanin.[122]

With forensic casework it has also been necessary to exhume individuals and determine the quantities of drugs that may have been used or administered. For example, morphine was found in serial hair sections taken from a Greek woman exhumed seven months post-mortem,[123] similarly morphine was found in hair from two Italian women exhumed seven months after burial[124] and cocaine and lidocaine were detected in scalp and pubic hair from a man and a woman exhumed after seven and two months burial respectively.[125]

The first analysis of drugs from ancient hair investigated coca use in 163 individuals representing 7 different cultures from South America. Of those tested, 76 tested positive for benzoylecgonine (BZE) a stable metabolic product of cocaine.[126,127] Attempts to characterise drugs in hair samples from Egyptian and Peruvian remains at the Museum für Völkerkunde, Munich, Germany during the early 1990s[128–135] have largely been discredited because the results were implausible in the light of archaeological knowledge. Their discoveries proved astonishing – cocaine and nicotine (both New World drugs) found in the hair of Old World Egyptian mummies, and hashish (an Old World drug) found in the hair of New World Peruvian mummies. The controversies that were generated by these results have been discussed widely both in the popular media[136,137] and the academic literature[138–140] [Table 14.1]. The claims of the Munich-based researchers, whilst widely disputed, withdrew considerable confidence from the emerging field of drug analysis in ancient hair.[67] The explanations of the presence of New World drugs in Egyptian mummies

Table 14.1 *The controversy surrounding cocaine in Egyptian mummy hair*

Problems	Source of Discussion
Absence of genus *Erythroxylun* (coca) and genus *Nicotiana* (tobacco) from Old World	139, 212, 213
Absence of *Cannabis sativa* from New World	139
Phytochemical studies demonstrate that cocaine occurs only in the New World species of *Erythroxylum*	213
The genus *Nicotiana* is essentially neotropical	213
Pre-Columbian contact between Egypt and the New World not possible	139, 213–215
Hashish is the principle form of the drug THC, as usually found in Europe	214
Lack of correlation between the drugs described and reference to a text concerning a remedy using poppy seeds	214
Practice not documented by the ancient Egyptians	139, 214
Lack of appropriate control samples	213

Proposed explanations	Source of Discussion
Instrument error, *e.g.* false-positives at high dilution using immunoassays	213–216
Substances related to cocaine exist in various plants	216
Use of cocaine and especially nicotine as nsecticide/fungicide	140, 212
Faked mummified material	212
Use of tropine-alkaloid-containing plants (*e.g. Belladonna* or *Hyoscyamus* sp.) in the mummification process	212
Instability of psychotropically active substances/oxidation	212
Contamination during curation history, *e.g.* nicotine from cigarette smoke	212–216
Presence of nicotine in plants other than the American tobacco plant	213, 216

ignored not only the post-excavation histories of these remains, but also the native origins of the plants from which the drugs are derived.[140]

More recently, the controversies surrounding drug testing in archaeological hair have subsided as further systematic studies emerge for a range of drug types[6,67,141] including nicotine[142] and alcohol.[143] The absence of evidence for drugs in the hair of nineteen Formative period mummies from Chile was interpreted as either demonstrating that those individuals in the survey group did not use drugs or that despite the wide range of known drugs analysed (cocaine, opiates, cannabis), the analyses did not consider some group-specific drugs derived from local or imported plants.[5] Further work in hair from three archaeological sites in N. Chile has produced evidence of the hallucinogenic substances harmaline, tetrahydroharmine, dimethyltriptamine and bufotenine that supports a wider knowledge of plant-based substances in pre-Columbian South America.[144]

14.7 Hair as a Source of Genetic Information

Whereas nuclear DNA may be expected to survive in most freshly shed hair where the root bulb is intact, mitochondrial (mt) DNA polymorphisms have proved particularly important in the analysis of forensic hair evidence derived from humans [145–147] and other animals [148,149] where little or no genomic DNA survives. mtDNA from the hair shaft was first validated for forensic casework in the mid 1990s,[150] shortly afterwards the potential for using ancient hair samples was discussed.[151] The first successful use of ancient DNA from hair helped to confirm the sex as female of the otherwise skeletal remains of an adult body from a cemetery at Sowinki, near Poznan, Poland.[152] Ancient DNA recovered from animal hair at Smith Creek Cave confirmed, *via* gene sequencing, the presence of desert bighorn sheep in the Great Basin 9800 years ago.[153] In casework, DNA from the hair and bones of British national Paul Wells, kidnapped, murdered and buried for almost two years in southern Kashmir, were matched to blood samples from his parents[154] and mtDNA has been successfully recovered from long-curated hair samples.[155]

The development of robust extraction protocols for ancient or degraded samples[156–158] has resulted in the more widespread use of mtDNA from hair. It has been encouraging to note that recovery and/or decontamination of mtDNA from hair has been more successful than from bone or teeth, and that the oldest authentic mtDNA from hair is on a par with the oldest DNA recovered from bone and teeth, with surprising quantities of DNA surviving.[158,159] However, as with all ancient DNA work there are issues of DNA damage[160] and therefore safeguards against contamination and independent replication are required to confirm authenticity.[161]

Researchers may also use archaeological samples to further our understanding of the genetic origins of the melanocortin 1 receptor (MC1R) gene responsible for red hair and normal variation in skin colour.[162] Yet disputes regarding ownership of ancient samples and ethical approval for the use of hair for genetic studies have proved controversial. This is made all the more problematic because access to modern reference samples of blood or hair is restricted by many countries for fear that genetic material will be misused, as in a study attempting to unravel links between Andamanese and Africans.[163]

14.8 Accelerator Mass Spectrometry Radiocarbon Dating of Hair

With the advent of accelerator mass spectrometry (AMS) radiocarbon dating, the ability to use very small samples of organic material, as little as 20 μg of carbon,[164] means that only a relatively small number of hairs are needed. Hair recovered from caves and rock shelters in Nevada, USA were the first archaeological hair samples to be dated.[165] More recently, AMS radiocarbon dates were obtained from three hairs of ancient bighorn sheep (*Ovis canadensis*) averaging 9880 ± 200 BP.[153] The oldest hair so far dated is putative bison hair recovered from permafrost that has been radiocarbon dated to more than 64,800 years, at least twice as old as hair recovered from a woolly mammoth (*Mammuthus primigenius*) found in Alaska,[166] but because it was recovered below a volcanic tephra dating to approximately 190,000 years it may be considerably older.[158] Samples have also been taken from European bog bodies for dating.[167] The uptake of ^{14}C into biological tissues as a direct result of nuclear testing in the 1950s and 1960s has recently been suggested as a potential means of dating more recent forensic material.[168,169]

14.9 Hair Grooming Practice and Cosmetic Treatment

The use of cosmetics and hair care products is not unique to modern society.[170] Traces of hair colourants such as henna have been identified from ancient samples dating back to roughly 3500 BC.[171] With the development of the pharmaceutical industry during the 19th century a whole host of products designed for use with hair began to be marketed (Figure 14.4).

Not all traces of substances found in hair relate to cosmetic treatment *in vivo*. Traces of materials clearly related to the process of artificial mummification were identified from samples of 21st Dynasty Egyptian mummy hair (1080–946 BC) using FTIR microscopy and these substances included dammar, tragacanth and myrrh.[172] Elemental mapping of hair from two Egyptian mummies using microfocus synchrotron techniques showed a marked heterogeneous distribution that was interpreted as being related to mummification and cosmetic treatments.[173]

Scanning electron microscopy was used to suggest that Lindow Man's hair was cut using some form of cutting tweezers or shears.[174] Furthermore, variation in

Figure 14.4 *The use of hair care products is not a modern phenomenon. This advert for 'Atkinson's Bears Grease' said to promote hair growth, making it 'beautifully soft and glossy' comes from The Leeds Mercury, Saturday May 6th, 1826*

ancient hairstyles has been studied in early Iron Age Danish bog bodies,[175] samples from Roman Poundbury, UK[176] and hair/human hair wigs from ancient Egypt.[171] Not unsurprisingly the presence of ectoparasites has been noted on ancient samples of hair in Egypt[177] and elsewhere throughout the world.[178,179]

14.10 Hair Stylistic Information and Symbolism

In addition to the scientific potential of archaeological hair, it is important to recognise the importance of hair to past societies. Hair and hair pieces have been used to denote sex, status and power, as in the sophisticated use of braids in hair from male individuals (Figure 14.5a) and hats made from hair in high status graves of the Chiribaya culture (*ca.* 1000 AD) from the lower reaches of the Osmore valley, Peru (Figure 14.5b). Furthermore, there are practical reasons for use of certain hair styles and hair pieces, as in human hair wigs, commonly used in ancient Egyptian society for reasons of hygiene and fashion.[177]

Figure 14.5 *Hair used to signify sex and status: (a) a male individual excavated from a pre-Inca cemetery site in the lower reaches of the Osmore Valley, Peru has a mass of braided hair, whilst (b) a head-piece made of woven human hair from the same cemetery site was used for status and display*

Hair was also used in ritual and magic including the ritual treatment of hair in funerary contexts.[180] The Incas believed hair had magical qualities and they performed cutting rituals on a child's first locks, often tying long hair into braids to prevent strands from falling out.[106] Hair-offerings in ancient Egyptian society are considered to have been an important element of popular religion or 'family magic'.[181,182] Similar caches of tubes made from bovine horn and stuffed with human hair have been excavated in Menorca.[183] Indeed the 'magical' importance of hair in England is evidenced by the discovery of a 5 inch long bundle of hair in a late 17th century 'Belarmine Witch Bottle'.[184] Human hair has also been used to construct artefacts, such as mats and bindings at Kulubnarti, Sudanese Nubia.[185]

14.11 Caveats in Using Archaeological Hair – Taphonomy, Contamination and Curation History

Although 'Taphonomy' is a term derived from geology, literally meaning the 'laws of burial', it has become associated with the much wider study of decay processes within the disciplines of archaeology and forensic science. Whilst hair, as a robust biomaterial, may be recovered from archaeological contexts that favour good organic preservation, it does not survive universally. As a consequence there are good reasons to suspect that surviving hair may be partially degraded. Research has therefore been directed towards understanding the condition of hair and how this may impact on the reliability of analytical data derived from hair. Similarly within the museum environment undocumented curation histories can raise questions over sample authenticity.[186]

To date, numerous different techniques have been applied to the assessment of hair condition including protein oxidation,[187] racemisation[188] and fluorescence microscopy.[189] Whilst techniques such as FTIR have been used to suggest the lack of structural change to hair from certain micro-environments over thousands of years,[190] molecular spectroscopy has also provided evidence of progressive conformational change, to both the disulfide linkages characteristic of alpha keratin and to the amide protein groups, as hair starts to degrade.[31,173,191]

Histology is perhaps one of the most widely applied techniques for condition assessment of archaeological biomaterials. Much of the literature on archaeological hair focusses on techniques used for gross morphological assessment, *e.g.* scanning electron microscopy,[192–199] with only limited reference to the more detailed infor-mation on the internal condition of the hair provided by whole fibre mounts[200] and transmission electron microscopy of cut cross-sections.[201,202] However, through detailed examination of the internal morphology of both archaeological[203,204] and experimentally-degraded samples[205] we have demonstrated that alteration is selective and may occur within the underlying cortex in otherwise outwardly well-preserved samples (Figure 14.6).[206,207] A histological index enables screening of samples.[208,209]

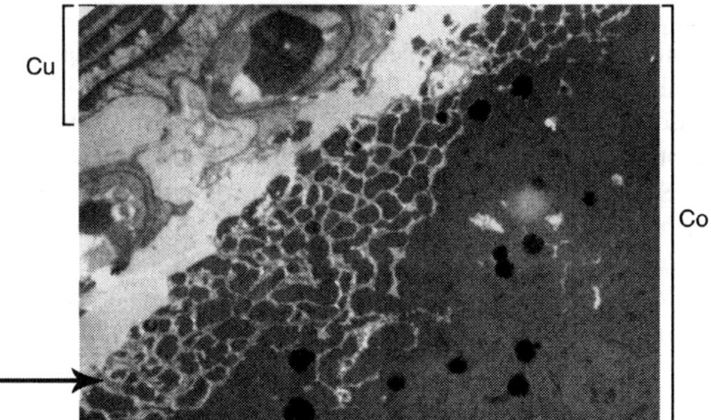

Figure 14.6   *A transmission electron micrograph of a hair sample buried for 12 months in loam at an experimental field site shows progressive breakdown (↑) of the cortex (Co) underlying the cuticle (Cu) that despite being degraded is still present*

14.12 Summary

Hair survives in diverse environments within the archaeological record and has found tremendous utility in bioarchaeology and forensic investigation as well as demonstrating considerable potential for future investigation. Hair can provide unique information due to the time-resolved nature of its formation and small sample sizes required for analyses. Throughout this chapter several interesting case studies have been used to illustrate the potential problems associated with the analysis of archaeological hair. For the most part, these are problems common to all archaeological material. Perhaps most serious is the direct application of analytical techniques devised for modern biomaterials to degraded or contaminated material without adequate sample preparation, or an appreciation of the potential differences in instrument sensitivity with degraded material. Importantly, while the study of single 'exceptionally-preserved' bodies[210] are fascinating in their own right, they are fundamentally weakened by the inability to place them in the context of an assemblage of other remains. Clearly, caution must be exercised when analysing degraded ancient samples[211] and it is important to gain an understanding of the environment to which samples have been exposed.[80] However, it is also clear that with due care it is possible to obtain useful information relating to diet, population movement, drug use, toxicology, dating, genetic, cosmetic, ritual and stylistic information from archaeological hair.

14.13 Acknowledgements

The author would like to acknowledge the support of the following colleagues at the University of Bradford: Mr R.C. Janaway, Professor D.J. Tobin, Dr H.I. Dodson, Professor A.M. Pollard, Professor H.G. Edwards and Professor M.P. Richards; Dr Sonia Guillen at Centro Mallqui (Peru) for permission to use Figures 14.1 and 14.5 and to thank the Wellcome Trust for funding continued work with archaeological hair through a Fellowship (grant 024661) and PhD funding (grant 053966).

14.14 References

1. V. Morell, *Science*, 1994, **265**, 741.
2. A.C. Aufderheide, M.A. Kelley, M. Rivera, L. Gray, L.L. Tieszen, E. Iversen, H.R. Krouse, A. Carevic, *Journal of Archaeological Science*, 1994, **21**, 515.
3. M. Trotter, *American Journal of Physical Anthropology*, 1943, **1**, 69.
4. S. Pain, *New Scientist*, 1998, **12 December**, 34.
5. H. Baez, M.M. Castro, M.A. Benavente, P. Kintz, V. Cirimele, C. Camargo and C. Thomas, *Forensic Science International*, 2000, **108**, 173.
6. D. Counsell, L. Lunt and E.K. Sutherland, *Chromatography and Separation Technology – CAST*, 2000, 6, May.
7. R. Bonnichsen, M.T. Beatty, M.D. Turner and M. Stoneking (ed), *What Can Be Learned From Hair? A Hair Record From The Mammoth Meadow Locus, Southwestern Montana*, Oxford Science Publications, Oxford, 1996.
8. A.S. Wilson, R.A. Dixon, H.I. Dodson, R.C. Janaway, A.M. Pollard, B. Stern, and D.J. Tobin, *Biologist*, 2001, **48**, 213.
9. T. Toribara and A. Muhs, *Arctic Anthropology*, 1984, **21**, 99.
10. D. Brothwell, *Pact 38*, 1993, **III.5**, 317.
11. A.C. Aufderheide, *The Scientific Study of Mummies*, Cambridge University Press, 2003.
12. G. Loussouarn, *British Journal of Dermatology*, 2001, **145**, 294.
13. H. Harding and G. Rogers, in *Forensic Examination of Hair*, J., Robertson (ed) Taylor and Francis, London, 1999, 1.
14. V.A. Randall and F.J. Ebling, *British Journal of Dermatology*, 1991, **124**, 146.
15. S. Mays, in *Human Osteology in Archaeology and Forensic Science*, M. Cox and S. Mays (ed), Greenwich Medical Media, London, 2000, 425.
16. M. Schwertl, K. Auerswald and H. Schnyder, *Rapid Communications in Mass Spectrometry*, 2003, **17**, 1312.
17. R. Bol and C. Pflieger, *Rapid Communications in Mass Spectrometry*, 2002, **16**, 2195.
18. A.M. Tsatsakis and M. Tzatzarakis, *Pure and Applied Chemistry*, 2000, **72**, 1057.
19. F.M. Brown, *Proceedings of the American Philosophical Society*, 1942, **85**, 250.
20. Y. Hayashiba, H. Iki, T. Fukuyama, T. Kita, K. Shintaku and Y. Furuya, *Igaku Kenkyu*, 1983, 17.
21. U. Korber-Grohne, *Journal of Archaeological Science*, 1988, **15**, 73.
22. A. Valente, *Journal of Zoology*, 1983, **199**, 271.
23. M.L. Ryder, *Journal of Archaeological Science*, 1984, **11**, 477.
24. M.L. Ryder, *Journal of Archaeological Science*, 1984, **11**, 99.
25. M. Goodway, *Journal of the American Institute of Conservation*, 1987, **26**, 27.
26. C.J. Dove and S.C. Peurach, in *To the Aleutians and Beyond*, B. Frohlich, A.B. Harper and R. Gilberg (ed), The National Museum of Denmark Ethnographic Series, Copenhagen, 2002, **Vol 20**, 51.

27. J.W. Hicks, *Microscopy of Hairs, a Practical Guide and Manual*, Federal Bureau of Investigation (FBI) Laboratory, Washington DC, 1977.
28. S., Seta, H. Sato and B. Miyake in *Forensic Science Progress*, A., Maehly and R.L. Williams (ed), Springer, Berlin, 1988, 47.
29. A.S. Wilson and M.T.P. Gilbert in *Introduction to Biological Human Identification*, T. Thompson and S. Black (ed), CRC Press, Boca Raton, in Press.
30. Anon, *Proceedings of the Society of Antiquaries of London*, 1891, **13**, 198.
31. H.G.M. Edwards, N.F.N. Hassan and A.S. Wilson, *The Analyst*, 2004, **129**, 956.
32. D. Pruner-Bey, *Journal of the Royal Anthropological Institute*, 1877, **6**, 71.
33. M. Trotter, *American Journal of Physical Anthropology*, 1938, **XXIV**, 105.
34. E. Rabino Massa and B. Chiarelli, *Journal of Human Evolution*, 1972, **1**, 259.
35. S. Titlbachova and Z. Titlbach, *ZAS*, 1977, **104**, 79.
36. D.B. Hardy, *Human Biology*, 1978, **49**, 277.
37. V.H. Mair, *Archaeology*, 1995, **28**, 28.
38. D. Brothwell and R. Spearman, in *Science in Archaeology*, 1st edn, D. Brothwell and E. Higgs (ed), Thames and Hudson, Bristol, 1963, 427.
39. P.W. Post and F. Daniels, *American Journal of Physical Anthropology*, 1969, **30**, 269.
40. H. Craig, *Geochimica et Cosmochimica Acta*, 1957, **12**, 133.
41. T.C. O'Connell, and R.E. Hedges, *American Journal of Physical Anthropology*, 1999, **108**, 409.
42. J. Yoshinaga, M. Minagawa, T. Suzuki, R. Ohtsuka, T. Kawabe, T. Inaoka and T. Akimichi, *American Journal of Physical Anthropology*, 1996, **100**, 23.
43. T.C. O'Connell and R.E.M. Hedges, *Journal of Archaeological Science*, 1999, **26**, 661.
44. T.C. O'Connell, R.E.M. Hedges, M.A. Healey and A.H.R.W. Simpson, *Journal of Archaeological Science*, 2001, **28**, 1247.
45. P. Iacumin, H. Bocherens, A. Mariotti and A. Longinelli, *Palaeogeography Palaeoclimatology Palaeoecology*, 1996, **126**, 15.
46. C.D. White, *Journal of Archaeological Science*, 1993, **20**, 657.
47. J.S. Williams and M.A. Katzenberg, *Paper presented at the V World Congress of Mummy Studies, Torino, Italy* 2004.
48. A.S. Wilson, M.P. Richards, R.C. Janaway, A.M. Pollard and D.J. Tobin, *Paper presented at the V World Congress of Mummy Studies, Torino, Italy* 2004.
49. C.D. White and H.P. Schwarcz, *American Journal of Physical Anthropology*, 1994, **93**, 165.
50. H.P. Schwarcz and C.D. White, *Journal of Archaological Science*, 2004, 31, 753.
51. C.D. White, F.J. Longstaffe and K.R. Law, *Palaeogeography Palaeoclimatology Palaeoecology*, 1999, **147**, 209.
52. J. Fernandez, H.O. Panarello and J. Schobinger, *Geoarchaeology*, 1999, **14**, 27.
53. J.C. Fernandez and H.O. Panarello, in *El Santuario Incaico Del Cerro Aconcagua*, J. Schobinger (ed), Universidad Nacional de Cuyo, Mendoza, 2001, 335.

54. H.O. Panarello, S.A. Valencio and J. Schobinger in, *IV South American Symposium on Isotope Geology,* A.N. Sial (ed), Institute de recherche pour le developpment, Salvador, 2003, 100.

55. Z.D. Sharp, V. Atudorei, H.O. Panarello, J. Fernandez and C. Douthitt, *Journal of Archaeological Science,* 2003, **30**, 1709.

56. P. Iacumin, S. Davanzo and V. Nikolaev, *Palaeogeography, Palaeoclimatology, Palaeoecology,* 2005, **218**, 317.

57. K. Spindler, *The Man in the Ice,* Weidenfeld and Nicolson, London, 1994.

58. R.J. Cano, F. Tiefenbrunner, M. Ubaldi, C. Del Cueto, S. Luciani, T. Cox, P. Orkand, K.H. Kunzel and F. Rollo, *American Journal of Physical Anthropology,* 2000, **112**, 297.

59. S.A. Macko, G. Lubec, M. Teschler-Nicola, V. Andrusevich and M.H. Engel, *Faseb Journal,* 1999, **13**, 559.

60. J.H. Dickson, K. Oeggl, T.G. Holden, L.L. Handley, T.C. O'Connell and T. Preston, *Philosophical Transactions of the Royal Society of London Series B–Biological Sciences,* 2000, **355**, 1843.

61. D.M. Roy, R. Hall, A.C. Mix, R. Bonnichsen, *American Journal of Physical Anthropology,* 2005, in press.

62. P. Manson and S. Zlotkin, *CMAJ: Canadian Medical Association Journal,* 1985, **133**, 186.

63. N. Miekeley, M.T. Dias Carneiro and C.L. da Silveira, *Science of the Total Environment,* 1998, **218**, 9.

64. R.J. Shamberger, *Biological Trace Element Research,* 2002, **87**, 1.

65. J.H. Burton and T.D. Price in *Biogeochemical Approaches to Paleodietary Analysis,* S.H. Ambrose and M.A. Katzenberg (ed), Kluwer Academic/ Plenum Publishers, New York, 2000, 159.

66. K.M. Hambidge, *American Journal of Clinical Nutrition,* 1982, **36**, 943.

67. L.W. Cartmell and C. Weems, *Chungara, Revista de Antropologia Chilena,* 2001, **33**, 289.

68. J. Sen and A.B.D. Chaudhuri, *Journal of the Indian Anthropological Society,* 1995, **30**, 259.

69. M.K. Sandford and G.E. Kissling, in *Investigations of Ancient Human Tissue,* M.K. Sandford, (ed), Gordon and Breach, 1993, 131.

70. M.K. Sandford, D.P. Van Gerven and R.R. Meglen, *Human Biology,* 1983, **55**, 831.

71. M.A. Katzenberg and M.K. Sandford, In *Proceedings of the 1ˢᵗ World Congress on Mummy Studies,* Tenerife, M. A. y. E. d., Ed., Cabildo de Tenerife, 1992, 543.

72. M.K. Sandford and M.A. Katzenburg, in *Proceedings of the 1ˢᵗ World Congress on Mummy Studies,* Tenerife, M. A. y. E. d., Ed, Cabildo de Tenerife, 1992, 535.

73. R.A. Benfer, J.T. Typpo, V.B. Graf and E.E. Pickett, *American Journal of Physical Anthropology,* 1978, **48**, 277.

74. P.P. Parnigotto, M. Folin, A. Marigo, R. Gerbasi, A. Martorana and E.R. Massa, *Journal of Human Evolution,* 1982, **11**, 591.

75. E. Gonzalez-Reimers, M. Arnay de la Rosa, V. Castro-Aleman and L. Galindo-Martin, *Human Evolution,* 1991, **6**, 159.

76. L. Capasso, R. Mariani-Constantini, S. Calvieri and L. Frati, *Journal of Palaeopathology,* 1997, **9**, 153.
77. G. Grupe and K.Z. Dorner, *Morph. Anthrop.,* 1989, **77**, 297.
78. R. Doi, L. Raghupathy, H. Ohno, A. Naganuma, N. Imura, and M. Harada, *Science of the Total Environment,* 1988, **77**, 153.
79. R. Casallas, N.F. Mangelson, M. Kuchar, C.W. Griggs, L.B. Rees and N. Iskander, Nuuk, Greenland, 2001.
80. I.M. Kempson, W.M. Skinner, P.K. Kirkbride, A.J. Nelson and R.R. Martin, *European Journal of Mass Spectrometry,* 2003, **9**, 589.
81. M.K. Sandford and G.E. Kissling, *American Journal of Physical Anthropology,* 1994, **95**, 41.
82. M. Wolfsperger, H. Wilfing, K. Matiasek and M. Teschler-Nicola, *International Journal of Anthropology,* 1993, **8**, 27.
83. A.Y. Du, N.F. Mangelson, L.B. Rees and R.T. Matheny, *Nuclear Instruments & Methods in Physics Research Section B– Beam Interactions with Materials and Atoms,* 1996, **109**, 673.
84. M. Wilhelm and H. Idel, *Zentralblatt fur Hygiene und Umweltmedizin,* 1996, **198**, 485.
85. D. Brothwell and G. Grime, in *Mummies in a New Millenium,* N. Lynnerup, C. Andreasen and J. Berglund (ed), Danish Polar Center, Copenhagen, 2003, 66.
86. I.M. Dale, *Proceedings of the Dorset Natural History and Archaeological Society,* 1981, **103**, 92.
87. D. Weiss, B. Whitten and D. Leddy, *Science,* 1972, **178**, 69.
88. A.C. Leslie, H. Smith, *Archives of Toxicology,* 1978, **41**, 163.
89. G.M. Egeland, R. Ponce, R. Knecht, N.S. Bloom, J. Fair and J.P. Middaugh, *International Journal of Circumpolar Health,* 1999, **58**, 52.
90. B. Wheatley and M.A. Wheatley, *Arctic Medical Research,* 1988, **47**, 163.
91. J.C. Hansen, T.Y. Toribara and A.G. Muhs, in *The Mummies from Qilakitsoq – Eskimos in the 15th Century*, J.P. Hart Hansen and H.C. Gullov (ed), Danish Polar Center, Copenhagen, 1989, 161.
92. J.C. Hansen G. and Asmund, in *Mummies in a New Millenium,* N. Lynnerup, C. Andreasen and J. Berglund (ed), Greenland National Museum and Archives and Danish Polar Center: Nuuk, Greenland, 2003, 69.
93. A. Wyttenbach, S. Bajo and E. Hug, *Journal of Radioanalytical Chemistry,* 1973, **15**, 9.
94. Anon, http://www.ornl.gov/ORNLReview/rev27–12/text/ansside6.html, n.d., 2002.
95. J.M.A. Lenihan, *The Lancet,* 1971, **2**, 1030.
96. L.M. Deppisch, J.A. Centeno, D.J. Gemmel and N.L. Torres, *Journal of the American Medical Association,* 1999, **282**, 569.
97. D. Derbyshire, in *Daily Telegraph,* London, 2001.
98. B. Besserglik, wysiwyg://22/http://www.smh.com.au/icon/0103/21/news8.html, 2001.
99. H. Smith, S. Forshufvud and A. Wassen, *Nature,* 1962, **194**, 725.
100. J.T. Hindmarsh and P.F. Corso, *Journal of the History of Medicine,* 1998, **53**, 201.

101. S. Kage, K. Kudo, A. Kaizoji, J. Ryumoto, H. Ikeda and N. Ikeda, *Journal of Forensic Sciences*, 2001, **46**, 830.
102. K.P. Sanders, M. Marshall, J. Oxley, J.L. Smith, and Egee, L. *Science and Justice*, 2002, **42**, 137.
103. J.H. Bock and D.O. Norris, *Journal of Forensic Science*, 1997, **42**, 364.
104. J.D. McMahan R.J. Dale, Office of History and Archaeology, Department of Natural Resources Anchorage, 1990.
105. J.D. McMahan and R.J. Dale, Office of History and Archaeology, Alaska Division of Parks and Outdoor Recreaction, Unalaska, 1995.
106. K. Douglas, *New Scientist*, 2001, **172**, 30.
107. A.R. Cortella, M.L. Pochettino, A. Manzo and G. Ravina, *Journal of Archaeological Science*, 2001, **28,** 787.
108. M.C. Ceruti, in *Mummies in a New Millenium* N. Lynnerup, C. Andreasen, and J. Berglund, (ed), Danish Polar Center, Copenhagen, 2003, 178.
109. J. Schobinger, *Natural History*, 1991, 62.
110. M.R. Moeller, *Therapeutic Drug Monitoring*, 1996, **18**, 444.
111. F. Tagliaro, F.P. Smith, Z. De Battisti, G. Manetto and M. Marigo, *Journal of Chromatography*, 1997, **B**, 261.
112. H. Sachs, *Forensic Science International*, 1997, **84**, 7.
113. Y. Gaillard, and G. Pepin, *American Clinical Laboratory*, 1997, **16**, 18.
114. Y. Nakahara, *Journal of Chromatography* 1999, *B*, 161.
115. P. Kintz, *Toxicology Letters*, 1998, **103**, 109.
116. Y. Gaillard, F. Vayssette, and G. Pepin, *Forensic Science International*, 2000, **107**, 361.
117. L. Rivier, *Forensic Science International*, 2000, **107**, 309.
118. D. Thieme, J. Grosse, H. and Sachs R.K. Mueller, *Forensic Science International*, 2000, **107**, 335.
119. D.A. Kidwell, E.H. Lee and S.F. DeLauder, *Forensic Science International*, 2000, **107**, 39.
120. R.E. Joseph, Jr. T.P. Su, and E.J. Cone, *Journal of Analytical Toxicology*, 1996, 20, 338.
121. M. Rothe, F. Pragst, S. Thor and J. Hunger, *Forensic Science International*, 1997, **84**, 53.
122. Y. Nakahara, K. Takahashi and R. Kikura, *Biological and Pharmaceutical Bulletin*, 1995, **18**, 1223.
123. A.M. Tsatsakis, M.N. Tzatzarakis, D. Psaroulis, C. Levkidis and M. Michalodimitrakis, *American Journal of Forensic Medicine and Pathology*, 2001, **22**, 73.
124. F. Mari E. Bertol, in *Proceedings of XXXV TIAFT Annual Meeting – Poster Abstracts*; *http://www.tiaft.org/tiaft97/proceedings/abstract/posters/113.html*, 1997.
125. M.G. Arado, I.V. Garrote, L. Laborde, A. Bosch, and L.A. Ferrari, in *Proceedings of the 2001 TIAFT Conference – Poster Abstracts, http://www.tiaft.org/tiaft2001/posters/p75.doc*, 2001.
126. L.W. Cartmell, A.C. Aufderheide, A. Springfield, C. Weems and B. Arriaza, *Latin American Antiquity*, 1991, **2**, 260.

127. A.C. Springfield, L.W. Cartmell, A.C. Aufderheide, J. Buikstra and J. Ho, *Forensic Science International,* 1993, **63**, 269.
128. S. Balabanova, F. Parsche and W. Pirsig, *Naturwissenschaften,* 1992, **79**, 358.
129. S. Balabanova, F. and Parsche, W. Pirsig, *Baessler-Archiv,* 1992, **40**, 87.
130. S. Balabanova, W. Pirsig, F. Parsche, and E. Schneider, in *Proceedings of the I^st World Congress on Mummy Studies* Organismo Autonomo de Museos y Centros, Cabildo de Tenerife, 1992, 465.
131. U. Hobmeier, and F. Parsche, *Homo,* 1994, **45**, S59.
132. F. Parsche, S. Balabanova, and W. Pirsig, *The Lancet,* 1993, **341**, 1157.
133. F. Parsche, and A. Nerlich, *Fresenius' Journal of Analytical Chemistry,* 1995, **352**, 380.
134. S. Balabanova, F. Parsche, G. Buhler, and W. Pirsig, *Homo,* 1993, **44**, 92.
135. Balabanova, F.W. Rosing, G. Buhler, W. Schoetz, G. Scherer and J. Rosenthal, *Homo* 1997, **48**, 72.
136. S. Connor, in *Sunday Times,* London, 1996, 6.
137. H. Pringle, *The Mummy Congress: Science, Obsession and the Everlasting Dead,* Theia, New York, 2001.
138. G. Hertting, T. Schafer, L.O. Bjorn, N.G. Biset, M.H. Zenk, N.D.P. McIntosh, and F. Parsche, *Naturwissenschaften,* 1993, **80**, 243.
139. N. Moore, D. Brothwell, and M. Spigelman, *The Lancet,* 1993, **341**, 1157.
140. P.C. Buckland and E. Panagiotakopulu, *Antiquity,* 2001, **75**, 549.
141. L. Cartmell, A.C. Aufderheide, L.E. Wittmers and C. Weems, in *Mummies in a New Millenium,* N. Lynnerup, C. Andreasen and J. Berglund, (ed), Danish Polar Center Copenhagen, 2003, 79.
142. L.W. Cartmell, A.C. Springfield and C. Weems, in *Studies on Ancient Mummies and Burial Archaeology,* F. Cardenas Arroyo and C. Rodriguez-Martin, (ed), Instituto Canario de Bioantropologia, Bogota, Colombia, 2001, 237.
143. L. Cartmell, A.C. Aufderheide, D. Caprara, J. Klein and G. Koren, *Paper Presented at the V World Congress of Mummy Studies,* Torino, Italy, 2004.
144. M.M. Castro, C. Camargo, M.A. Benavente, H. Baez, S. Salas and E. Aspillaga, in *Mummies in a New Millenium,* N. Lynnerup, C. Andreasen and J. Berglund, (ed), Danish Polar Center, Copenhagen, 2003, 75.
145. J. Huhne, H. Pfeiffer, K. Waterkamp and B. Brinkmann, *International Journal of Legal Medicine,* 1999, **112**, 172.
146. F. Rousselet and P. Mangin, *International Journal of Legal Medicine,* 1998, **111**, 292.
147. M. Allen, A.S. Engstrom, S. Meyers, O. Handt, T. Saldeen, A. von Haeseler, S. Paabo and U. Gyllensten, *Journal of Forensic Sciences,* 1998, **43**, 453.
148. P. Savolainen and J. Lundeberg, *Journal of Forensic Sciences,* 1999, **44**, 77.
149. P.M. Schneider, Y. Seo and C. Rittner, *International Journal of Legal Medicine,* 1999, **112**, 315.
150. M.R. Wilson, J.A. Dizinno, D. Polanskey, J. Replogle and B. Budowle, *International Journal of Legal Medicine,* 1995, **108**, 68.
151. R. Bonnichsen and A.L. Schneider, *The Sciences,* 1995, 26.
152. A. Krzyszowski, M. Stochaj, R. Pawlowski and A. Welz, *Prseglad Antropologiczny,* 1996, **59**, 97.

153. R. Bonnichsen, L. Hodges, W. Ream, K.G. Field, D.L. Kirner, K. Selsor and R.E. Taylor, *Journal of Archaeological Science,* 2001, **28**, 775.
154. P. Swami, *Frontline* 2000, 17.
155. I.A. Lebedeva, E.E. Kulikov, N.V. Ivanova and A.B. Poltaraus, *Sudebno-Meditsinskaia Ekspertiza,* 2000, **43**, 9.
156. L.E. Baker, W.F. McCormick and K.J. Matteson, *Journal of Forensic Sciences,* 2001, **46**, 126.
157. K. Takayanagi, H. Asamura, K. Tsukada, M. Ota, S. Saito and H. Fukushima, *International Congress Series,* 2003, **1239**, 759.
158. M.T.P. Gilbert, D.J. Tobin and A.S. Wilson, in *Molecular markers, PCR, Bioinformatics and Ancient DNA – Technology, Troubleshooting & Applications,* G. Dorado, (ed), Science Publishers, New York, in press.
159. M.T.P. Gilbert, A.S. Wilson, M. Bunce, A.J. Hansen, E. Willerslev, B. Shapiro, T.F. Higham, M.P. Richards, T.C.O'Connell, D.J. Tobin, R.C. Janaway, A. Cooper, *Current Biology,* 2004, **14:12**, R463.
160. M.T.P. Gilbert, A.J. Hansen, E. Willerslev, L. Rudbeck, I. Barnes, N. Lynnerup and A. Cooper, *American Journal of Human Genetics,* 2003, **72**, 48.
161. M.T.P. Gilbert, L. Menez, R.C. Janaway, D.J. Tobin, A. Cooper, A.S. Wilson *Forensic Science International,* 2005, in press.
162. K. Aoki, *Annals of Human Biology,* 2002, **29**, 589.
163. M. Mukerjee, *Scientific American,* 1999, 24.
164. D.L. Kirner, R. Burkey, R.E. Taylor and J.R. Southon, *Nuclear Instrumentation and Methods in Physics Research,* 1997, **B123**, 214.
165. R.E. Taylor, P.E. Hare, C.E. Prior, D.L. Kirner, L. Wan and R.R. Burky, *Radiocarbon,* 1995, **37**, 319.
166. J.M. Gillespie, *Science,* 1970, **170**, 1100.
167. J. van der Plicht, W.A.B. van der Sanden, A.T. Aerts, H.J. Streurman *Journal of Archaeological Science,* 2004, **31**, 471.
168. M.A. Geyh, *Radiocarbon,* 2001, **43**, 723.
169. E.M. Wild, K.A. Arlamovsky, R. Golser, W. Kutschera, A. Priller, S. Puchegger, W. Rom, P. Steier and W. Vycudilik, *Nuclear Instruments and Methods in Physics Research Section B– Beam Interactions with Materials and Atoms,* 2000, **172**, 944.
170. P. Walter, P. Martinetto, G. Tsoucaris, R. Breniaux, M.A. Lefebvre, G. Richard, J. Talabot and E. Dooryhee, *Nature,* 1999, **397**, 483.
171. J. Fletcher, in *Ancient Egyptian Materials and Technology,* P.T. Nicholson and I. Shaw (ed), Cambridge University Press, Cambridge, 1999, 495.
172. R. Kellner, C. Minich, N. Iskander and M.M. Khater, *Proceedings of SPIE (International Society for Optical Engineering),* 1989, **1145**, 310.
173. L. Bertrand, J. Doucet, P. Dumas, A. Simionovici, G. Tsoucaris and P. Walter, *Journal of Synchrotron Radiation,* 2003, **10**, 387.
174. D. Brothwell and K. Dobney, in *Lindow Man, The Body in the Bog,* I.M. Stead, J.B. Bourke and D. Brothwell, (ed), British Museum Publications, London, 1986, 66.
175. E. Munksgaard, *Aarborger for Nordisk Oldkyndigmed og Histoirie,* 1976, 5.

176. D.J. Fletcher, *Postgraduate Medicine,* 1982, **72**, 79–81, 84, 87.

177. A.J. Fletcher, *Egyptian Archaeology,* 1994, 5, 31.

178. E. Martinson, Y.A. K.J. Reinhard, J.E. Buikstra, K.D. de la Cruz *Mem Inst Oswaldo Cruz,* 2003, **98**, 195.

179. A.V. Trofimov, Y.A. Yakovlev, A.I. Bobrova in *Northern Archaeological Congress Abstracts,* Ural State University (ed) Academkniga Publishers, Ekaterinburg, 2002, 180.

180. J. Fletcher and D. Montserrat, in *Proceedings of the 7th International Congress of Egyptologists,* C.J. Eyre, (ed), Peeters, Leuven, 1998, 401.

181. G.J. Tassie, *Papers from the Institute of Archaeology,* 1996, **7**, 59.

182. G.J. Tassie, *Papers from the Institute of Archaeology,* 2000, **11**, 27.

183. H. Wellman, *Journal of Conservation and Museum Studies,* 1996, 1999, http://www.ucl.ac.uk/archaeology/conservation/jcms/issue1/wellman.html.

184. A. Massey, *Current Archaeology,* 2000, **169**, 34.

185. N.K. Adams, *Ancient Textile News,* 1999, **28**, 20.

186. A.S. Wilson, M.P. Richards, M.T.P. Gilbert *Journal of the German Society of Dermatology,* 2004, **2**, 549.

187. G. Lubec, M.R. Zimmerman, M. TeschlerNicola, V. Stocchi and A.C. Aufderheide, *Free Radical Research,* 1997, 26, 457.

188. G. Lubec, M. Weninger and S.R. Anderson, *FASEB Journal,* 1994, **8**, 1166.

189. G.J. Smith, *Antiquity,* 1993, **67**, 117.

190. G. Lubec, G. Nauer, K. Seifert, E. Strouhal, H. Porteder, J. Szilvassy and M. Teschler, *Journal of Archaeological Science,* 1987, **14**, 113.

191. A.S. Wilson, H.G.M. Edwards, D.W. Farwell and R.C. Janaway, *Journal of Raman Spectroscopy,* 1999, **30**, 367.

192. A. Conti-Fuhrman and E. Rabino Massa, *Journal of Human Evolution,* 1972, **1**, 487.

193. M.W. Hess, G. Klima, K. Pfaller, K.H. Kunzel and O. Gaber, *American Journal of Physical Anthropology,* 1998, **106**, 521.

194. H. Hino, T. Ammitzboll, R. Moller and G. Asboehansen, *Journal of Cutaneous Pathology,* 1982, **9**, 25.

195. E. Rabino Massa, M. Masali and A.M. Conti Fuhrman, *Journal of Human Evolution,* 1980, **9**, 133.

196. E. Rabino Massa, *Arch. It. Anat. e Embriol.,* 1976, **81**, 301.

197. J.A. Swift, *Nature,* 1972, **238**, 161.

198. D.H. DeGaetano, J.B. Kempton and W.F. Rowe, *Journal Forensic Science,* 1992, **37**, 1048.

199. W.F. Rowe, in *Forensic Taphonomy – The Post-mortem Fate of Human Remains,* W.D. Haglund and M.H. Sorg (ed), CRC Press, London, 1997, 337.

200. W. Widy and B. Andreas-Ludwicka, *Przeg. Derm.,* 1970, **LVII**, 159.

201. D.J.P. Ferguson, *Microscopy and Analysis,* 1992, 5.

202. D.A. Birkett, C.L. Gummer and R.P.R. Dawber, in *Science in Egyptology,* R.A. David, (ed), Manchester University Press, Manchester, 1986, 367.

203. A.S. Wilson, R.A. Dixon, H.G.M. Edwards, D.W. Farwell, R.C. Janaway, A.M. Pollard and D.J. Tobin, *Chungara, Revista de Antropologia Chilena,* 2001, **33**, 293.

204. A.S. Wilson, R.C. Janaway, A.M. Pollard, R.A. Dixon and D.J. Tobin, in *Human Remains, Conservation, Retrieval and Analysis,* E. Williams (ed), British Archaeology Reports, Oxford, 2001, **S934**, 119.
205. A.S. Wilson, H.I. Dodson, R.C. Janaway, A.M. Pollard and D.J. Tobin in *Mummies in a New Millenium,* N. Lynnerup, C. Andreasen, J. Berglund (ed), Greenland National Museum and Archives, Danish Polar Center, Nuuk, Greenland, 2003, 63.
206. A.S. Wilson, in *Principles of Forensic Taphonomy: Applications of Decomposition Processes in Recent Gravesoils,* M. Tibbett and D.O.Carter (ed), Humana Press, New Jersey, in press.
207. A.S. Wilson, R.C. Janaway, D.J. Tobin, *Journal of Investigative Dermatology Symposium Procedings,* 1999, **4**, 353.
208. A.S. Wilson, H.I. Dodson, R.C. Janaway, A.M. Pollard, D.J. Tobin, *Journal of the German Society of Dermatology,* 2004, **2**, 515.
209. M.T.P Gilbert, R.C. Janaway, D.J. Tobin, A. Cooper, A.S. Wilson *Forensic Science International,* 2005, in press.
210. R.P. Ambler, S.A. Macko, B. Sykes, J.B. Griffiths, J. Bada and G. Eglinton, *Philosophical Transactions of the Royal Society of London Series B–Biological Sciences,* 1999, **354**, 75.
211. A. Coghlan, *New Scientist,* 1999, 16th October, 24.
212. T. Schafer, *Naturwissenschaften,* 1993, **80**, 243.
213. N.G. Bisset and M.H. Zenk, *Naturwissenschaften,* 1993, **80**, 244.
214. P. McIntosh, *Naturwissenschaften,* 1993, **80**, 245.
215. D. Brothwell and M. Spigelman, *The Lancet,* 1993, **341**, 1157.
216. L.O. Bjorn, *Naturwissenschaften,* 1993, **80**, 244.

A Perspective on Future Directions

DESMOND J. TOBIN

In preparation for this perspective I entered the four lonely letters of the word H-A-I-R into Google$^{\text{®}}$ and within 0.11 seconds 60,800,000 possible hits came rushing forth. Fitting then, that by stripping away the 'H' and so yielding a similarly important substance, generates only <4 times more hits! Even 'hair analysis' yields >3 million hits – clearly, hair is important to us humans.

The hair fibre is the macroscopic end point of a bewilderingly complex *menage-à-trois* involving ectodermal, mesodermal and neuroectodermal components – the nature of which is largely set during our development *in utero* but which is dramatically revisited in the adult during the unique tissue remodelling events required for hair follicle cycling. Given the hair follicle's extraordinary dynamics, it has of late become a temptress with considerable powers of seduction for a growing and disparate band of scientists. These have been enticed to exploit this veritable mini-organ's highly accessible 'one-stop-shop' of answers to many of science's unyielding enigmas. This combined force of effort has, over the last ten years, yielded a flurry of new and often challenging data covering a diverse range of scientific specialties. In this way, the status of the hair follicle has been raised significantly as a model tissue for study in science today.

Not only does hair immediately identify us as a mammal, the symbolism of hair has featured intensely throughout human civilisation. Its length, colour, texture, shape or whether it still grows or falls out has fascinated us all for eons. Literature from every human society extols the hair fibre's virtues and deprecates its vices as both sets of traits are woven through with myth and magic. Hair has been used to explain and justify the anger or delight of the gods, justify or signpost our fertility, and even regulate our crop harvests. Its perceived contribution to personal strength underlies, and often justifies, religious prohibitions to its cutting and shaving in men. Its cropping and coverage in women of some religious faiths is said to regulate its sexual potency, while shaving widely represents a most harrowing of public female humiliations. Finally, removal of the hair *via* 'scalping' was believed to extract a person's soul. Clearly, we have invested much in our scalp hair, and so it is perhaps surprising that its systematic study has only recently become a serious scientific pursuit.

For the cutaneous biologist, the explosion in hair biology research can be traced to the approval by the FDA (US) of the first proven hair growth promoter – Upjohn's (now Pharmacia's) minoxidil in the early 1980s. This advance can be credited with opening the door for many a budding biologist (myself included) to scientific hair follicle biology research. The last 20 years therefore has finally seen the putting to bed of age-old spurious 'old wives tales' regarding hair and hair growth.

When I embarked on this project I set for myself the goal to bring together into one volume diverse strands (no pun intended... it is almost impossible to avoid them – there are so many...indeed one wishes often to just...eh...curl up and dye) of scientific endeavour where the hair follicle or its hair fibre has become a central bioresource. Clearly, much of this progress is fuelled by our increasing knowledge of the biology of the hair follicle and its fibre – a unique physicochemical amalgam – with regard to its development, formation, engineering and pigmentation *etc*. As our only major trait which we can alter phenotypically with relative ease and at relatively minor cost, the market place for hair growth and hair appearance modulators is great.

However, just how far we are prepared to expose our systemic biology and physiology to potentially hazardous hair 'treatments' is likely to become an issue of increasing concern. Our 'reliance' on 'appearance enhancers' is likely to continue spiralling upwards, as we spend more and more of our time on earth beyond the first flush of youth. Moreover, more of the world's populations will become economically developed and so join in on this pursuit. Thus, we need to be surer than ever before that the products we use will not harm us. Having said that, the hair follicle (and fibre) should remain its own robust watchdog. Its unique growth dynamics neatly partition post-biogenic change, due to endogenous and exogenous factors (*e.g.* pollutants, drugs, environmental toxins *etc.*) from what the body 'saw' during formation of the hair itself.

Clearly, hair analysis (*e.g.* for trace metals *etc.*) as a sub-speciality has still some way to go before it can generate wholly reliable information – but many of these limitations have been identified and will be overcome. Concerns also apply to the cosmetics and cosmeceutical industries, where not only improved safety profiles are needed for specific product categories, but perhaps more importantly, greater innovations are urgently needed in basic product development terms. The dramatic advances in the power and sophistication of molecular and genetic technologies have also positioned the hair centrally to yield enormously valuable information in such diverse fields as individualisation, forensics and genealogical profiling (*e.g.* within maternal lineages using mtDNA). There will continue to be scientists who excel in getting more and more useful information from less and less tissue. Downsides, and there may be many, will be the perception that we may have less personal control of our body's tissues and altered right of refusal, as scientific wizards profile us on the basis of a single hair strand. Upsides however, will permit us to test the quality of our environments in ways that will become more and more convincing and so will provide ordinary people with significant tools to engage in the monitoring and regulation of the activities of governments and big business. Exploiting the 'no further biogenic change' characteristic of hair fibre post-formation will continue to enable significant new dietary information to be gleaned from hair and this should provide an important adjunct in the critically important fields of nutrition, dietetics, forensics *etc.* Here again, we will need to reach a consensus on the development and implementation internationally of acceptable standards for analysis and assessment of hair.

When after reaching a grand old age we finally shed this mortal coil, for those of us not too keen on a firey end, it may be comforting to know that archeologists, our modern day head-hunters, may fuss about our ruined remains attempting to untangle the complexities of our lost lives, lifestyles and loves. Recent advances in our understanding of how hair degrades may yield important information to help guide future generations on their own merry quests to the tomb.

Though not a focus in this volume, I add the following if only not to end on a morbid note. The extraordinary regenerative power of the hair follicle, with its stem cell reservoirs for epithelial, connective, neural, muscle and perhaps even hematopoeitic tissues, may contribute substantially to the exploitation of adult stem cell technologies for human health. Thus, from our first imaginings to our last breath, the hair follicle and its fibre is likely to play an increasingly important role in our lives, not only medically, legally, cosmetically, but also in getting to understand the world about us. We ignore the news coming down the fibre wires at our peril! 'Wisdom is the comb given to a man after he has lost his hair' (*Irish saying*).

Subject Index